Advances in Science, Technology & Innovation

IEREK Interdisciplinary Series for Sustainable Development

Advances in Science, Technology & Innovation (ASTI) is a series of peer-reviewed books based on important emerging research that redefines the current disciplinary boundaries in science, technology and innovation (STI) in order to develop integrated concepts for sustainable development. It not only discusses the progress made towards securing more resources, allocating smarter solutions, and rebalancing the relationship between nature and people, but also provides in-depth insights from comprehensive research that addresses the **17 sustainable development goals (SDGs)** as set out by the UN for 2030.

The series draws on the best research papers from various IEREK and other international conferences to promote the creation and development of viable solutions for a **sustainable future and a positive societal** transformation with the help of integrated and innovative science-based approaches. Including interdisciplinary contributions, it presents innovative approaches and highlights how they can best support both economic and sustainable development, through better use of data, more effective institutions, and global, local and individual action, for the welfare of all societies.

The series particularly features conceptual and empirical contributions from various interrelated fields of science, technology and innovation, with an emphasis on digital transformation, that focus on providing practical solutions to **ensure food, water and energy security to achieve the SDGs.** It also presents new case studies offering concrete examples of how to resolve sustainable urbanization and environmental issues in different regions of the world.

The series is intended for professionals in research and teaching, consultancies and industry, and government and international organizations. Published in collaboration with IEREK, the Springer ASTI series will acquaint readers with essential new studies in STI for sustainable development.

ASTI series has now been accepted for Scopus (September 2020). All content published in this series will start appearing on the Scopus site in early 2021.

Federico Lucci · Domenico M. Doronzo ·
Jasper Knight · Anna Travé · Stefan Grab ·
Amjad Kallel · Sandeep Panda · Helder I. Chaminé ·
Jesús Rodrigo-Comino · Sami Khomsi ·
Santanu Banerjee · Broder Merkel ·
Haroun Chenchouni
Editors

Selected Studies in Geomorphology, Sedimentology, and Geochemistry

Proceedings of the 3rd Conference of the Arabian Journal of Geosciences (CAJG-3)

Springer

Editors
Federico Lucci
Dipartimento di Scienze della Terra e Geoambientali
Università degli Studi di Bari "Aldo Moro"
Bari, Italy

Jasper Knight
University of the Witwatersrand
Johannesburg, South Africa

Stefan Grab
School of Geography, Archaeology and Environmental Studies
University of the Witwatersrand
Wits, South Africa

Sandeep Panda
Gujarat Biotechnology University
Gandhinagar, Gujarat, India

Jesús Rodrigo-Comino
Departamento de Análisis Geográfico Regional y Geografía Física
Facultad de Filosofía y Letras
Campus Universitario de Cartuja
University of Granada
Granada, Spain

Santanu Banerjee
Indian Institute of Technology Bombay
Mumbai, India

Haroun Chenchouni
Higher National School of Forests
Khenchela, Algeria

Domenico M. Doronzo
Istituto Nazionale di Geofisica e Vulcanologia
Sezione di Napoli
Naples, Italy

Anna Travé
Department of Mineralogy, Petrology and Applied Geology
Faculty of Earth Sciences
Universitat de Barcelona (UB)
Barcelona, Spain

Amjad Kallel
ENIS
University of Sfax
Sfax, Tunisia

Helder I. Chaminé
School of Engineering (ISEP)
Polytechnic of Porto
Porto, Portugal

Sami Khomsi
Georesources Lab
CERTE
University of Carthage
Tunis, Tunisia

Broder Merkel
Geology Department
Technical University Bergakademie
Freiberg, Germany

ISSN 2522-8714 ISSN 2522-8722 (electronic)
Advances in Science, Technology & Innovation
IEREK Interdisciplinary Series for Sustainable Development
ISBN 978-3-031-43743-4 ISBN 978-3-031-43744-1 (eBook)
https://doi.org/10.1007/978-3-031-43744-1

This Springer imprint is published by the registered company Springer Nature Switzerland AG
The registered company address is: Gewerbestrasse 11, 6330 Cham, Switzerland

Paper in this product is recyclable.

About the 2nd Springer Conference of the *Arabian Journal of Geosciences* (CAJG-3), Online November 2–5, 2020

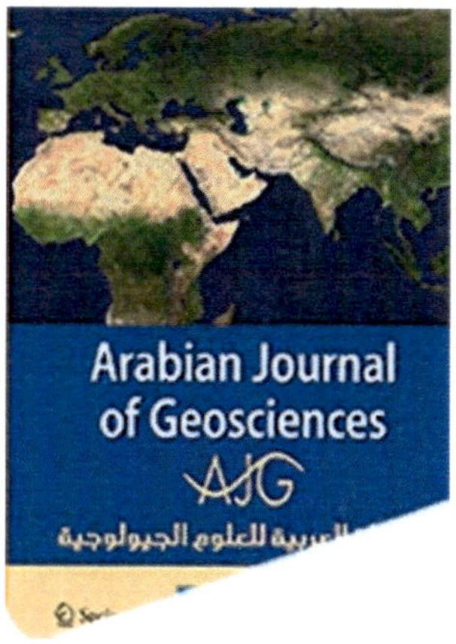

The *Arabian Journal of Geosciences* (AJG) is a Springer journal publishing original articles on the full range of Earth sciences in partnership with the Saudi Society for Geosciences. The journal focuses on, but is not limited to, research themes which have regional significance for the Middle East, the Euro-Mediterranean, Africa, Asia, and some other regions of the world. The journal receives on average 3500 submissions a year and accepts around 800 papers for publication in its 24 annual issues (acceptance rate around 20%). It benefits from the participation of an editorial team of 15 topical chief editors and 100 international associate editors who generously help in evaluating and selecting the best papers.

In 2008, Prof. Abdullah Al-Amri, in close partnership with Springer, founded the *Arabian Journal of Geosciences (AJGS)*. In 2018, the journal celebrated its tenth anniversary. To mark the event, the founder and the editor-in-chief of the AJGS organized the 1st Conference of the *Arabian Journal of Geosciences* (CAJG) in close collaboration with Springer on November 12–15, 2018. The conference was an occasion to endorse the journal's long-held reputation and brought together 450 authors from 70 countries, who work in the wide-ranging fields of Earth sciences. The dynamic four-day conference in a stimulating environment in Hammamet, Tunisia, provided attendees with opportunities to share their latest unpublished findings and learn about the latest geosciences studies. The event also allowed attendees to meet and talk to the journal's editors and reviewers.

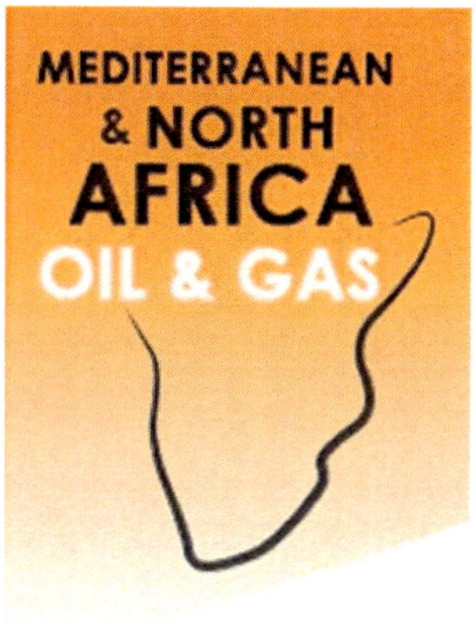

In a continuation of the successful 1st CAJG and 2nd CAJG, this year's conference aims to bring geoscientists from all over the world to present and discuss their most recent findings. The 3rd CAJG will continue to publish the newest findings in its proceedings and some special issues by Springer, with a clear mission to drive greater North–South (Europe-Africa) scientific cooperation and to open doors to new and enriching collaborations with geoscientists based in Asia and the Americas. Research studies from the wide-ranging fields of Earth sciences can be submitted for evaluation by the Conference Scientific Committee. In particular, the 3rd CAJG will devote a special session to studies focusing on unraveling the undiscovered oil and gas resources in the Mediterranean and North Africa. Many international experts will be invited to take part in the discussion, in particular we are expecting a massive participation of researchers from North Africa Research Group (NARG) of Manchester University (UK) and other scientists based in North Africa.

Conference Tracks

The scientific committee of the 3rd CAJG invites research papers on all cross-cutting themes of Earth sciences, with a main focus on the following 15 conference tracks:

- Track 1. Atmospheric Sciences, Meteorology, Climatology, Oceanography
- Track 2. Biogeochemistry, Geobiology, Geoecology, Geoagronomy
- Track 3. Earthquake Seismology and Geodesy
- Track 4. Environmental Earth Sciences
- Track 5. Exploration and Theoretical Geophysics, Seismic and Well-Logging Methods, Mathematical Geosciences
- Track 6. Geo-Informatics and Remote Sensing
- Track 7. Geochemistry, Mineralogy, Petrology, Volcanology
- Track 8. Geological Engineering, Geotechnical Engineering
- Track 9. Geomorphology, Geography, Soil Science, Glaciology, Geoarcheology, Geoheritage
- Track 10. Hydrology, Hydrogeology, Hydrochemistry
- Track 11. Marine Geosciences, Historical Geology, Paleoceanography, Paleoclimatology
- Track 12. Numerical and Analytical Methods in Mining Sciences and Geomechanics
- Track 13. Petroleum and Energy Engineering, Petroleum Geochemistry
- Track 14. Sedimentology, Stratigraphy, Paleontology, Geochronology
- Track 15. Structural Geology, Tectonics and Geodynamics, Petroleum Geology

About the 2nd Springer Conference of the *Arabian Journal of Geosciences* (CAJG-3), Online November 2–5, 2020

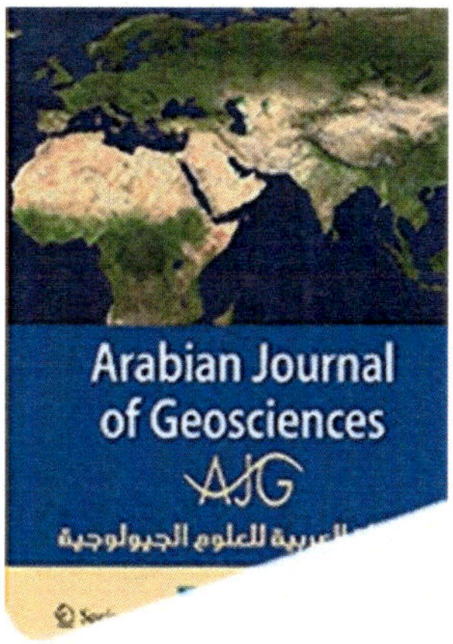

The *Arabian Journal of Geosciences* (AJG) is a Springer journal publishing original articles on the full range of Earth sciences in partnership with the Saudi Society for Geosciences. The journal focuses on, but is not limited to, research themes which have regional significance for the Middle East, the Euro-Mediterranean, Africa, Asia, and some other regions of the world. The journal receives on average 3500 submissions a year and accepts around 800 papers for publication in its 24 annual issues (acceptance rate around 20%). It benefits from the participation of an editorial team of 15 topical chief editors and 100 international associate editors who generously help in evaluating and selecting the best papers.

In 2008, Prof. Abdullah Al-Amri, in close partnership with Springer, founded the *Arabian Journal of Geosciences (AJGS)*. In 2018, the journal celebrated its tenth anniversary. To mark the event, the founder and the editor-in-chief of the AJGS organized the 1st Conference of the *Arabian Journal of Geosciences* (CAJG) in close collaboration with Springer on November 12–15, 2018. The conference was an occasion to endorse the journal's long-held reputation and brought together 450 authors from 70 countries, who work in the wide-ranging fields of Earth sciences. The dynamic four-day conference in a stimulating environment in Hammamet, Tunisia, provided attendees with opportunities to share their latest unpublished findings and learn about the latest geosciences studies. The event also allowed attendees to meet and talk to the journal's editors and reviewers.

In a continuation of the successful 1st CAJG and 2nd CAJG, this year's conference aims to bring geoscientists from all over the world to present and discuss their most recent findings. The 3rd CAJG will continue to publish the newest findings in its proceedings and some special issues by Springer, with a clear mission to drive greater North–South (Europe-Africa) scientific cooperation and to open doors to new and enriching collaborations with geoscientists based in Asia and the Americas. Research studies from the wide-ranging fields of Earth sciences can be submitted for evaluation by the Conference Scientific Committee. In particular, the 3rd CAJG will devote a special session to studies focusing on unraveling the undiscovered oil and gas resources in the Mediterranean and North Africa. Many international experts will be invited to take part in the discussion, in particular we are expecting a massive participation of researchers from North Africa Research Group (NARG) of Manchester University (UK) and other scientists based in North Africa.

Conference Tracks

The scientific committee of the 3rd CAJG invites research papers on all cross-cutting themes of Earth sciences, with a main focus on the following 15 conference tracks:

- Track 1. Atmospheric Sciences, Meteorology, Climatology, Oceanography
- Track 2. Biogeochemistry, Geobiology, Geoecology, Geoagronomy
- Track 3. Earthquake Seismology and Geodesy
- Track 4. Environmental Earth Sciences
- Track 5. Exploration and Theoretical Geophysics, Seismic and Well-Logging Methods, Mathematical Geosciences
- Track 6. Geo-Informatics and Remote Sensing
- Track 7. Geochemistry, Mineralogy, Petrology, Volcanology
- Track 8. Geological Engineering, Geotechnical Engineering
- Track 9. Geomorphology, Geography, Soil Science, Glaciology, Geoarcheology, Geoheritage
- Track 10. Hydrology, Hydrogeology, Hydrochemistry
- Track 11. Marine Geosciences, Historical Geology, Paleoceanography, Paleoclimatology
- Track 12. Numerical and Analytical Methods in Mining Sciences and Geomechanics
- Track 13. Petroleum and Energy Engineering, Petroleum Geochemistry
- Track 14. Sedimentology, Stratigraphy, Paleontology, Geochronology
- Track 15. Structural Geology, Tectonics and Geodynamics, Petroleum Geology

About the Conference Steering Committee

General Chair

Abdullah Al-Amri

Founder and Editor-in-Chief,
Arabian Journal of Geosciences,
King Saud University, Saudi Arabia

Conference Supervisor

Nabil Khélifi

Senior Publishing Editor, MENA program,
Journal Publishing Manager,
Arabian Journal of Geosciences,
Springer, a part of Springer Nature, Germany

Advisory Co-chair

Walter D. Mooney

Guest of Editorial Board,
Arabian Journal of Geosciences,
United States Geological Survey Western Region, USA

Advisory Co-chair

Giovanni Bertotti

Guest of Editorial Board,
Arabian Journal of Geosciences,
Geoscience and Engineering, Delft University of Technology, The Netherlands

Scientific Committee Co-chair

François Roure

Chief Editor—Track 15,
Arabian Journal of Geosciences,
IFP—Energies Nouvelles, France

Scientific Committee Co-chair

Biswajeet Pradhan

Chief Editor—Track 6,
Arabian Journal of Geosciences,
University of Technology Sydney, Australia

Local Organizing Co-chair

Mohamed Soussi

Former Associate Editor,
Arabian Journal of Geosciences,
Tunis El Manar University, Tunis, Tunisia

Local Organizing Co-chair

Mounir Medhioub

Guest of Editorial Board,
Arabian Journal of Geosciences,
University of Sfax, Sfax, Tunisia

Publications Co-chair

Beatriz Bádenas

Chief Editor—Track 14,
Arabian Journal of Geosciences,
University of Zaragoza, Zaragoza, Spain

Publications Co-chair

Marina Rabineau

Chief Editor—Track 11,
Arabian Journal of Geosciences,
University of Brest, Brest, France

Program Co-chair

Amjad Kallel

Chief Editor—Track 4,
Arabian Journal of Geosciences,
ENIS, University of Sfax, Tunisia

Program Co-chair

Sami Khomsi

Georesources Lab,
CERTE, University of Carthage
Tunis, Tunisia

Proceedings Editorial Manager

Mourad Amer

Editor of Springer/IEREK ASTI Series,
Guest of Editorial Board of AJGS,
IEREK, Alexandria, Egypt

Communication Chair

Zakaria Hamimi

Associate Editor,
Arabian Journal of Geosciences,
Benha University, Benha, Egypt

Public Relations Chair

Faïez Gargouri

Director of the Higher Institute of Computer Science and Multimedia,
President of the AIG,
University of Sfax, Tunisia

Conference Manager

Mohamed Sahbi Moalla

Journal Coordinator,
Arabian Journal of Geosciences,
ISET, University of Sfax, Tunisia

Preface

The proceedings of the 3rd Springer Conference of the *Arabian Journal of Geosciences* (CAJG-3) which was held online in November 2020 consists of four volumes titled:

Proceedings Volume 1: New Prospects in Environmental Geosciences and Hydrogeosciences
Proceedings Volume 2: Advances in Geophysics, Tectonics, and Petroleum Geosciences
Proceedings Volume 3: Recent Research on Geomorphology, Sedimentology, and Geochemistry
Proceedings Volume 4: Research developments in Geotechnics, Geo-Informatics and Remote Sensing

These volumes are based on the accepted conference papers for either oral/poster presentations or selected for online publication during the CAJG-3.

The first volume offers a broad range of recent studies that discuss the latest advances in geoenvironmental and hydrogeosciences from diverse backgrounds including climate change, geoecology, biogeochemistry, water resources management, and environmental monitoring and assessment. It shares insights of experienced scientists from, but not limited to, research institutes worldwide, focused the Mediterranean and Middle East regions on how the understanding of ecological, climatological, oceanic, and hydrological processes is the key for improving practices in environment management, including the eco-responsibility, scientific integrity, and social and ethical dimensions.

The second volume includes a series of research methods that are nowadays in use for measuring, quantifying, and analyzing the targeted geological domains. Remote sensing with high-resolution satellite imagery, seismology, geochemistry, theoretical geophysics and related profiles with tomographic images, earthquake geology with times series radar interferometry and related geodetic-GPS campaigns, and well-logging contribute to scrutinizing the Earth tectonic architecture. Fundamental questions that address the genesis and evolution of our planet are built upon data collection and experimental investigations under physical constitutive laws which are the conditions for a successful scientific research. These multidisciplinary approaches combined with the geodynamics of tectonic provinces and investigations of potential zones of natural resources (petroleum reservoirs) provide the basis for the economic development.

The third volume presents an updated unique view in conjugating field studies and modeling to better quantify the process-product binomial unusual in geosciences. Earth systems require a comprehensive understanding of processes and dynamics of geology, morphotectonics, sedimentology, stratigraphy, and geochemistry. In the geomorphology section, several studies deal with topics related to fault slip and incision rates, soil science, landslides and debris flows, coastal processes, and geoarcheology and geoheritage. Under the sedimentology section, researchers present recent advances in stratigraphy, environmental, tectonic and diagenetic processes, together with evolutionary, biostratigraphic and paleoenvironmental significance of paleontology are presented. Additionally, this section also contains papers on marine geosciences, from molecular proxies related to climate to geophysical surveys. Last but not least, the third section on geochemistry focuses on sedimentary geochemistry and mineralogical characterization, magmatic and metamorphic processes and products, and the origin and exploration of mineral deposits. This volume resumes the current situation related to the above-mentioned topics mainly in the Mediterranean realm, testifying the geological importance of this area and surroundings.

The fourth volume contains research studies in the fields of three sections; (1) geological and geotechnical engineering, (2) geomechanical studies based on numerical and analytical methods, and (3) geo-informatics and remote sensing. The content of these research studies would provide new scientific knowledge for further understanding on landslides, new stabilization techniques, importance of geophysics for engineering geology investigations as well as new empirical approaches for easily predicting some physical and hydrogeomechanical properties of geomaterials.

The proceedings of the CAJG-3 is of interest to all researchers, practitioners, and students in all fields of geosciences, including the environmental geosciences.

Bari, Italy	Federico Lucci
Naples, Italy	Domenico M. Doronzo
Johannesburg, South Africa	Jasper Knight
Barcelona, Spain	Anna Travé
Wits, South Africa	Stefan Grab
Sfax, Tunisia	Amjad Kallel
Gandhinagar, India	Sandeep Panda
Porto, Portugal	Helder I. Chaminé
Granada, Spain	Jesús Rodrigo-Comino
Tunis, Tunisia	Sami Khomsi
Mumbai, India	Santanu Banerjee
Freiberg, Germany	Broder Merkel
Khenchela, Algeria	Haroun Chenchouni
July 2021	

Contents

About the Editors

Federico Lucci Dipartimento di Scienze della Terra e Geoambientali, Università degli Studi di Bari Aldo Moro, Italy.

Federico Lucci holds a Ph.D. degree in Earth Sciences (2010) from Università degli Studi Roma Tre (Italy). Theses and specialties related to the degrees are petrography, multivariate statistics in Earth Sciences and igneous petrology-structural geology. Since February 2022, he is Researcher of petrography and petrology at the University of Bari 'Aldo Moro,' where is dealing with the teaching course petrography of metamorphic rocks. Since 2008, he was co-tutor of numerous bachelor, master and Ph.D. theses, the bulk of them dealing with the evolution of crystalline basements. Petrographic fieldwork and laboratory activity were always part of the work in each thesis. His research is focused on modeling and timing of magmatic complexes as key-constraint for understanding regional tectonics and geodynamic scenarios, through a multidisciplinary approach that integrates field survey, rock fabric characterization, mineral chemistry, geochemistry, geochronology, inverse and forward thermobarometry and FC-AFC-Mixing modeling. He has done extensive fieldwork in Alpine and Hercynian orogens (Rif Chain, Corsica, Sardinia, Pyrenees, Central Iran), Gondwanan orogens (Central Iran), Pacific-Cocos convergence zone (Trans Mexican Volcanic Belt), intraplate volcanic chains (Cameroon Line). His investigations are related to the definition of the magmatic heat source in areas of geothermal interests (Eastern Trans Mexican Volcanic Belt), understanding the granulite-granite connection (Rif Chain, Pyrenees and Central Iran), unraveling the Mesozoic geodynamic scenario of the Neo-Tethys realm of Iran through its magmatic record and its Cadomian basement, understanding the processes of crust-forming and growth through anatexis processes (Rif Chain in Morocco; Sanandaj-Sirjan Zone of Iran), and understanding the existing relationships between ore deposits, magmatism and crust structure (Iran). He serves as Reviewer for international ISI peer-reviewed journal since 2015. In 2018, he joined *Arabian Journal of Geosciences* (Springer Nature) as an Associate Editor for evaluating submission in the fields of mineralogy, petrology, geochemistry and volcanology. In early 2018, he joined the *Geochemistry—Chemie Der Erde* (ELSEVIER) Journal as Associate Editor responsible for evaluating submissions in the field of igneous and metamorphic petrology and geochemistry.

Domenico M. Doronzo Istituto Nazionale di Geofisica e Vulcanologia, Italy.
Dr. Domenico M. Doronzo holds a B.Sc., a M.Sc., and a Ph.D. degree (in 2011) in Earth Sciences from Università degli Studi di Bari Aldo Moro, Italy. Theses and specialties related to the degrees are physical volcanology, experimental and computational fluid dynamics, petrology, and natural hazards. Then, he has worked in volcanology and sedimentology, fluid dynamics and combustion, environmental sciences, and rock physics in the USA, Mexico, Spain, and Italy. He is currently a contract Researcher at Istituto Nazionale di Geofisica e Vulcanologia, Italy. Particularly, he has received the Rittmann Medal (in 2014) from Associazione Italiana di Vulcanologia and Istituto Nazionale di Geofisica e Vulcanologia, which is assigned to the best young Italian volcanologist. His research interests focus on integrating theory, field, numerical modeling, experiments, and laboratory to study geological processes and products in volcanic areas (Vulcano, Vesuvio, Campi Flegrei, Etna, Colli Albani, Tenerife, Altiplano Puna, Colima) from fluid dynamic and natural hazard perspectives. Particularly, he studies pyroclastic energy currents, explosive volcanic eruptions, pre-eruptive conditions, debris flows, flow-building interactions, sand and dust storms, turbidity currents, man-made environmental phenomena and georesources. He has compiled the chapter on Campi Flegrei volcano for the European Catalogue of Volcanoes and has reviewed the 79 AD 'Pompeii' eruption of Vesuvius. He has edited the 1st CAJG proceedings on Petrogenesis and Exploration of the Earth's Interior, and 2nd CAJG on Recent Research on Geomorphology, Sedimentology, Marine Geosciences and Geochemistry. He is Chief Editor of the *Arabian Journal of Geosciences* responsible for evaluating submissions in the fields of geochemistry, mineralogy, petrology and volcanology.

Jasper Knight University of Witwatersrand, South Africa.
Professor of Physical Geography, University of the Witwatersrand, Johannesburg, South Africa.
I am a geoscientist with research interests in the spatial and temporal variability in landscape system responses to climate and environmental changes during the late Pleistocene and Holocene, looking at sediments and geomorphology. I focus thematically on glaciers, rivers, coasts and mountains. I focus geographically on the UK and Ireland, northwest USA, Australia, European Alps, various locations in Asia, and across both southern and northern (Saharan) Africa. I am the editor of three major international journals (Sedimentary Geology, Journal of Maps, Land Degradation and Development). My teaching and student supervision focuses on the interconnections between the physical and human environments, landscape change, and environmental resources of the past, present and future.

Anna Travé Universitat de Barcelona, Barcelona. Spain, Spain.
I graduated in Geological Sciences in 1987 and received my Ph.D. in 1992 with the thesis 'stratigraphy, Sedimentology and Geochemistry of Stromatolites of the South Pyrenean Eocene Basin'. During my Ph.D., I did three stays in the isotope laboratory in Liverpool (England) as well as other stays in the USA, England, France and Italy for work meetings. In 1993, I participated in a field campaign in the Salt Range (Pakistan) with the UNESCO-INTAS international correlation program 'Early Paleogene Benthos.' During 1994 and 1995, I had a postdoctoral position of the European Community at the University of Montpellier where I began my research in the field of fluid circulation and its relationship with tectonics. From 1996 to 2001, I get a postdoctoral contract at the University of Barcelona and several projects financed by the ITGE ('Instituto Geológico and Minero of Spain'), REPSOL, NORSK-HYDRO, ELF AQUITAINE and BP, to work on carbonate deposits.) At this time, I also participate in international correlation field campaigns, within the UNESCO-INTAS program, in Ukraine and Jamaica. Since 2002, I am professor at the University of Barcelona where in May 2019 I won the position of Full Professor in Sedimentary Petrology. The two main lines of research I have developed are: (1) sedimentology, petrology and sedimentary geochemistry and, (2) fluid geochemistry and the relationship between fluid and tectonic migration. I have participated in 20 R&D projects financed in public calls and 28 R&D contracts of special relevance with companies and/or administrations. I have presented 75 papers in international congresses and I have published more than 63 articles in international journals. I have been evaluator of on-demand research projects by the Research Council of the Katholieke Universiteit Leuven (Belgica), of research projects by ANECA, and of the Ramon y Cajal and Juan de La Cierva contracts. For years I have been a reviewer of articles from international journals such as *Tectonophysics*, *Chemical Geology*, *Oil and Gas Science and Technology*, Revue from the Institut français du pétrole, Geological Society of America GSA Bulletin, *Sedimentary Geology* and *Geologica Acta*. I have been a member of the Editorial Committee of Geologica Acta since 2004. The last big contracts with companies have been with EXXONMOBIL and Maersk Oil Qatar. From 2011 to 2021, I have supervised seven finished Ph.D. theses.

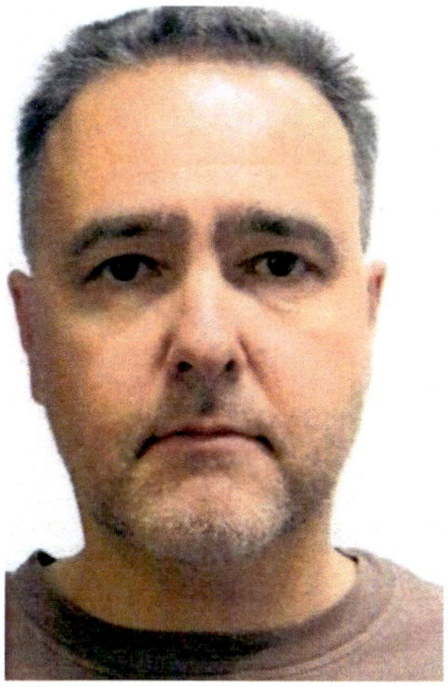

Stefan Grab School of Geography, Archaeology and Environmental Studies, University of the Witwatersrand, South Africa.
Swiss by birth but grew up in Pietermaritzburg, South Africa. Studied geomorphology at the University of Natal with specialization in periglacial geomorphology. Today works in both geomorphology and historical climate change sectors. Research regions include southern and eastern Africa, the mid-Atlantic and Iran (Middle East).

Amjad Kallel ENIS, University of Sfax, Sfax, Tunisia.
Dr. Amjad Kallel is currently an Associate Professor of Environmental Geology at the Sfax National School of Engineers at the University of Sfax, Tunisia. He holds a B.Eng. in Georesources and Environment (1998) from the University of Sfax (Tunisia) and an M.Sc. degree and a Ph.D. degree in Georesources and Environment (2004) from Hokkaido University (Japan). He joined Venture Business Laboratory (VBL) at Akita University, Japan (2005–2006), as a researcher focusing on refining and recycling technologies for the recovery of rare elements from natural and secondary sources. On his return to Tunisia, he worked at the University of Gabes from 2006 to 2011, where he contributed to the elaboration of teaching programs at the Higher Institute of Water Sciences and Technologies of Gabes. Since 2011, he has joined the Sfax National School of Engineers. There, he has also been involved in various research projects related to Environmental Geology and Environmental Geotechnics. He has co-organized many prestigious workshops, seminars, and international conferences. In 2016, he joined the *Arabian Journal of Geosciences* (Springer) and the *Euro-Mediterranean Journal for Environmental Integration* (Springer) as Chief Editor and Managing Editor, respectively.

Dr. Sandeep Panda is currently working as an Assistant Professor at the Department of Industrial Biotechnology, Gujarat Biotechnology University, Gujarat, India. His main research areas include (Bio) hydrometallurgical approaches for metal extraction from primary and secondary resources, Bio-desulphurization of Coal, Bio and Chemical approaches for Mine Water treatment and application of eco-friendly approaches for sustainable mineral-metal waste recycling and management. He has been involved as Principal/Co-Principal Investigator and as Team Member in several R&D and industrial research projects (at both national and international levels) since 2009. As of 2023, he has published 55 International journal papers; four book chapters and five Editorial books of high quality and impact that have received nearly 1950 citations (h-index—25). One of his articles published in the journal *NATURE* related to Tech-Metal supply is receiving much attention and appreciation from global leaders working in his area. His research works have invited the attention of several press and media. He has been featured in the WORLD RANKING OF THE TOP 2% SCIENTISTS in the Domain of Mining and Metallurgy (Sub Domain—Environmental Sciences) as per the Stanford University list of top-cited scientists in 2020, 2021, 2022. He is an active member of several prestigious professional bodies and technical/scientific committee member in several reputed international conferences. He has served as a reviewer in over 40 reputed International Journals and is currently serving as: (1) Associate Editor of the *Arabian Journal of Geosciences* and *Euro-Mediterranean Journal of Environmental Integration* (Springer Nature) and Editorial Board Member of: (2) *Frontiers in Microbiology* and *Frontiers in Environmental Sciences* (Frontiers Publications),

(3) *Mineral Processing* and *Extractive Metallurgy Review Journal* (Taylor & Francis Group) and (4) Editorial Advisory Board of *Detritus Journal* (IWWG).

Helder I. Chaminé School of Engineering (ISEP), Polytechnic of Porto, Portugal.
Helder I. Chaminé is a skilled Geologist and Professor of engineering geosciences at the School of Engineering (ISEP) of the Polytechnic of Porto, with over 32 years of experience in multidisciplinary geosciences research, consultancy and practice. He studied geological engineering and geology (B.Sc., 1990) at the Universities of Aveiro and Porto (Portugal), respectively. He received his Ph.D. in geology at the University of Porto in 2000 and spent his postdoctoral research in applied geosciences at the University of Aveiro (2001–2003). In 2011, he received his Habilitation (D.Sc.) in geosciences from the University of Aveiro. Before joining the academy, he worked for over a decade on international projects for mining, geotechnics, groundwater industry and/or academia related to geodynamics and regional geology, hard rock hydrogeology and water resources, engineering geosciences and applied geomorphology, rock engineering and georesources. His research interests span fundamental to applied fields: GIS mapping techniques for applied geology, structural geology and regional geology, engineering geosciences and rock engineering, slope geotechnics, mining geology and hydrogeomechanics, hard rock hydrogeology, exploration hydrogeology, urban groundwater and hydromineral resources. He is interested in mining geoheritage, history of cartography, military geosciences and higher-education dissemination, skills, and core values. Presently, he is Head of the Laboratory of Cartography and Applied Geology (LABCARGA | ISEP), Senior Researcher at Centre GeoBioTec | U.Aveiro and Centre IDL | U.Lisbon, as well as belongs to the executive board of the M.Sc. + B.Sc. Geotechnical and Geoenvironmental Engineering Programmes (OE + EUR-ACE Label) and the Department of Geotechnical Engineering (ISEP). He currently belongs to the board of the Technical Commission of Environmental Geotechnics from SPG. He was a Board Member of APGeom—Portuguese Association of Geomorphologists (2009–2013), SPG—Portuguese Geotechnical Society (2016–2020), APG—Portuguese Association of Geologists (2020–2021), and AIH-GP—Portuguese Chapter of the International Association of Hydrogeologists (2019–2022). He was a consultant and or responsible for over 75 projects of applied geology, hydrogeomechanics, slope geotechnics, mining geology, exploration hydrogeology, hard rock hydrogeology, water resources, urban groundwater and applied mapping (Mozambique, Portugal, and Spain). He has co-authored over 210 publications in indexed journal articles, conference proceedings/full papers, chapters, technical and professional papers. He co-edited over 15 special volumes and is presently involved in editing themed issues for some international journals (e.g., *Environmental Earth Sciences*—Springer, *Springer Nature Applied Sciences*, *Discover Water*—Springer, *Arabian Journal of Geosciences*—Springer,

Water—MDPI). As a referee for several international journals in applied mapping, geosciences, geotechnics, hydrogeology, water resources, and geohazards, he has a broad activity. He served as Invited Expert Evaluator of Bologna Geoscience Programme for DGES (Portugal) and Scientific Projects Evaluation for NCST, 2017–2019 (Kazakhstan), and NRF | RISA, 2019 (South Africa), as well as Coordinator of 'Geology on Summer | Ciência Viva' Programme at ISEP (2005–2019) for geoscience dissemination. He has also been active in teaching and supervising many Ph.D., M.Sc. and undergraduate students. He has been on the editorial board, among others, of the *Arabian Journal of Geosciences* (SSG + Springer), *Hydrogeology Journal* (IAH + Springer), *Euro-Mediterranean Journal for Environmental Integration* (Springer), *Springer Nature Applied Sciences* (Springer), *Mediterranean Geoscience Reviews* (Springer), *Discover Water* (Springer), *Geotechnical Research* (ICE), *Geosciences* (MDPI), *Revista Geotecnia* (Portugal) and *Geología Aplicada a la Ingeniería y al Ambiente* (Argentina). He integrates as Moderator or Session Chair in several conferences, workshops, and meetings. He was the Scientific Chair of the 1st International Conference on Georesources, Geomaterials, Geotechnologies and Geoenvironment, 4GEO (Porto, Portugal, November 7–8, 2019). He is currently on the organizing and scientific committees of the 3rd International Workshop on Natural Hazards—NATHAZ'22 (Terceira Island, Azores, May 2022), supported by Springer.

Jesús Rodrigo-Comino University of Granada, Granada, Germany.
Dr. Jesús Rodrigo-Comino, graduated in Geography, currently works as Assistant Professor at the University of Granada. He carried out a thesis with an extraordinary degree award with the work: 'The soils of the province of Malaga: Revision according to the classification criteria of FAO-WRB (2006).' He holds a Master's Degree in Territorial Planning and Geographic Information Systems (2013) from the University of Granada, whose final work was divided into three national publications and a monograph. During his predoc stage, he obtained three scholarships for doctoral studies: DAAD (German Academic Exchange Service), La Caixa Foundation and FPU (Ministry of Education, Spain). During this period, he completed his doctoral thesis between the University of Trier (Germany -2 years-) and Malaga (2 years). Doctor in Geography (2018) from the University of Malaga with a doctoral thesis consisting of eight international publications, international mention and outstanding cum laude: 'Actual geomorphological processes in sloping vineyards. A comparison between Ruwer-Mosel Valley (Trier, Germany) and Montes de Málaga (Málaga, Spain).' Currently, he is also preparing a second doctorate in Geomatics engineering at the Polytechnic University of Valencia. His research career consists of four full monographs (Nova, Springer, etc.) and one edited alone (Elsevier), 171 indexed publications (Scopus)/130 JCR, leading international collaborations with research teams from Iran, China, Kazakhstan, Sudan, India, Brazil, Croatia, Iraq or the USA.

Regular reviewer in more than 140 international indexed journals (e.g., *Scientific Reports*, *Science of the Total Environment*, *PlosOne*, *Catena*, *Geoderma*, *Agriculture*, *Ecosystem and Environment or Earth-science Review*), member of the panel of two doctoral theses, evaluator of projects for the Ministries of Science of Chile, Peru, the USA, Serbia, Switzerland, Kazakhstan or Poland, and postgraduate scholarships for DAAD. He has organized several scientific meetings and congresses (e.g., Action Cost, V Biohydrology, Fire in the Earth, etc.), sessions at international conferences (EGU, TerraEnvision, Conference of the *Arabian Journal of Geosciences*, etc.), oral presentations and conferences magisterial (Germany, Bulgaria, Norway, etc.). Editor-in-chief of the indexed journal (Scopus and ESCI; Q2) *Air Soil and Water Research* (SAGE). In addition, he is an associate editor at *Scientific Reports* (Nature-Springer), *Hydrological Science Journal* (Taylor & Francis), *Arabian Journal of Geosciences* (Springer), *Euro-Mediterranean Journal for Environmental Integration* (Springer), and *Journal of Mountain Science* (Springer). He has participated as a researcher in R&D&I projects on social issues related to housing or the census, or transfer and knowledge at European level such as the INTERREG Smart-Light HUB project (light pollution) or COST FIRElinks (fires). He has been invited to give lectures on agriculture, sustainable management and erosion in several Universities. He has supervised five final degree projects and three completed master's degrees in Germany. He has taught regulated and certified teaching at the Universities of Granada, Valencia, Malaga, León, Oviedo and Trier (in German) on development, geomorphology, Geographic Information Systems, remote sensing and statistical techniques. He was recently awarded with a Leonardo Grant (BBVA Foundation).

Sami Khomsi Georesources Lab, CERTE, University of Carthage, Tunis, Tunisia.

Sami Khomsi is a structural-petroleum geologist. He got his Ph.D. at Tunis University in 2004 and his Habilitation à Diriger des Recherches (HDR) at the same university in 2015. Since 2011, he works in the Department of petroleum geology and sedimentology, and in the Geoexploration Department in the Faculty of Earth Sciences, King Abdulaziz University, Jeddah, Saudi Arabia. He worked also in different universities in Tunisia and France. Past employment within academia (research and teaching) and industry in the exploration activities: involved in many international programs, regional exploration assets, regional synthesizes, sites survey, prospects evaluation and professional expert training (field trip leader for professionals, coaching of geoscientists) in basin analysis involving basin modeling, structural analysis and balanced cross sections. He collaborated with major oil companies in Eastern Maghreb on field studies, structural analysis of fold–thrust structures, subsurface reservoir characterizations and seismic interpretations. He has also experience in geophysical methods involving fast processing and interpretation of seismic reflexion sections in terms of structural geology-investigation-evaluation of plays, characterization of fractured reservoirs and subsurface mapping. His main research interests: geometry, kinematics and dynamics of fold–thrust systems and foreland basins in North Africa, Middle East and the Mediterranean realm, fractured reservoir analogues, applying structural analysis and subsurface techniques in the characterizations of drillable structures and structural evaluation of traps using regional cross sections at large-scale as well as structural subsurface based on regional seismic reflection profiles in different basins. https://www.scopus.com/authid/detail.uri?authorId=16175410900.

Santanu Banerjee Indian Institute of Technology Bombay, Mumbai, India.
Santanu Banerjee currently works as a Professor at the Department of Earth Sciences, Indian Institute of Technology Bombay. He does research in sedimentology and petroleum geology. His research areas include authigenic iron silicates, microbially induced sedimentary structures, basin analysis, provenance interpretations and alternate potash fertilizer. He is on the Editorial board of *Journal of Palaeogeography*, *Minerals*, *Arabian Journal of Geosciences* and *Journal of Earth Systems Science.*

Broder Merkel TUBAF, Freiberg, Germany.
• Dissertation (Ph.D.) Dr. rer.nat. as Hydrogeologist (TU München) • Habilitation (Dr. habil) Christian-Albrechts Universität Kiel • Full Professor for Hydrogeology at TUBAF (1993–2015) Hydrogeology, hydrochemistry, modeling chemical thermodynamics and reactive transport, geo-statistics, GIS and remote sensing • Dean of the Faculty Geoscience, Geotechnics and Mining • DFG liaison officer of TUBAF • Vice-rector of TUBAF • Head of the Scientific Diving Center of TUBAF • Fieldwork in Bolivia, China, Chile, Czech Republic, Iraq, Iran, Israel, Italy, Jordan, Hungary, Mexico, Mozambique, Namibia, Palestine, Russia, Spain, USA • Organization of seven International Conferences Uranium Mining and Hydrogeology.

Haroun Chenchouni Higher National School of Forests, Khenchela, Algeria.
Dr. Haroun Chenchouni is an associate professor and research scientist (Ecologist) at the Higher National School of Forests (Khenchela, Algeria). He is a former associate professor at the University of Tebessa (Algeria). He holds a M.Sc. (Magister) in Dryland Ecology from the University of Ouargla (Algeria) and a doctorate degree in Ecology and Environment from the University of Batna. He graduated as an engineer in Plant Ecology and Forest Ecosystems from the Department of Biological Sciences (University of Batna, Algeria). His research interests are fairly broad; he uses statistical modeling approaches to understand how natural environments, mainly climatic and edaphic factors, and anthropogenic perturbations

influence biological interactions, shape trends in population dynamics, and influence community diversity. He uses various biological models to investigate biological interactions and community ecology of arid and semiarid ecosystems of North Africa. At various universities in Algeria, he teaches forest ecology, biostatistics, and ecological modeling. He has published more than 100 peer-reviewed publications and internationally recognized research papers. He is also involved in national and international research projects. In 2017, he joined the Arabian Journal of Geosciences (AJGS) as an associate editor. Then in 2019, he was assigned as a chief editor of Topic 2 (biogeochemistry, geobiology, geoecology, geoagronomy) to handle submissions dealing with various fields of biogeosciences, geoecology, climate change, plant and soil science, agricultural and forest environment, and environmental sciences.

Geochemistry, Mineralogy, Petrology, Volcanology

Documenting the Partial Melting of Gabbroic Rocks: A Natural Example from the Alvand Plutonic Complex (Northern Sanandaj-Sirjan Zone, West Iran)

Federico Lucci, Adel Saki and Mirmohammad Miri

Abstract

The Alvand Complex in Hamedan area (NW-Iran) represents a key area to document the partial melting of a gabbroic protolith. In this area, where the hornblende-gabbros are in contact with coeval to younger olivine-gabbros and pyroxenites, they are characterized by the diffuse presence of trondhjemite leucosomes veins. Equilibrium batch melting modelling indicates trondhjemites as the products of 2–5% degree of partial melting of the hornblende-gabbro. Thermobaric estimates indicate T 720 °C and P 3.5 kbar for the anatectic event.

Keywords

Gabbro partial melting · Trondhjemite · Alvand complex · Sanandaj-Sirjan zone · Iran

1 Introduction

The anatexis (or partial melting) is widely accepted as one of the most important processes in the formation and growth of continental crust (Gao et al., 2016), due to the generation of up to great volumes of intermediate to felsic melts (Gao et al., 2016; Rossetti et al., 2020). Even if a consistent number of experimental studies explored the partial melting of mafic rocks (Gao et al., 2016), the documented evidence in nature is still sparse. In this view, the Jurassic Alvand plutonic complex in the Hamedan area (Fig. 1a; northern Sanandaji-Sirjan Zone, western Iran) represents an important case study, where migmatization of mafic rocks is documented (Saki et al., 2020). In the northern sector of the Alvand complex (Cheshmeh-Ghasaban, Fig. 1b), hornblende-gabbros are in contact with olivine-gabbros and pyroxenites (Saki et al., 2020) and exhibit felsic leucosomes in the shape of leucocratic drops and pods, veins, layers, and net-structure veinlets (Fig. 1c).

2 Methods

Fifteen samples from the Cheshmeh-Ghasaban gabbros and from leucosome veins were selected for optical petrographic investigations and whole rock analyses. The samples were analysed using a Perkin Elmer Sciex ELAN 6000 ICP/MS instrument in Activation Laboratories (Canada).

3 Results

The Cheshmeh-Ghasaban gabbros show a primary assemblage of labradoritic plagioclase (30–40%), Ca-hornblende (up to 35%), augitic clinopyroxene (15–25%), biotite (up to 15%), rare orthopyroxene (<5%), and rare quartz (<5%). The leucosomes are dominated by oligoclase plagioclase (up to 65%), quartz (up to 35%), and rare K-feldspar (<10%). Rare crystals of clinopyroxene are identified. Remnants of hornblende, biotite, and pyroxenes from the protolith are also observed.

The gabbro protolith shows SiO_2 50.4 wt%, with Al_2O_3 17.3 wt%, FeO* 6.7 wt%, CaO 12.3 wt%, low TiO_2 1.1 wt%, and low MgO 7.5 wt%. The Na_2O (2.1 wt%) content is higher than K_2O (0.8 wt%). The relatively low MgO content excludes this gabbro are primitive melts (Azizi

F. Lucci
Dipartimento di Scienze, Università degli Studi Roma Tre, Roma, Italy

F. Lucci (✉)
Dipartimento di Scienze della Terra e Geoambientali, Università degli Studi di Bari “Aldo Moro”, Bari, Italy
e-mail: federico.lucci@uniba.it

A. Saki · M. Miri Department of Geology, Faculty of Earth Sciences, Shahid Chamran University of Ahvaz, Ahvaz, Iran

F. Lucci et al. (eds.), *Selected Studies in Geomorphology, Sedimentology, and Geochemistry*, Advances in Science, Technology & Innovation, https://doi.org/10.1007/978-3-031-43744-1_1

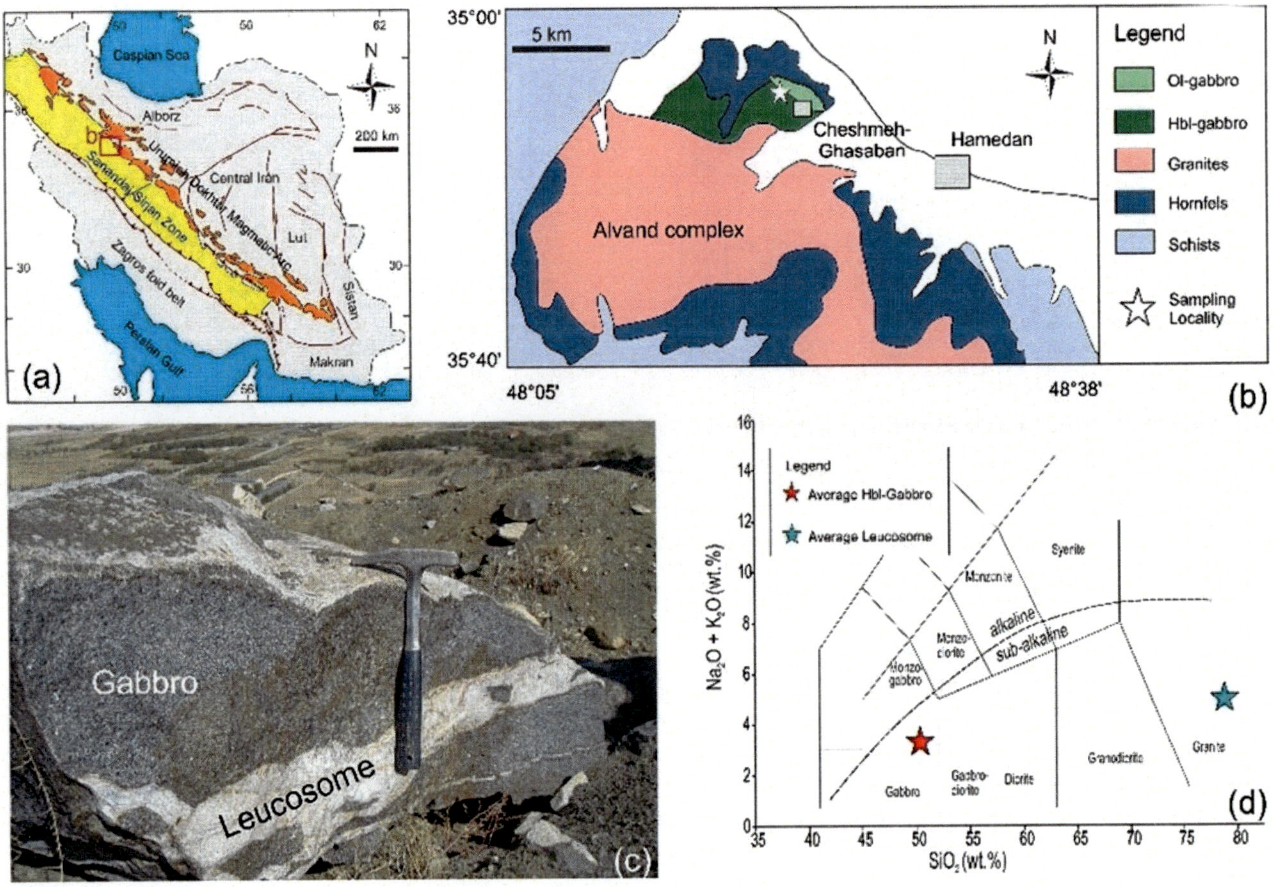

Fig. 1 **a** geological map of Iran (Saki et al., 2020) showing the location of the Alvand complex (red square). **b** geological map of the Alvand Complex (Saki et al., 2020). The white star indicates the sampling locality. **c** Field evidence of mafic migmatites with leucosome veins and patches in the hornblende-bearing gabbro. **d** Total alkali versus silica (TAS) diagram indicative of average protolith gabbro and average leucosome compositions

et al., 2018). Gabbro shows Sr 515 ppm and Y 15 ppm. The leucosomes have very high-SiO_2 (78.1 wt%) felsic composition associated with low MgO (<0.3 wt%), low FeO* (<1.1 wt%), low TiO_2 (<0.2 wt%), and low CaO (ca. 2 wt%). They show Al_2O_3 content of ca.12–13 wt%, and they are strongly enriched in Na_2O (up to 5 wt%) with respect to K_2O (<1.3 wt%). The Sr, Y, and Zr average contents are 198 ppm, 8 ppm, and 62 ppm, respectively. On the TAS diagram, samples plot in the fields of "Gabbro" and "Granite" (Fig. 1d), respectively. Concerning leucosomes, based on the estimated modal abundances, together with the normative An-Ab-Or components, they are furthermore classified as trondhjemites (Fig. 2a), comparable to those from Trondheim Region (Nilsen et al., 2003).

4 Discussion

Equilibrium (or Modal) batch melting (EBM) model is used to test the hypothesis that trondhjemitic leucosomes represents the product of in situ partial melting of the Cheshmeh-Ghasaban hornblende-gabbro. The EBM major-element mass balance modelling (Lucci et al., 2016; Moghadam et al., 2020; White et al., 2009) was calculated using our preliminary bulk compositions integrating mineral chemistry from literature (Saki et al., 2020). It shows that trondhjemite melts reflect ca. 2% partial melting of the gabbro protolith with a residual assemblage of plagioclase, hornblende, and clinopyroxene ($Pl_{45}+Hbl_{40}+Cpx_{15}$). Based on EBM major-element solution, we approached the EBM modelling in the Sr-Y system. Phase/element partition coefficients for a gabbroic source were selected from literature through the GERM K_D database (GERM Homepage, 2020). The results obtained for trace elements are consistent with those from major-elements indicating that trondhjemite leucosomes were produced by 2–5% partial melting of the hornblende-gabbro protolith.

Further constraints on the partial melting event are derived from (i) Zircon-saturation thermometry (T_{Zr}) (Watson & Harrison, 1983) for leucosome and (ii) Ti-in-hornblende thermometry (T_{Hbl}) (Otten, 1984) and Al-in-barometry (P_{Hbl}) (Schmidt, 1992) for amphiboles in

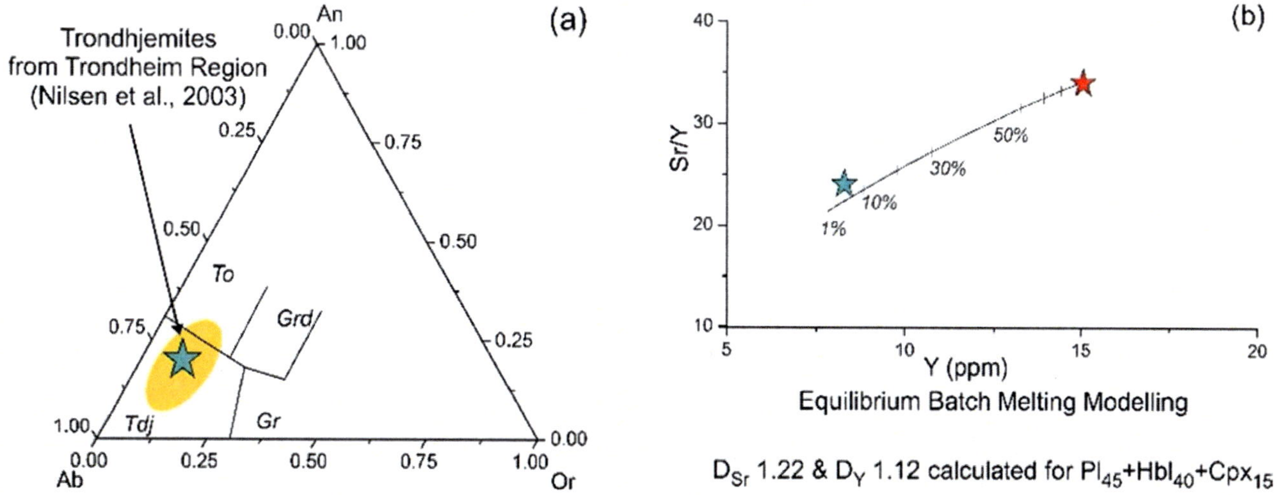

Fig. 2 **a** Normative An-Ab-Or diagram. **b** Results from equilibrium batch melting modelling. Percentages of melt produced are indicated

mesosomes (i.e. metagabbro) at contact with leucosomes (Saki et al., 2020). Obtained thermobaric estimates indicate that the gabbro partial melting occurred at ca. 720°C ($T_{Zr}=T_{Hbl}$) in the upper crust (P_{Hbl} ca. 3.5 kbar). Since leucosomes are found only where hornblende-gabbro are in contact with the coeval to younger (Yang et al., 2018) olivine-gabbros and pyroxenites, these latter ones are suggested as a possible heat source that triggered the partial melting event (Lucci et al., 2021; Saki et al., 2022).

5 Conclusions

The main outcomes of this work are: (i) documenting the partial melting of a hornblende-gabbro and the genesis of trondhjemite melts in Hamedan area of Iran, (ii) quantify the partial melting degree through EBM modelling, and (iii) defining the P–T conditions of the partial melting event.

Funding This research is part of the corresponding author Postdoc programme at the Roma Tre University. This research was supported by PRIN2017 Project 20177BX42Z_001 (Intraplate deformation, magmatism, and topographic evolution of a diffuse collisional belt: Insights into the geodynamics of the Arabia-Eurasia collisional zones) funded by MIUR-Italy. The research was also supported by Shahid Chamran University of Ahvaz (grants no. EG98.44295 and SCU. EG98.44295). FL gratefully acknowledges the grant to Department of Science, Roma Tre University (MIUR-Italy Dipartimenti di Eccellenza, ARTICOLO 1, COMMI 314–337 LEGGE 232/2016).

References

Azizi, H., Nouri, F., Stern, R. J., Azizi, M., Lucci, F., Asahara, Y., Zarinkoub, M. H., & Chung, S. L. (2018). New evidence for Jurassic continental rifting in the northern Sanandaj-Sirjan zone, western Iran: The Ghalaylan seamount, southwest Ghorveh. *International Geology Review*. https://doi.org/10.1080/00206814.2018.1535913

Gao, P., Zheng, Y. F., & Zhao, Z. F. (2016). Experimental melts from crustal rocks: A lithochemical constraint on granite petrogenesis. *Lithos, 266*, 133–157.

GERM Homepage. https://earthref.org/GERM/. Accessed August 26, 2020.

Lucci, F., Rossetti, F., White, J. C., Moghadam, H. S., Shirzadi, A., & Nasrabady, M. (2016). Tschermak fractionation in calc-alkaline magmas: The Eocene Sabzevar volcanism (NE Iran). *Arabian Journal of Geosciences*, *9*(10):573, 1–12.

Lucci, F., Saki, A., Miri, M., Rabiee, A., & White, J. C. (2021). Genesis of trondhjemite by low-pressure low-melt fraction anatexis of hornblende-gabbro at Alvand Plutonic complex (Hamedan, NW Iran): Insights from geochemical modelling. *Arabian Journal of Geosciences, 14*(17), 1–20.

Moghadam, H. S., Li, Q. L., Griffin, W. L., Stern, R. J., Ishizuka, O., Henry, H., Lucci, F., O'Reilly, S. Y., & Ghorbani, G. (2020). Repeated magmatic buildup and deep "hot zones" in continental evolution: The Cadomian crust of Iran. *Earth and Planetary Science Letters, 531*, 115989.

Nilsen, O., Sundvoll, B., Roberts, D., & Corfu, F. (2003). U-Pb geochronology and geochemistry of trondhjemites and a norite pluton from the SW Trondheim Region, Central Norwegian Caledonides. *Norges Geologiske Undersokelse, 441*, 5–16.

Otten, M. T. (1984). The origin of brown hornblende in the Artfjället gabbro and dolerites. *Contributions to Mineralogy and Petrology, 86*(2), 189–199.

Rossetti, F., Lucci, F., Theye, T., et al. (2020). Hercynian anatexis in the envelope of the Beni Bousera peridotites (Alboran Domain, Morocco): Implications for the tectono-metamorphic evolution of the deep crustal roots of the Mediterranean region. *Gondwana Research, 83*, 157–182.

Saki, A., Miri, M., & Oberhansli, R. (2020). High temperature—Low pressure metamorphism during subduction of Neo-Tetysbeneat the Iranian plate: evidence for mafic migmatite formation in the Alvand Complex (western Iran). Mineralogy and Petrology (in press).

Saki, A., Lucci, F., Miri, M., & White, J. C. (2022). Trondhjemite leucosomes generated by partial melting of a hornblende-gabbro (Alvand plutonic complex, Hamedan, NW Iran). *International Geology Review, 64*(5), 597–630.

Schmidt, M. W. (1992). Amphibole composition in tonalite as a function of pressure: An experimental calibration of the Al-in-hornblende barometer. *Contributions to Mineralogy and Petrology, 110*(2–3), 304–310.

Watson, E. B., & Harrison, T. M. (1983). Zircon saturation revisited: Temperature and composition effects in a variety of crustal magma types. *Earth and Planetary Science Letters, 64*(2), 295–304.

White, J. C., Parker, D. F., & Ren, M. (2009). The origin of trachyte and pantellerite from Pantelleria, Italy: Insights from major element, trace element, and thermodynamic modelling. *Journal of Volcanology and Geothermal Research, 179*(1–2), 33–55.

Yang, T. N., Chen, J. L., Liang, M. J., Xin, D., Aghazadeh, M., Hou, Z. Q., & Zhang, H. R. (2018). Two plutonic complexes of the Sanandaj-Sirjan magmatic-metamorphic belt record Jurassic to early cretaceous subduction of an old Neotethys beneath the Iran microplate. *Gondwana Research, 62*, 246–268.

The Twihinate–Lamlaga Carbonatite Complex: A Potential Source of REE, Nb, and Phosphates at the NW Margin of the West African Craton (Dakhla Province, Morocco)

Cheikh Elwali Malainine, Muhammad Ouabid, Fleurice Parat, Otmane Raji, Jean-Louis Bodinier, Nasrrddine Youbi, Moulay Ahmed Boumehdi, Jean-Marie Dautria, Carlos J. Garrido and Hicham El Messbahi

Abstract

Carbonatite-alkaline complexes and their weathering products are the worldwide primary source of strategic resources such as rare earth elements (REE) and igneous phosphates. These commodities are indispensable for many high-technology and industrial applications, which make carbonatite-alkaline complexes important targets for geological exploration. In this work, we focus on the Twihinate–Lamlaga carbonatite complex of the Cretaceous age in the Dakhla province at the NW margin of the West African Craton. We discuss the petrogenetic implications of carbonatite and associated REE-Nb-P mineralization using field, petrological, and geochemical data. Two distinct calcio-carbonatites (white and grey) have been identified. These carbonatites are associated with apatite-rich rocks and iron oxide–silica breccia rocks, which cap almost totally the complex. White carbonatite is formed of Sr- and light REE-rich calcite and apatite with abundant phlogopite. The grey carbonatite appears as evolved facies and locally intrudes the white carbonatite. It is composed of calcite, heavy REE-rich apatite, Fe-oxide/hydroxides with minor quartz, and REE-Nb-bearing minerals (mainly monazite-(Ce), synchysite-(Ce), and pyrochlore). However, the iron oxide rocks and silica breccia (mostly quartz) contain the highest REE and Nb concentrations, associated with Th-poor monazite (<150 ppm) and pyrochlore. Besides, the apatite-rich rocks have up to 29 wt.% P_2O_5 and host appreciable heavy REE and Y concentrations. The magmatic evolution and further hydrothermal/weathering processes were likely behind the observed critical metal and phosphate enrichment.

Keywords

Carbonatite · Twihinate–Lamlaga carbonatite complex · REE · Nb and apatite deposits · Dakhla province · Morocco

C. E. Malainine (✉) · M. Ouabid · O. Raji · J.-L. Bodinier
Mohammed VI Polytechnic University, Geology and Sustainable Mining Institute, Benguerir, Morocco
e-mail: cheikh-elwali.malainine@um6p.ma

C. E. Malainine · N. Youbi · M. A. Boumehdi
Faculty of Sciences Semlalia, Department of Geology, Cadi Ayyad University, Marrakech, Morocco

F. Parat · J.-L. Bodinier · J.-M. Dautria
University of Montpellier, Geoscience Montpellier, Montpellier, France

C. J. Garrido
Instituto Andaluz de Ciencias de La Tierra (IACT), CSIC–UGR, Granada, Spain

H. E. Messbahi
Sidi Mohamed Ben Abdallah University, Polydisciplinary Faculty, Taza, Morocco

1 Introduction

Carbonatites are the primary igneous rocks hosting significant strategic metal concentrations, especially rare earth elements (REE) and Nb, which are vital for emerging industries and green technologies (Jones et al., 1995; Mariano, 1989a; Verplanck et al., 2016). The magmatic and weathering processes within the carbonatite complexes often lead to economically valuable accumulations of phosphate deposits, which have been mined to produce fertilizers and critical metals, e.g. in Brazil, Russia, South Africa,

F. Lucci et al. (eds.), *Selected Studies in Geomorphology, Sedimentology, and Geochemistry*, Advances in Science, Technology & Innovation, https://doi.org/10.1007/978-3-031-43744-1_2

Zimbabwe, and Finland (Mariano, 1989b; O'Brien et al., 2015; Pufahl & Groat, 2017; Walter et al., 1995). These aspects make carbonatite-related deposits highly sought to satisfy the growing need for fertilizers and strategic metals.

In the southern part of Morocco, three main carbonatite occurrences have been reported recently: (1) the Paleoproterozoic magnesio-carbonatites of Gleibat Lafhouda complex are the best exposed and known to host iron oxide-apatite ores (Benaouda et al., 2020; Malainine et al., 2022a; Montero et al., 2016); (2) the Aghracha carbonatite of unknown age consists of 0.5–1.5 m-thick dykes within the Ediacaran basement, and (3) the Twihinate–Lamlaga Cretaceous carbonatites, the subject of this study. Restricted investigations of the National Office of Hydrocarbons and Mines (ONHYM) have reported significant REE and Nb ores related to the Twihinate–Lamlaga carbonatites (ONHYM, 2016). Therefore, this contribution attempts to better understand the petrogenetic implications of the Twihinate–Lamlaga carbonatite complex and related mineralization and discuss their metallogenetic potential.

2 Geological Setting

The Twihinate and Lamlaga Cretaceous carbonatites (ca. 104 ± 4 Ma) intruded into the Siluro–Devonian granitic gneisses (ca. 421–410 Ma) on the western part of the Oulad Dlim massif (Montero et al., 2016). The Twihinate intrusion is characterized by an annular structure (~3 km in diameter) with a central core surrounded by a thick ring (Fig. 1a). At 5 km further to the North from the Twihinate, outcrops the Lamlaga structure with its annular shape. Both structures are dominated by massif hard iron oxide cap rocks and brownish silica breccia (Malainine et al., 2022b). The carbonatite outcrops were observed only in the central core of the Twihinate structure, whereas they were concealed in the Lamlaga structure, as confirmed through core drilling (ONHYM, 2016). Our field and petrographic observations show that the carbonatites occur as grey and white facies. They are associated with apatite-rich rocks that appear as an accumulation of altered apatite within the carbonatite body.

3 Materials and Methods

About 50 rock samples were collected from both Twihinate and Lamlaga structures for petrographic studies. Whole-rock geochemistry was determined by X-ray fluorescence at the Instituto Andaluz de Ciencias de la Tierra (IACT, Granada, Spain) and inductively coupled plasma mass spectrometry (ICP-MS) (Agilent 7700x) at the Géosciences Montpellier laboratory (France). In situ major and trace element compositions of the main mineral phases were performed using the Cameca SX100 electron probe microanalyzer and Laser Ablation–ICP-MS at the Géosciences Montpellier laboratory.

4 Results

4.1 Petrography

The petrographic characteristics of the different facies from the Twihinate–Lamlaga carbonatite complex are summarized in the following table:

Facies	Major phases	Accessory minerals	Ore bearing minerals
White carbonatite	Calcite (90–95 vol.%) Apatite (~4 vol.%) Phlogopite (~2 vol%)	Iron oxides	–

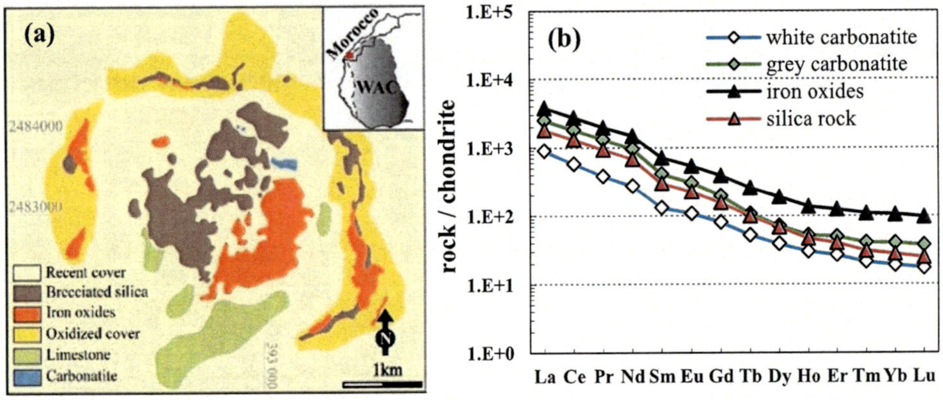

Fig. 1 Simplified geological map of Twihinate structure showing the principal facies encountered at the Twihinate–Lamlaga carbonatite complex **a**, and their chondrite-normalized REE patterns **b**

Facies	Major phases	Accessory minerals	Ore bearing minerals
Grey carbonatite	Calcite (>70 vol.%) Apatite (1–8 vol. %) Iron oxides (~10%)	Quartz, barite, pyrite, K-feldspars	Pyrochlore, synchysite, and monazite
Iron oxides	Goethite (75–85 vol. %) Hematite (10–20 vol.%) Quartz (~5 vol.%)	Apatite, magnetite, calcite	Pyrochlore, synchysite, and monazite
Silica breccia	Quartz (~90 vol.%) Iron oxides (5–10 vol.%)	Calcite, apatite, and barite	Pyrochlore, synchysite, and monazite
Apatite-rich rock	Apatite (up to 80 vol.%) Iron oxides (~10 vol.%)	Calcite and quartz	Pyrochlore

4.2 Whole-Rock and Mineral Chemistry

All the studied carbonatites of the Twihinate–Lamlaga complex are calcio-carbonatites, as stated previously (Benaouda et al., 2020; Montero et al., 2016). Their chondrite-normalized REE patterns show high REE concentration (1000–10,000 times chondrite) with significant REE enrichment of grey carbonatite relative to white carbonatite (Fig. 1b). However, the associated iron oxide rocks show the highest REE and Nb concentrations (up to 1.2 wt.% REE and 0.5 wt.% Nb), whereas the silica rocks display quite elevated REE contents (up to 3884 ppm REE). Overall, all these rock types have rather similar and parallel REE profiles and present light REE enrichment over the heavy REE with $La_N/Yb_N \sim 25–78$ (Fig. 1b). On the other hand, apatite-rich rocks are characterized by appreciable Nb and Y, and heavy REE concentrations (up to ~0.2 wt.% HREE+Y and 0.22 wt.% Nb).

Trace element compositions of apatite and calcite, as major phases within Twihinate–Lamlaga carbonatites, show distinct variations. In detail, apatite in the white carbonatite facies shows typical chemical features of worldwide carbonatitic apatite with light REE, Sr, and Mn enrichments relative to that of grey carbonatite. Calcite within white carbonatite is REE-Sr-Ba-rich without remarkable anomaly, whereas REE-depleted with well-expressed negative Ce anomaly within the grey carbonatite.

The REE and Nb mineralization has been observed in the studied carbonatites and the associated rocks as well. The ore-bearing minerals occur as fine-grained aggregates scattered in the rock and filling quartz-rich veins. They are dominated by monazite-(Ce) and synchysite-(Ce), which are the most REE-bearing (~51–58 wt.% light REE_2O_3), and pyrochlore (Nb_2O_5 ~48–60 wt.%) as the principal Nb-bearing mineral.

5 Discussions and Conclusions

Our petrography and whole-rock–mineral chemistry data show that the Twihinate–Lamlaga white carbonatite seemed to be formed during the early magmatic stage. The mineral paragenesis is dominated by light REE-rich calcite and apatite, which present a range of trace elements similar to the magmatic worldwide carbonatites (Belousova et al., 2002). On the other hand, the grey carbonatite is characterized by heavy REE-rich apatite and appears more evolved and enriched in REE linked mostly to monazite-(Ce) and synchysite-(Ce), attesting of their late stage evolution which often leads to such high REE enrichment in carbonatite complexes (Chakhmouradian & Wall, 2012).

The associated iron oxide and silica breccia cap rocks, as well as the apatite-rich rocks, present quite similar geochemical features with carbonatites which likely attest their strong genetic link (Fig. 1b). In terms of petrogenetic implications, hydrothermal to weathering processes are suspected of inducing primary leaching of carbonatite carbonates and their progressive replacement by oxide/hydroxides (Malainine et al., 2023). Apatite, as a resistant mineral to weathering, subsequently results in residual apatite accumulations. This might lead to the formation of cap rocks similar to those observed in the Twihinate–Lamlaga carbonatite complex (Lottermoser, 1990).

References

Belousova, E. A., Griffin, W. L., O'Reilly, S. Y., & Fisher, N. I. (2002). Apatite as an indicator mineral for mineral exploration: Trace-element compositions and their relationship to host rock type. *Journal of Geochemical Exploration, 76*(1), 45–69. https://doi.org/10.1016/S0375-6742(02)00204-2

Benaouda, R., Kraemer, D., Sitnikova, M., Goldmann, S., Schwarz-Schampera, U., Errami, A., Mouttaqi, A., & Bau, M. (2020). Discovery of high-grade REE-Nb-Fe mineralization associated with calciocarbonatite in south Morocco. *Ore Geology Reviews* 124, 103631. https://doi.org/10.1016/j.oregeorev.2020.103631

Benaouda, R., Kraemer, D., Sitnikova, M., Goldmann, S., Freitag, R., Bouali, A., Mouttaqi, A., El Haloui, R., Essaadaoui, M., & Bau, M. (2020). Thorium-poor monazite and columbite-(Fe) mineralization in the Gleibat Lafhouda carbonatite and its associated iron-oxide-apatite deposit of the Ouled Dlim Massif, South Morocco. *Gondwana Research, 77*, 19–39. https://doi.org/10.1016/j.gr.2019.06.011

Chakhmouradian, A. R., & Wall, F. (2012). Rare earth elements: Minerals, mines, magnets (and more). *Elements, 8*(5), 333–340. https://doi.org/10.2113/gselements.8.5.333

Jones, A. P., Wall, F., & Williams, C. T. (1995). *Rare earth minerals: Chemistry, origin and ore deposits* (Vol. 7). Springer Science & Business Media.

Lottermoser, B. G. (1990). Rare-earth element mineralisation within the Mt. Weld carbonatite laterite, Western Australia. *Lithos, 24*(2), 151–167. https://doi.org/10.1016/0024-4937(90)90022-S

Malainine, C.-E., Raji, O., Ouabid, M., Khouakhi, A., Bodinier, J. L., Laamrani, A., El Messbahi, H., Youbi, N., & Boumehdi, M. A. (2022a). An integrated ASTER-based approach for mapping carbonatite and iron oxide-apatite deposits. *Geocarto International, 37*, 6579–6601. https://doi.org/10.1080/10106049.2021.1953617

Malainine, C.-E., Raji, O., Ouabid, M., Bodinier, J.-L., & El Messbahi, H. (2022b). Prospectivity mapping of carbonatite-associated iron oxide deposits using an integration process of ASTER and Sentinel-2A multispectral data. *International Journal of Remote Sensing, 43*, 4951–4983. https://doi.org/10.1080/01431161.2022.2121189

Malainine, C.-E., Ouabid, M., Raji, O., Bodinier, J.-L., Parat, F., El Messbahi, H., Dautria, J.-M., Jourani, E.-S., Yazami, O.K., & Garrido, C.J. (2023). Origin of Cretaceous alkaline annular structures in the peri-cratonic terranes of the Reguibat Rise in Morocco: New constraints from carbonatite–silicate pairs and subsolidus processes. *Chemical Geology, 632*, 121536. https://doi.org/10.1016/j.chemgeo.2023.121536

Mariano, A. N. (1989). Nature of economic mineralization in carbonatites and related rocks. In *Carbonatites: Genesis and evolution* (pp. 149–176). Unwin Hyman.

Mariano, A. N. (1989a). Economic geology of rare earth minerals. *Mineralogist Society America, 21*, 309–337.

Montero, P., Haissen, F., Mouttaqi, A., Molina, J. F., Errami, A., Sadki, O., Cambeses, A., & Bea, F. (2016). Contrasting SHRIMP U-Pb zircon ages of two carbonatite complexes from the peri-cratonic terranes of the Reguibat shield: Implications for the lateral extension of the West African Craton. *Gondwana Research, 38*, 238–250. https://doi.org/10.1016/j.gr.2015.12.005

O'Brien, H., Heilimo, E., & Heino, P. (2015). The Archean Siilinjärvi carbonatite complex. In *Mineral deposits of Finland*, (pp. 327–343). Elsevier. https://doi.org/10.1016/B978-0-12-410438-9.00013-3

ONHYM. (2016). Annular structure of Twihinate (REE, Nb, Fe, U; southern provinces, Morocco). Office National des Hydrocarbures et des Mines.

Pufahl, P. K., & Groat, L. A. (2017). Sedimentary and igneous phosphate deposits: Formation and exploration: An invited paper. *Economic Geology, 112*(3), 483–516. https://doi.org/10.2113/econgeo.112.3.483

Verplanck, P. L., Mariano, A. N., & Mariano, A. (2016). Rare earth element ore geology of carbonatites. In *Rare earth and critical elements in ore deposits*. Society of Economic Geologists. https://doi.org/10.5382/Rev.18.01

Walter, A.-V., Nahon, D., Flicoteaux, R., Girard, J. P., & Melfi, A. (1995). Behaviour of major and trace elements and fractionation of REE under tropical weathering of a typical apatite-rich carbonatite from Brazil. *Earth and Planetary Science Letters, 136*(3–4), 591–602. https://doi.org/10.1016/0012-821X(95)00195-I

Petrogenesis of Pyroxenite Layers in Orogenic Peridotite of Beni Bousera (Internal Rif, Morocco)

Zahra Mourabit, Abdelhalim Tabit, Ahmed Algouti, Abdellah Algouti and Fatiha Hadach

Abstract

The Beni Bousera orogenic peridotites (northern Morocco) are associated with a variety of pyroxenite layers of variable volumes. These pyroxenites' mineralogy and their chemical compositions are better explained by means of a complex multi-stage model implying in the possibility that it is formed as (1) high-pressure cumulates or (2) residues from the re-fusion of gabbroic layers. The pyroxenite layers are chemically of basaltic composition. Mineralogically, they have modal compositions varying from olivine websterite to garnet clinopyroxenite. There are two different groups of genetically distinct pyroxenites. Pyroxenite type I is Fe-rich, with low contents of Al, Si, Ti, and Cr. These rocks may result from the re-crystallization processes of peridotite/gabbro complexes under high-pressure eclogite or granulite facies conditions. These rocks are thought to have initially crystallized at lower pressure from mafic liquids injected into crustal domains. Pyroxenite type II equilibrated at high temperatures 1000 °C to 1400 °C. They are characterized by enrichment in Al and Fe/(Fe+Mg) ratios which, however, remain very low. Further chemical signatures are negative Eu anomalies and low isotopic ratios of Sr and Nd to deduce that the pyroxenite type II may be the result of a recycled oceanic lithosphere.

Keywords

Beni Bousera · Pyroxenite · Mafic layers · Websterite · Clinopyroxenite

Z. Mourabit (✉) · A. Tabit · A. Algouti · A. Algouti · F. Hadach
2GRNT Laboratory, Geology Department, Faculty of Sciences Semlalia Marrakech, Cadi Ayyad University, Marrakech, Morocco
e-mail: zahra.mourabit@gmail.com

1 Introduction

A better knowledge of the structural, petrographic, and geochemical characteristics of ultramafic massifs and enclaves in alkaline basalts and in kimberlites are fundamental to understanding the history of the upper mantle. The petrology has attempted to decipher the information conveyed by magmas of mantle origin, and the mineralogical and chemical characteristics of their source regions. The lherzolitic massifs are the case of the Beni Bousera massif outcropping in the internal Rif (North of Morocco) (Kornprobst, 1969, 1974). They are supposed to be fragments of the upper mantle set up technically on the continent. Given their size, they allow to understand "in situ" the structure of a part of the earth: "the upper mantle". It could display the structural relationships between different types of rocks (lherzolites, harzburgites, dunites, and pyroxenites). These different petrographic facies bear witness to the processes which marked the mantle evolution of massifs, from their separation from the asthenosphere to their incorporation into the lithosphere. The various pyroxenites associated with peridotites in the form of regular layers, but often are boudinage structure and parallel to sub-parallel to the general foliation, are interpreted as being cumulates derived either from liquids resulting from the partial fusion of the host peridotite (Dickey, 1970; Kornprobst, 1969), or from the fusion of an altered and subducted oceanic lithosphere (Pearson et al., 1993). In the recent years, the origin and significance of pyroxenites in lherzolitic massifs or in the form of enclaves in alkaline basalts and in kimberlites have been the subject of several studies and rigorous debates due to their importance in mantle dynamics (Allègre & Turcotte, 1986; Dickey, 1970; Kornprobst, 1969, 1970; Nixon et al., 1986). This contribution attempts to discuss the pyroxenites petrogenesis in the orogenic peridotites of Beni Bousera.

F. Lucci et al. (eds.), *Selected Studies in Geomorphology, Sedimentology, and Geochemistry*, Advances in Science, Technology & Innovation, https://doi.org/10.1007/978-3-031-43744-1_3

2 Materials and Methods

As part of this work, the chemical analysis of pyroxenites involved many samples. These rocks' petrology is firstly interested in their mineralogical diversity, and then in the chemical composition evolution of the main mineral phases: clinopyroxene. Petrographically, clinopyroxene is the common mineral of all pyroxenites, except orthopyroxene where it exists, but in small quantities. Clinopyroxenes were analyzed with an electronic microprobe (Camebax - Clermont II) under the following physical conditions: potential difference U = 15 kV, the current intensity I = 12.5 nA, counting time (10–20) seconds. The total Fe is given in the form of FeO for all the minerals except for the spinel, for which, the proportions of Fe_2O_3 are calculated by stoichiometry (Camebax. B.R.G.M).

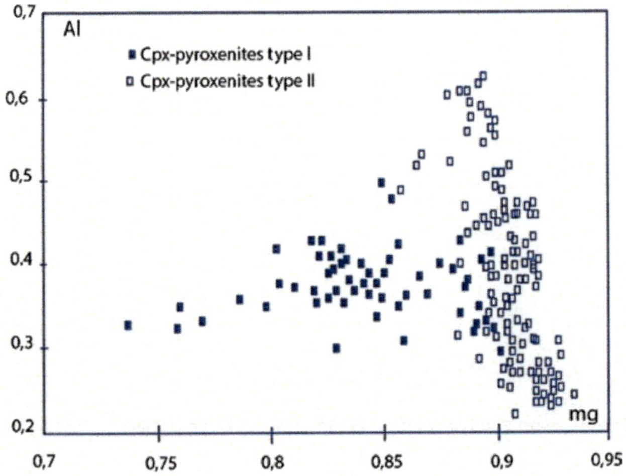

Fig. 2 Al Vs mg = (Mg/Mg + Fe) ratio of Clinopyroxene

3 Results

On the diagram (Ca–Mg–Fe) by Poldervaart and Hess, (195 na1), all the clinopyroxenes of the pyroxenite are diopside rich in Ca and Al. There are differences in composition between clinopyroxene in type I pyroxenites and clinopyroxene in type II pyroxenites. This is an enrichment of Ca in the clinopyroxenes of pyroxenites type II illustrated by the slightly high percentage of wollastonite, while the clinopyroxenes of pyroxenites type I exhibit enrichment in Fe + Mn, resulting in a slight enrichment of ferrosilite (Fig. 1). In Fig. 2, Al content versus (Mg/Mg + Fe) ratio in the clinopyroxenes varies from 0.20 to 0.65 (in cations by structural formula on 6.0_{2-}) of pyroxenites type II, whereas, in pyroxenite type I, it is almost constant. It varies on average from 0.30 to 0.50. There is a clear distinction between the two types of pyroxenites, well underlined by the mg ratio which is high in the clinopyroxenes of pyroxenites type II compared to those of type I.

The clinopyroxenes of spinel pyroxenites are Na-poor than those of garnet and corundum pyroxenites. The distribution of Ti versus Al content (Fig. 3) shows two trends, one is around 0.012 and 0.038, the other varies around 0 and 0.012. This characteristic of evolution is not verified in all the clinopyroxenes and reflects the petrological nature of the precursors from which these rocks originate. Ca content is rather high in all pyroxenites. In pyroxenite type I, it varies from 0.5 to 14wt%, while in pyroxenites type II varies from 6 to 18wt%. The pyroxenite type II shows a positive correlation of calcium content and the FeO/MgO ratio. This correlation could reveal a cumulative effect acquired by the parent magma leading to Ca enrichment.

The REE distribution in pyroxenites type II (Kornprobst et al., 1990; Pearson et al., 1993) and type I (spinel websterite, this work) displays a wide variation (Fig. 4). The pyroxenites type I normalized after Sun and McDonough (1989) shows LREE-rich patterns. It exhibits CeN contents

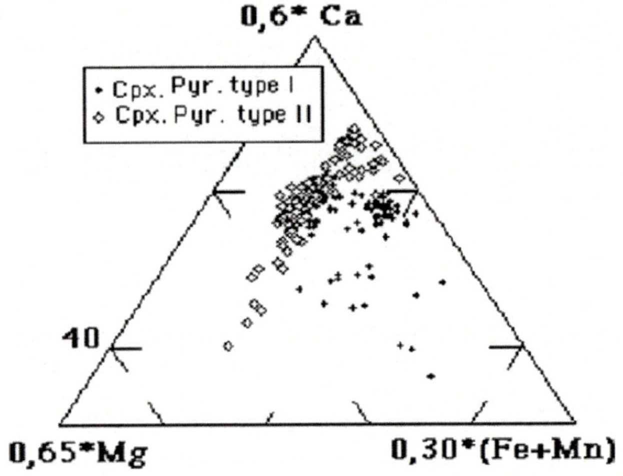

Fig. 1 Classification of Clinopyroxene of Beni Bousera pyroxenite

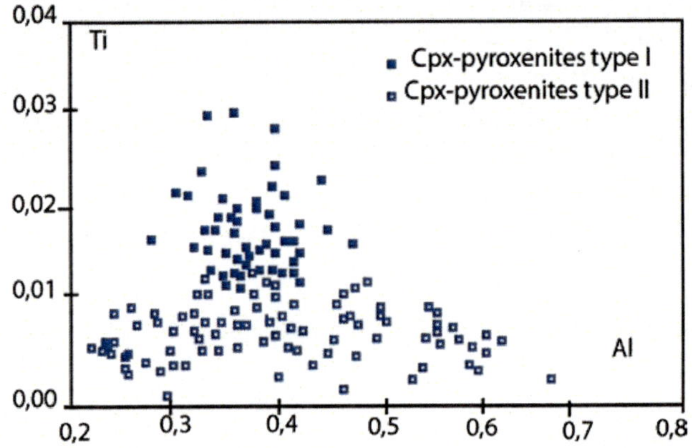

Fig. 3 Variation of Ti as a function of Al in the cpx (in atoms by structural formula)

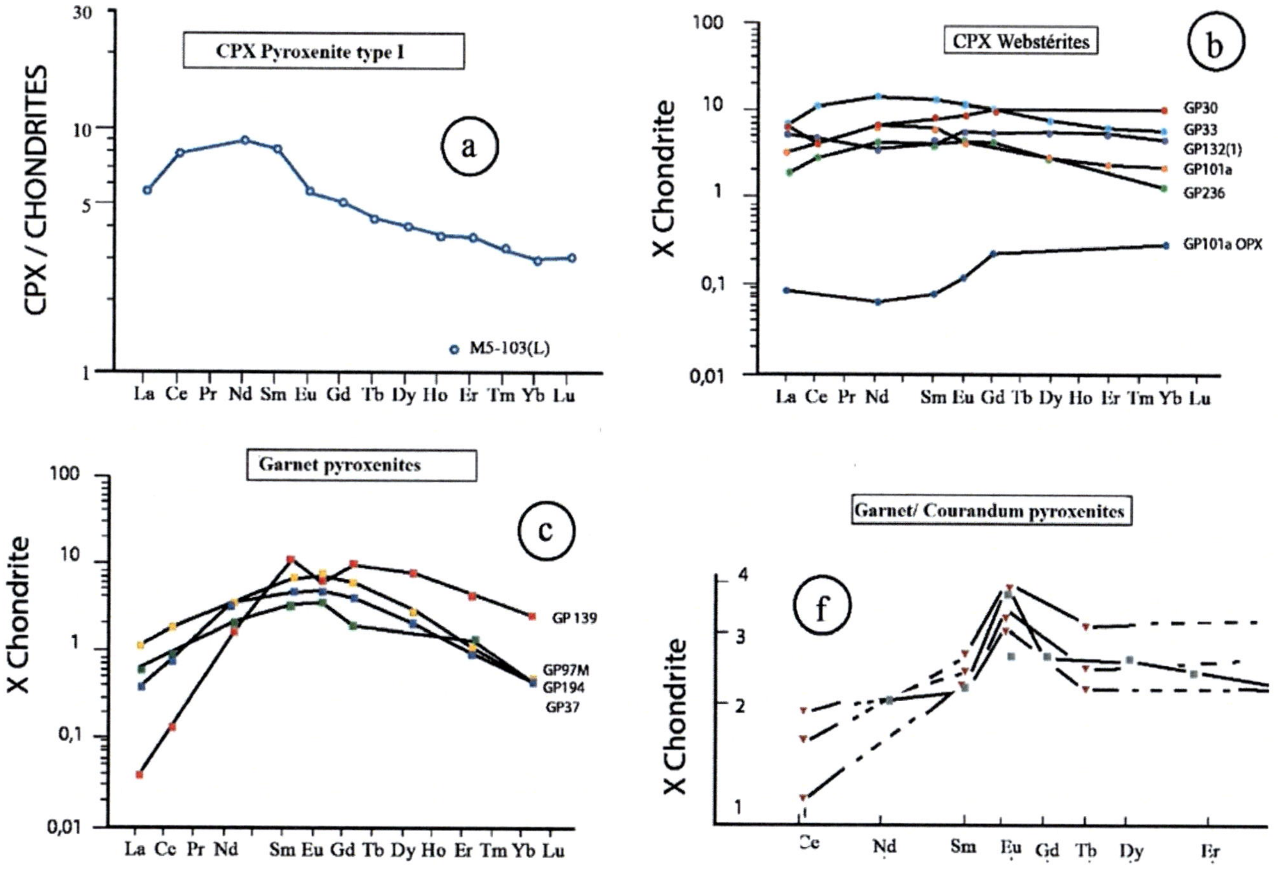

Fig. 4 REE contents of pyroxenites, REE contents of spinel websterite from this work normalized to chondrite. b-f-spectra are given after Pearson et al., (1993), normalized to chondrite

of the order of 2.65 and YbN of the order of 1; while the ratio (Ce/Sm) N is equal to 1.55. According to Kornprobstet al. (1990) in Corundum bearing garnet pyroxenites, the concentrations of CeN vary from 0.903 to 1.516, while the concentration of YbN varies from 1.99 to 3.23, and the (Ce/Sm) N ratio varies from 0.47 to 0.75. In the graphite bearing garnet pyroxenites (Pearson et al., 1993), CeN concentration is noted and it varies from 0.05 to 3.3; that of YbN varies from 0.56 to 24.27 and (Ce/Sm) N ratio varies from 0.016 to 0.91. In general, these rocks exhibit a high chemical diversity. It is underlined by wide variations in the contents of REE of different pyroxenites' facies.

The spectrum of incompatible elements of pyroxenite type I is normalized to chondrites (Sun & McDonough, 1989) for all the elements, except for the Sc which is still normalized to chondrites according to McDonough and Frey (1990) (Fig. 5). It shows strong enrichment in light incompatible elements such as Th and U against negative

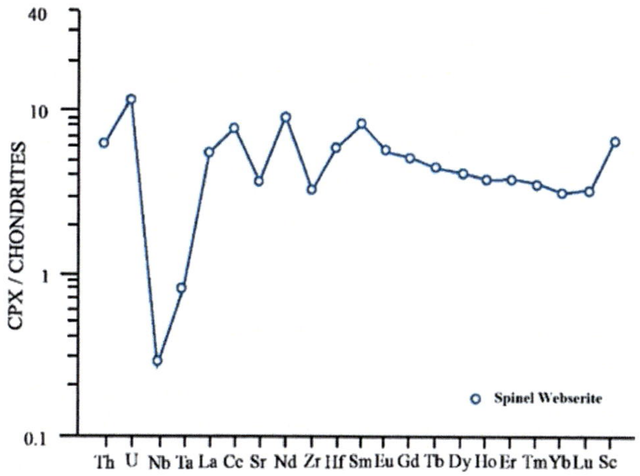

Fig. 5 Th, Nb, Ta, Sr, Zr, Hf, and REE of pyroxenite type I, normalized to chondrites (McDonough & Frey, 1990) for Sc. and (Sun & McDonough, 1989) for the other elements

anomalies in Sr and Zr. Thus, we note strong positive anomalies of the most incompatible elements (Th and U), as well as of Nd and Sm. However, incompatible elements at high charge, such as Nb, Zr, and Sr, show more pronounced negative anomalies except for Hf which shows a slight positive anomaly.

4 Discussion

The analyzed data of clinopyroxene shows a relative enrichment in Al and Ti, where Al content varies widely within the pyroxenite layer from the core to the rims. Moreover, the mg ratio vs Al in the clinopyroxene of pyroxenites type II is considerably not varying, this justifies the nature of the precursor of these rocks, supposed to be a gabbro. For which, the fractionation of plagioclase and of a sodium clinopyroxene was crucial subsequently giving rise to these rocks by isochemical metamorphism (Kornprobst et al., 1990).

The difference of major elements, Al, Ti, and Na between, clinopyroxene of pyroxenite type II and those of pyroxenite Type I, suggests two distinct origins for each type of pyroxenite. The mineralogical and experimental studies of these rocks (Dickey, 1970; Kornprobst, 1969, 1970) propose that the pyroxenites are formed by fractionation of the main phases clinopyroxene/garnet at high pressure and temperature, from a basic liquid of picric nature. Subsequently, the REE confirms that the pyroxenites are derived from peridotite with the possibility of the existence of certain pyroxenite layers having a multi-origin stage (Loubet & Allègre, 1979, 1982). The U and Th enrichment and the negative anomaly in Sr together argue that this rock seems to be crystallized among the last percolation stages "strongly incompatible character of Th" (Godard, 1993). Moreover, several spectra of pyroxenite type II exhibit significant anomalies of Eu. These anomalies are either positive or negative, although these rocks do not contain plagioclase. These Eu anomalies involve varying sources of plagioclase fractionation in a subducted oceanic crust that testifies to the important fractionation of this mineral as a cumulate in the history of these rocks. According to Pearson et al., (1993), the retromorphosed diamond-bearing in the form of graphite in pyroxenite type II suggests a crystallization of these pyroxenites well outside the stability field of plagioclase. After El Atrassi et al., (2011), the diamond relics bearing in graphite aggregate pyroxenites type II reveals evidence of partial melting and alternative metamorphic origin of these diamonds' relics. However, isotopic data rejects any genetic link between host peridotites and pyroxenites (Kornprobst et al., 1990; Nixon et al., 1986; Pearson et al., 1991; Polvé, 1983; Polvé & Allègre, 1980). They assume that pyroxenites are attested to a subducted oceanic lithosphere (Allègre and Turcotte, 1986; Kornprobst et al., 1990; Pearson et al., 1993; Polvé and Allègre, 1980; Varas et al., 2018). According to Allègre and Turcotte (1986), the pyroxenite layers represent a subducted oceanic lithosphere, metamorphosed, and variably thinned by convection and diffusion in the asthenosphere for a period equivalent to 1 billion years. This idea was supported and demonstrated by Kornprobst et al. (1990) as a source of pyroxenites type II in this massif.

5 Conclusion

We deduce that the great correlation between the chemistry of major elements and modal mineralogy, clinopyroxene, e.g., in addition to the systematic mineralogical zonation of certain layers suggests that the pyroxenites of the Beni Bousera massif are derived either from a liquid by fractional crystallization or, by re-crystallization at high depths of a precursor obtained by fractional crystallization at low pressure "Gabbro". Pyroxenites type I, the mineralogy is chemically characterized indicating that pyroxenites are produced by fractional crystallization mechanism at high depth from a picritic melt (Dickey, 1970; Kornprobst, 1969, 1970). Thus, petrological and experimental (Kornprobst, 1969, 1970) and geochemical (Javoy, 1977; Loubet & Allègre, 1982) data indicate those pyroxenites has been interpreted as the result of clinopyroxene-fractionation from garnet, at high pressure, during the differentiation of a picritic melt. Pyroxenites type II shows abnormal mineralogical characters compared to all the pyroxenites layers in lherzolite massifs. It is characterized by clinopyroxenes highly Al-rich and garnet pyrope-rich. The Al and Ca-rich suggests that these pyroxenites re-crystallized from a gabbroic protholite rich in anorthite as in the case of hyper-aluminous eclogites containing corundum and/or kyanite from crustal metamorphic units.

References

Allègre, C. J., & Turcotte, D. L. (1986). Implications of a two-component marble-cake mantle. *Nature, 263*, 123–127.

Dickey, J. S. (1970). Partial fusion products in alpine type peridotites: Serrania de la Ronda and other examples. *Minerals Social America Special Paper, 3*, 33–49.

El Atrassi, F., Brunet, F., Bouybaouene, M., Chopin, C., & Chazot, G. (2011). Melting textures and microdiamonds preserved in graphite pseudomorphs from the Beni Bousera peridotite massif, Morocco. *European Journal of Mineralogy, 23*(2), 157–168.

Godard, M. (1993). *Modélisation des interactions chimiques liquide/solide: Application à la circulation des magmas mantelliques.* Ph.D. diss., Montpellier 2.

Javoy, M. (1977). Stable isotopes and geothermometry. *Journal of the Geological Society, 133*(6), 609–636.

Kornprobst, J. (1974). *Contribution à l'étude pétrographique et structurale de la zone interne du Rif (Maroc septentrional).* Notes et Mémoire. Services Géology. Maroc, 251, 256.

Kornprobst, J. (1969). Le massif ultrabasique des Beni-Bouchera (Rif Interne, Maroc): Étude des péridotites de haute température et de haute pression. *Contributions Mineralogy and Petrology, 23*, 283–322.

Kornprobst, J. (1970). Les péridotites et les pyroxénolites du massif ultrabasique des Beni Bouchera. *Contributions Mineralogy Petrology, 29*, 290–309.

Kornprobst, J., Piboule, M., Roden, M., & Tabit, A. (1990). Corundum-bearing garnet clinopyroxenites at Beni Bousera (Morocco): Original plagioclase-rich gabbros recrystallized at depth within the mantle? *Journal of Petrology, 31*(3), 717–745.

Loubet, M., & Allègre, C. J. (1979). Element studies in the alpine type peridotite of Beni Bousera (Morocco). *Geochemical and Journal, 13*, 69–75.

Loubet, M., & Allègre, C. J. (1982). Trace elements in orogenic lherzolites reveal the complex history of the upper mantle. *Nature, 298*(5877), 809–814.

McDonough, W. F., & Frey, F. A. (1990). Rare earth elements in upper mantle rocks. In B. R. Lipin & G. A. McKay (Eds.), *Geochemistry and mineralogy of rare earth elements* (Vol. 21, pp. 99–145). Mineralogical Society of America.

Nixon, P. H., Davies, G. R., Slodkevich, V. V., & Bergman, S. C. (1986). Graphite pseudomorphs after diamonds in the eclogite-peridotite massif of Beni Bousera, Morocco and a review of anomalous diamond occurrences. In *4th International kimberlite conference, Perth, extended abstracts* (pp. 412–414).

Pearson, D. G., Davies, G. R., & Nixon, P. H. (1993). Geochemical constraints on the petrogenesis of diamond facies pyroxenites from the Beni Bousera peridotite massif, North Morocco. *Journal of Petroleum, 34*, 125–172.

Pearson, D. G., Davies, G. R., Nixon, P. H., Greenwood, P. B., & Mattey, D. P. (1991). Oxygen isotope evidence for the origin of pyroxenites in the Beni Bousera peridotite massif, north Morocco: Derivation from subducted oceanic lithosphere. *Earth Planetary, Science Letters, 102*, 289–301.

Poldervaart, A., & Hess, H. (1951). Pyroxenes in the crystallization of basaltic magma. *The Journal of Geology, 59*(5), 472–489.

Polvé, M. (1983). Les isotopes du Nd et du Sr dans les lherzolites orogéniques: Contribution à la détermination de la structure et de la dynamique du manteau supérieur. Thèse de Doctorat, Paris VII.

Polvé, M., & Allègre, C. J. (1980). Orogenic lherzolite complexes studied by ^{87}Rb-^{87}Sr: A clue to understand the mantle convection processes? *Earth Planetary Science Letters, 51*, 71–93.

Sun, S., & McDonough, W. F. (1989). Chemical and isotopic systematics of oceanic basalts: Implications for mantle composition and processes. In: A. D. Saunders, & M. J. Norry, (Eds.), *Magmatism in the ocean basins* (Vol. 42, pp. 313–345). Geological Society Special Publication.

Varas, M., Garrido, G., Marchesi, C., Bosch, D., & Hidas, K. (2018). Genesis of ultra-high-pressure garnet pyroxenites inorogenic peridotites and its bearing on the compositional heterogeneity of the earth's mantle. *Geochimica Et Cosmochimica Acta, 232*, 303–328.

Petrology and Mineral Chemistry of the Mobarak Abad Gabbro in Central Alborz, Iran

Ali A. Parchegani, Mohammad A. Arian, Mohammad Foudazi and Azam Hossein Nejad

Abstract

The Cenozoic Mobarak Abad Gabbro is a West-Northwest striking dike, approximately 6 km in length and 0.5 km in width, which outcrops several kilometres west of the Mosha-Fasham thrust, 75 km North-East of Tehran in the Mobarak Abad region of Iran. This gabbro is intruded into the Middle Eocene Karaj Formation, which is dominantly composed of volcaniclastic units including tuff, shale, and breccia. To better constrain the petrogenesis of the igneous rock in the study area, the petrology and mineralogy of the dike were investigated using electron microprobe analysis (EMPA) in conjunction with X-ray fluorescence whole rock analysis, optical mineralogy, and petrography. The petrographic study indicates that the dike ranges in composition and includes olivine gabbro, gabbro, diorite, and synodiorite. Following the Middle Eocene Alpine-Himalayan orogeny, a hotspot developed in the Central Alborz zone and was associated with widespread magmatic activity in the region, including the Damavand Volcano, the Qasr Firoozeh Granodiorite, the Lavasan Nepheline syenite, and the Karaj Dam Gabbro. Based on the geochemical studies, it is concluded that the magma that formed the Mobarak Abad Gabbro is a primary magma owing to the elevated Cr (>1000 ppm) and Ni (>400 ppm) and the presence of hyalosiderite olivine which is Mg-dominant. Geochemical studies presented here suggest that the source magma for the Mobarak Abad Gabbro was water-rich, calc-alkaline, and ranged from metaluminous to peraluminous. Assimilation of calcareous rocks by the source magma is suggested to account for the elevated concentration of Sr and contributed to the emplacement of the basaltic magma at high temperatures (700 °C to 1300 °C).

Keywords

Petrology · Central Alborz · Mobarak Abad · Karaj formation · Gabbro · Dike · Calc-alkaline magma

1 Introduction

The study area is located between 51°46′ to 52° longitude and 35°46′ to 35°48′ latitude in the region to the South of the central Alborz Mountain Range, Iran, and forms a part of the Central Alborz structural zone. In this area, there are several deformed mafic intrusive masses exposed beside the Haraz Road, close to Ardineh village and at Jafarabad. The Mobarak Abad intrusive body, which is considered to be the largest mass in this area, has been injected into one of the branches of the Mosha-Fasham thrust between Ardineh village and Mobarak Abad. The intrusive body takes the form of a large 6-km by 500-m-long dike with a West-Northwest, East-Southeast strike, occurring on the slopes of the Qaravol and Lar mountains. The maximum width of the dike on the west side is 800 m and the minimum width on the east side is 200 m. The Mobarak Abad intrusive mass has been intruded into the Karaj formation (middle Eocene), so it is younger than the Middle Eocene, but in the northern and western parts, it is in intrusive contact with Paleozoic sediments. The Middle Eocene sequence in this region includes coherent lava, pyroclastic units, fossiliferous limestone, dolomite, and evaporitic rocks (gypsum). The area has been strongly deformed with part of

A. A. Parchegani (✉) · M. A. Arian · M. Foudazi
Earth Science Faculty, Azad University, Tehran, Iran
e-mail: a.parchegani@uq.edu.au

A. H. Nejad
Earth Science Faculty, Science and Research Branch, Azad University, Tehran, Iran

A. A. Parchegani
W.H. Bryan Mining and Geology Research Centre, Sustainable Minerals Institute, The University of Queensland, Brisbane, QLD 4072, Australia

F. Lucci et al. (eds.), *Selected Studies in Geomorphology, Sedimentology, and Geochemistry*, Advances in Science, Technology & Innovation, https://doi.org/10.1007/978-3-031-43744-1_4

the melanocratic gabbro being intensely fractured and deformed. However, less deformed melagabbro boulders can be found in the region forming fresher cores surrounded by onion skin weathering. This is the first study to evaluate the genesis of the Mobarak Abad dike. This paper investigates the geochemical, mineralogical, and petrogenetic properties of these masses through integrated field observations, petrographic studies, as well as whole rock X-ray fluorescence and electron microprobe analysis of the exposed rocks.

Fieldwork was carried out to observe field relationships and to sample plutonic rocks in the study area. In total, 60 samples of the freshest rocks (with the least amount of weathering) were sampled. Thin sections were prepared and petrographic studies were carried out in transmitted light. After microscopic studies, 15 samples were selected for detailed XRF chemical analysis and mineralogical analysis by electron microprobe (EMPA) of olivine, pyroxene, and plagioclase.

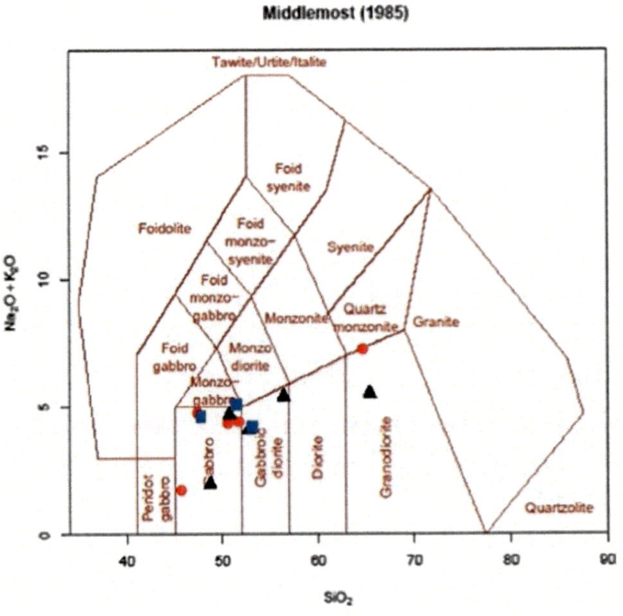

Fig. 1 $(Na_2O + K_2O)/SiO_2$ diagram for classification of plutonic igneous rocks (Middlemost, 1985). According to the diagram, the region's plutonic rocks are located in the gabbro, gabbrodiorite, and granodiorite areas

2 Geology and Petrology of Intrusive Masses

The geology of the study area lies between the Mosha-Fasham and Rudehen faults and has been affected by deformation associated with these faults. Based on the chronostratigraphic classification, the formations present in this area, which have common contacts with the intrusions of the region, include the Carboniferous Mobarak Formation the pre-Jurassic Shemshak Formation, the Paleocene-Eocene Fajan Formation, and the Middle Eocene Karaj Formation.

The igneous rocks in the region were classified using chemical discrimination diagrams based on the alkali-silica diagrams of Middlemost (1985). According to this diagram (Fig. 1a), the rocks of the region are classified as gabbro, gabbrodiorite, and granodiorite or tonalite (Figs. 1a and 2b), which are described below. Based on the mineralogy of the rocks and their frequency, the rocks of the Mobarak Abad mass are classified as melagabbro, leucogabbro, diorite, and synodiorite (Fig. 2 or 1b). In hand specimen, the colour of the melagabbro is dark, and alteration is only weakly observed. Melagabbro has an intergranular to granular texture and comprises augite-type clinopyroxene, bytownite and labradorite-type plagioclase, and hyalosiderite-type olivine. Apatite, uralite, and iron oxide are the sub-minerals of these rocks (Fig. 3). Olivine has been (partially) altered to serpentinite, augite has been altered to uralite, and plagioclase sosoritization and biotite alteration to chlorite is commonly observed in these rocks. In hand specimen, the weathered leucogabbro is light grey in colour but is dark grey in fresh, broken surfaces. Leucogabbro has a granular texture and contains pyroxenephenocrysts (augite) and also plagioclase (labradorite). The effects of alteration in these rocks are abundant (Fig. 3). In hand specimen, the diorite is weathered to a light grey colour but appears grey in fresh

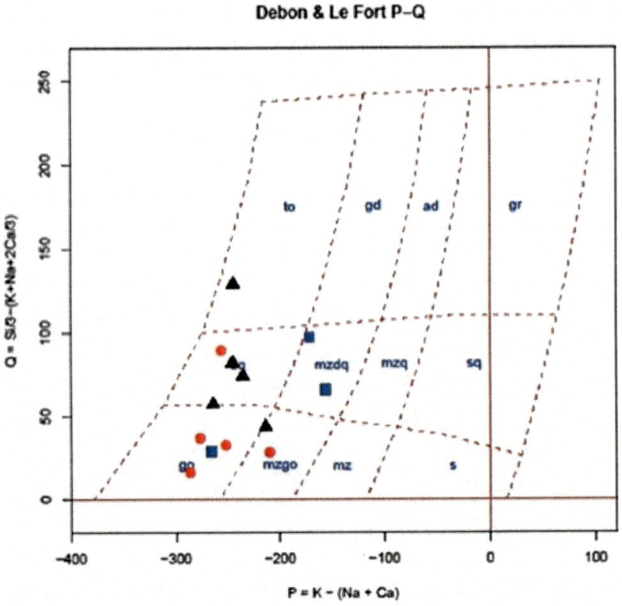

Fig. 2 De Bon and Le Fort (1983) Q/P diagram for classification of the area's plutonic igneous rocks by cationic method. Based on this diagram, the samples of rocks in the region are located in the gabbro (go), monzogabbro (mzgo), quartz diorite (dq), quartz monzodiorite (mzdq), and tonalite (to) fields

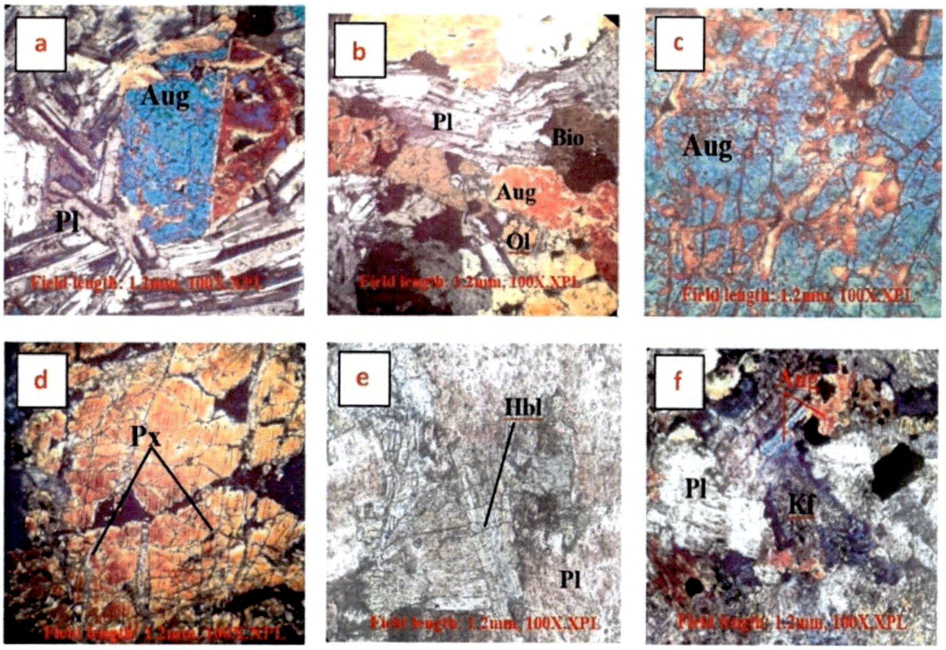

Fig. 3 a: Intergranular texture in melagabbro with the presence of plagioclase and augite crystals, and **b** Granular texture is found in melagabbro and minerals such as augite, plagioclase, olivine, and biotite. **c** Granular texture in leucogabbro with augite crystals and **d** Granular texture in leucogabbro and massive augite crystal. Granular texture in diorite with hornblende and plagioclase crystals. **e** and **f** Granular texture in synodiorite with augite, potassium feldspar, and plagioclase crystals

broken surfaces. Mineralogically, the leucogabbro comprises andesine (based on Michelle Levy's extinction angle method), hornblende, augite, and alkali feldspar (Fig. 3). In hand specimen, weathered surfaces are light grey and dark grey in fresh broken surfaces. The rock is porphyritic phenocrysts plagioclase, augite, hornblende, and alkali feldspar phenocrysts in a fine-grained ground mass (Fig. 3).

3 Electron Microprobe Analysis (EMPA)

To determine the chemical composition of the minerals that form the plutonic igneous rocks in the study area, olivine, pyroxene, and plagioclase were analysed using electron microprobe analysis (EMPA). Olivine in the samples is Mg-rich 67.8% of the magnesium type of hyalosiderite and is chemically located at the chrysolite border. Pyroxenes are rich in magnesium and are oxidised due to the high percentage of the main elements such as Fe, Ca, and Mg. Plagioclase analysed by EPMA in this study plots in the fields of bytownite and labradorite (Fig. 4).

4 Geochemistry

The amount of the main cation oxides relative to silica was investigated using X-ray fluorescence whole rock chemical analysis (Fig. 5). The amount of Al_2O_3 in the rocks of the area varies from about 10% in gabbro to 25% in diorite. In intermediate compositions, with the increase in the percentage of plagioclase, the amount of aluminium also increases, which is due to the presence of Al_2O_3 in the structure of plagioclase. The content of Fe_2O_3 in samples analysed varies from about 2.2% in synodiorite to 11.8% in gabbro. The

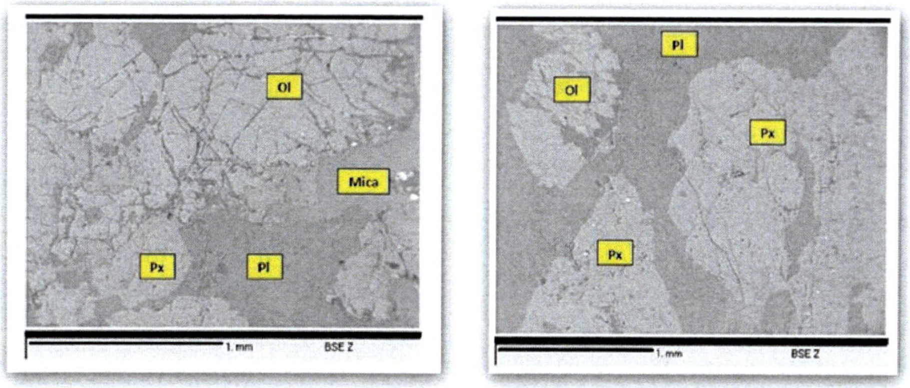

Fig. 4 (BSE) Electron microscope image of the minerals in the sample

amount of this element from mafic compositions to felsic compositions follows a downward trend and approaches zero with a steep slope. The concentration of iron in the ferromagnesian minerals of olivine and augite is the main reason for this. Iron is also present in the magnetite structure, and the small percentage in more felsic samples is related to the presence of magnetite. The weight percentage of CaO varies from about 3% in synodiorite to 14% in gabbro and basic compositions. The percentage curve of this oxide decreases from mafic compositions to felsic compositions with a regular downward trend. The main reason for this decline in Ca is in the structure of calcium-rich augite and plagioclase that are formed in the early stages of crystallisation, thus making the remaining calcium-poor magma. The amount of Na_2O varies from about 1.2% in gabbro and mafic compositions to 6.4% in synodiorite and more felsic compositions. This upward trend indicates that in the early stages of magma crystallisation, the amount of this oxide was low, but with the crystallisation of calcium plagioclase in magma, the percentage of Na_2O increased. It is noteworthy that the amount of this oxide in orthoclase increases up to 3 times compared to melagabbro. The amount of K_2O changes from about 1% in melagabbro to 2.5% in the diorite. As can be seen in Fig. 5, the curve is irregular and reaches its maximum in the middle compositions. The amount of MgO varies from about 2% in synodiorite to 17% in melagabbro. Like iron, this element participates in the composition of olivine and augite minerals, so its amount in the mafic compositions is much higher than more felsic compositions, and the trend of MgO diagram is declining. The gradual reduction of magnesium from mafic to felsic rocks may be due to the crystallisation and degradation of olivine and magnesium oxide. The frequency of TiO_2 is about 0.3 to 1.3% in the rocks of the area and its value is observed with an irregularity. Titanium oxide is a chemical component of augite and causes a distinctive purple coloration in these minerals. The MnO content varies from about 0.05 to about 0.6% and shows no systematic correlation with SiO_2 which may reflect variable hydrothermal alteration, which has caused modified primary manganese content some rocks, so the amount of manganese oxide in the rocks is not very dependent on magmatic changes.

5 Geochemical Spider Diagrams

Spider diagrams of the geochemical composition of rocks in the study area, normalised to oceanic crust (mid oceanic ridge basalt, MORB), show that elements including Th, K, Ba, Sr, and Rb are enriched up to 100 times. In contrast, elements such as Y and Ti show depletion (Fig. 6). Also, elements such as Nb, Zr, and Ce show enrichment in some samples and depletion in others (Fig. 6). Normalisation to chondrite composition, more than 800 times enrichment of uranium and thorium, is observed and all elements show enrichment relative to the normalised chondrite values (Fig. 7).

In the case of potassium, it shows enrichment compared to MORB types (Fig. 6) and depletion compared to continental crust. Strontium also indicates the same, with the exception of synodiorite; it appears that the entry of Sr into

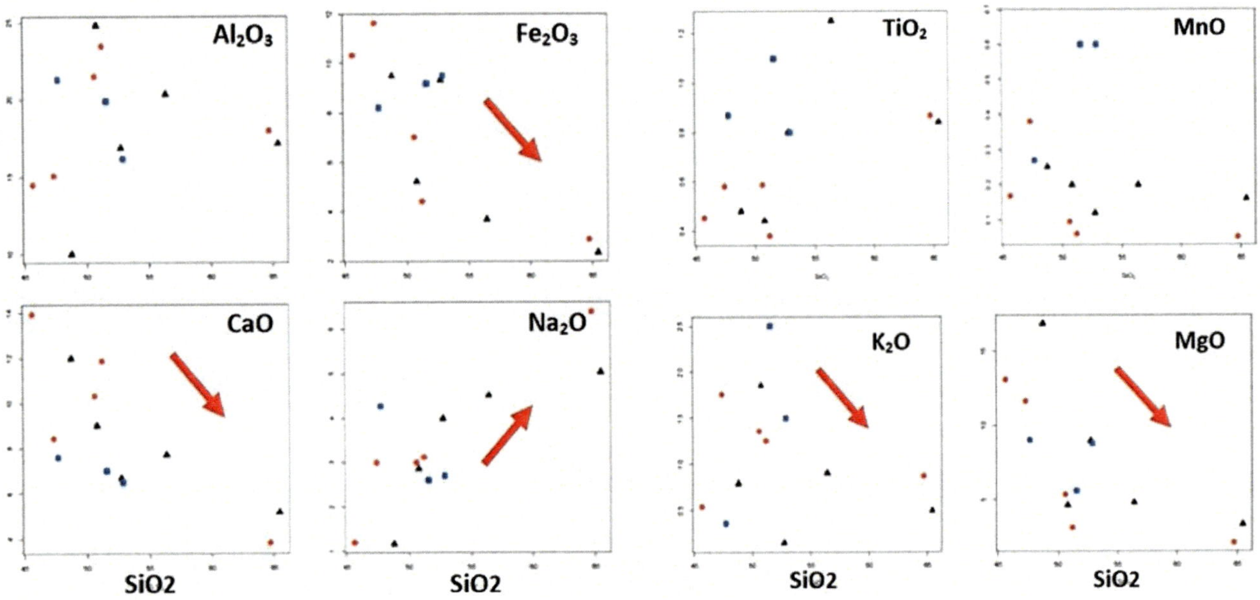

Fig. 5 Diagrams of chemical changes of the main oxides relative to silica in the plutonic igneous rocks, samples have been collected from three different outcrops in the area under study. To see the chemical composition differences, multiple symbols have been used

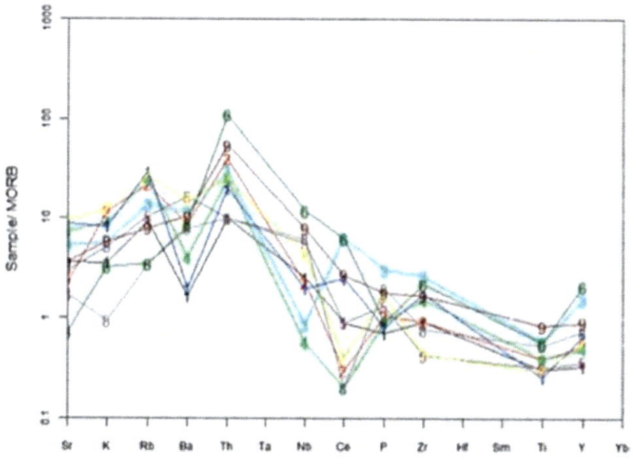

Fig. 6 Spider diagrams relative to the oceanic crust

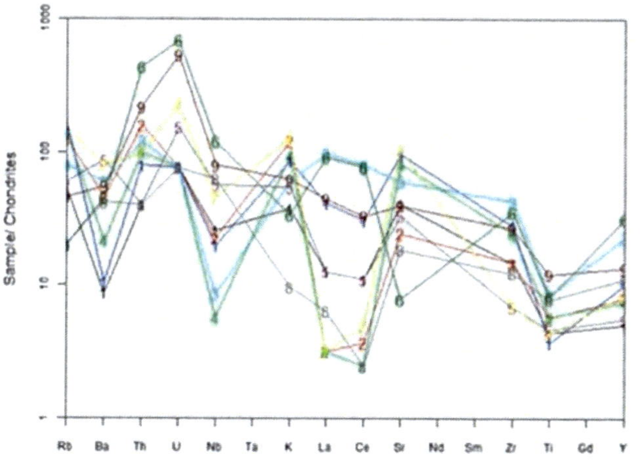

Fig. 7 Spider diagram relative to chondrite (Thompson, 1982)

the structure of the mineral plagioclase in diorite has caused a decrease in the percentage of this element in more felsic compositions. In most spider diagrams, lithophilic elements such as Rb, K, Ba, and Sr indicate enrichment relative to normalised values. The mobility of these elements is very high and they are concentrated in the continental crust and their concentration is in magma. It can indicate magma contamination with a continental crust. In melagabbro, the (Mg × 100) / (Mg + Fe) ratio is 76, between 68 and 73 in primary basalt and 88 to 89 in mantle peridotite (Roeder, 1970). Thus, melagabbro is considered a primary magma and cannot be the result of a low rate of partial melting of the mantle, because this ratio does not exceed 73% in primary melt at a rate of less than 25%, unless a significant portion of the peridotite has participated in the melting. In the other words, the resulting magma would be expected to have a tholeiitic composition and its differentiated compositions follow that of the tholeiitic series evolution. This scenario is not supported by the inferred magma compositions of Mobarak Abad, which are closer to the calc-alkaline series than the tholeiitic. An alternative interpretation is that the above high ratio (Mg × 100)/(Mg + Fe) in melagabbro is due to density segregation of olivine crystals (Vaziri and Ahmadi, 2010). In this case, leucogabbro can be considered the result of flotation and separation of plagioclase from mafic minerals, and the residual light magma in the upper part of the mass has led to the formation of diorite in the northern margin and synodiorite in the higher parts of the mass.

An interpretation consistent with the observed characteristics is that the primary magma had a moderate composition between leucogabbro and melagabbro before segregation and was derived from melting of the metasomatised peridotite. The presence of hydrated minerals such as amphibole (0.4%) and biotite (1 to 5.4%) in the basic rocks of Mobarak Abad supports the interpretation that the primary magma is water bearing which can also be interpreted from the fact that magma has cut off its solidus curve before reaching the Earth's surface. The frequency of hygromagmaphile elements in the segregated rocks suggests that the Mobarak Abad intrusives were derived from partial melting of enriched mantle (Vaziri and Ahmadi, 2010). Regarding the Alborz region, the lack of a continuous ophiolite chain along this route does not support the interpretation of a palaeosubduction zone between the oceanic crust and Central Iran continental crust mentioned by previous studies. However, the presence of a hotspot in central Alborz following the Eocene severe volcanism in this region is possible, and this thermal anomaly could be a continuation of the same process that produced the central Alborz volcano (Vaziri and Ahmadi, 2010).

6 Conclusions

Based on microscopic studies and electron microprobe analysis, the main minerals that constitute the Mobarak Abad dike are augite, bytownite and labradorite plagioclase, hyalosiderite olivine (decomposing into serpentine), and biotite. The main minerals that constitute leucogabbro include augite and labradorite. Diorites are also characterised by andesine, hornblende, and augite, and the main minerals that make up the synodiorite include andesine, augite, hornblende, and orthoclase. There is no depletion in the chondrite normalised REE spider diagrams. In the spider diagram, the rare elements are divided into three categories according to the diagonal: the first group of rare elements such as Th, K, Ba, Sr, and Rb is enriched, which sometimes reaches 100 times. The second group is elements such as Y and Ti that show depletion. Finally, a group of elements, such as Nb, Zr, and Ce, which in some cases show enrichment and in others show depletion. Formation of a water-rich calc-alkaline basic magma and the elements of a hygromagmaphile nature from an ordinary mantle is not possible unless the mantle is enriched locally or in a region of fluids. The main cause of enrichment is usually water, trace elements, or mantle's convection (Cox 1979). The basic calc-alkaline magma has been severely depleted of olivine minerals and has formed rocks of gabbroic composition. The gabbro is dark-coloured olivine melagabbro and leucogabbro without olivine, bright-colour, and a higher percentage of augite. Diorites in the suite are interpreted to have formed by a mechanism of buoyant segregation of plagioclase. Consumption of iron and magnesium in olivine and clinopyroxene causes magma to be depleted of these elements and is associated with the presence of calcium in platelets. As a result, the remaining magma is rich in silica and alkalis and forms alkali feldspar-bearing synodiorite.

After the Middle Eocene and the formation of the Alpine-Himalayan mountains, there was a hotspot in the Central Alborz zone, which can be traced to the magmatic activity in Central Alborz, including the Damavand Volcano and large intrusive bodies including the Qasr Firoozeh Granodiorite, the Lavasan Nepheline syenite, Monzonite, the Karaj Dam Gabbro, and the Mobarak Abad Gabbro. The movement of faults can, on the one hand, cause uplift and create an adiabatic phenomenon in the mantle, and on the other hand, they can be associated with local extension, which in both cases allows melting and the production of basic magmas (Vaziri and Ahmadi, 2010). In Mobarak Abad area, the existence of numerous faults and tectonic activity has caused the rocks in the mantle to melt by decreasing pressure at a constant temperature, and the decrease in density compared to the surrounding rocks. Secondly, these faults have provided a conduit for this formed magma to reach the earth's surface. Based on observations and studies, it is concluded that this magma is water-rich calc-alkaline magma (based on the diagrams of Irvin & Baragar, 1971), which is rich in aluminium. It appears possible that the passage of this magma through calcareous sequences, as well as the high temperature of the basaltic magmas, may have led to digestion of these calcareous rocks, leading to the observed high level of some trace elements such as Cr (>1000 ppm) and Ni (>400 ppm) in the dike.

References

Allenbach, P. (1966). Geology and petrography of Damavand (Central Alborz). *Iran, 63*, 114–118.

Assereto, R. (1966). Geological map of upper Djadjrud and Lar valleys (central Elburz, Iran), scale 1:50,000 with explanatory.

Berberian, M. (1976). Contribution to the seismotectonics of Iran (Part II). *Geo, Surv, of Iran, 39*, 518–523.

De Bon, F., & Le Fort, P. (1983). A chemical—Mineralogical classification of common plutonic rocks and associations, 73–135.

Clark, A. (1990). *Pocket guide to minerals gallery book.*

Cox, K. G., et al. (1979). *The Interpretation of Igneous Rocks*. Allen and Unwin, London, 450 p.

Dellenbach, J. (1964). *Contribution a I etude geologique de region situess al'est de Tehran (Iran)*. University of Strasbourg.

Fredrick, D. (1997). *Rocks and minerals: A portrait of the natural world.* Smith mark Publishers.

Irvin, T. N., & Baragar, W. R. A. (1971). A guide to the chemical classification of the common volcanic rocks, Canadian. *Journal of Earth, 8*, 523–548.

Mackenzie, W. S., & Adams, A. E. (1995). A color atlas of rocks and minerals in thin section.

Maniar, P. D., & Piccoli, P. M. (1989). Tectonic discrimination of granitoids. *Geological Society of America Bulletin, 101*, 635–643.

Middlemost, E. A. K. (1994). Naming materials in the magma/igneous rock system. *Earth-Science Reviews*.

Midelemost, E. A. K. (1985). Naming materials in the magma/igneous rock system *Longman Ground*, 73–86.

Roeder, E. (1970). *Petrology and geochemical data*. Evolution, presentation, interpretation, Longman scientific and Technical, U.

Stocklinn, J. (1974). Northern Iran: Alborz mountains. Mesozoic-Cenozoic orogenic belt, data for orogenic studies. *Geological Social, London., 4*, 213–234.

Thompson, R. N. (1982). Magmatism of the British Tertiary Volcanic Province. *Scottish Journal of Geology, 18*, 49–107.

Vaziri, M. H., & Ahmadi, A. (2010). *Petrology and petrography of igneous rock* (pp. 178–208).

Wilson, M. (1989). Partial melting processes in the earth's upper mantle. *Igneous Petrogenesis*, 37–44.

Morphometry, Petrography, and Geochemistry Analysis of Pillow Lava in Karangsambung (Indonesia): Implication for Geologic Setting of Karangsambung

Dina Gunarsih, Mirzam Abdurrachman and Idham Andri Kurniawan

Abstract

The Karangsambung area is characterized by pillow lava outcrops with different dimensions located at Muncar River (MUR), Mandala River (MAR), and Susu River (SUR). Pillow size indicates the original conditions when the pillow was formed. This research aims to understand the formation conditions of pillow lava and its implication for geologic setting in the Karangsambung area, based on their morphometric, geochemical, and petrographic data, and on the crystal size distribution of the hosted plagioclase. Pillow lavas at MAR are large, with a cooling time between 9.7 and 10.3 months, and a cooling rate between 7.90 and 9.09×10^{-10} K/s. At SUR, the size of pillow lavas is medium, with a cooling time between 5.1 and 10.3 months, and a cooling rate between 33.10 and 36.42×10^{-10} K/s. At MAR, pillow lavas are smaller, with a cooling time between 11.7 and 12 months, and a cooling rate between 1.76 K/s and 1.80×10^{-10} K/s. These results indicate different geological conditions of the pillow lava in Karangsambung. We suggest that pillow lavas at MUR were formed on the most-gentle slope, the depth of the water column was very deep, the cooling time was slower, and the cooling rate decreased. At SUR, pillow lavas were formed at a steeper slope and the depth of the water column was shallower than that of the pillow lava at MUR, but it had the fastest cooling time, and the cooling rate increased. The pillow lavas at MAR were formed at a very steep slope, the depth of the water column was very shallow, leading to slower cooling time and cooling rate. It can be concluded that there were three geological conditions of the formation of pillow lava in this area: (1) the most-gentle slope and very deep water column depth, (2) steep slope and shallower water column depth, (3) very steep slope and very shallow water column depth.

Keywords

Karangsambung · Pillow lava · Petrography · Geochemistry · Crystal size distribution

1 Introduction

Karangsambung area (Central Java, Indonesia) is characterized by the occurrence of some outcrops of pillow lava with different dimensions. These pillows are located at Muncar River (MUR), Mandala River (MAR), and Susu River (SUR). Many studies have been conducted in Karangsambung, but the presence of different pillow lavas in some locations has never been studied. Pillow size indicates the original conditions when the pillow was formed. Previous studies stated some factors that control the dimensions of the pillow lava, including the depth of the water column, slope, cooling time, cooling rate, and lava viscosity (Walker, 1992; Schnur, 2007).

Pillow lavas at MUR, MAR, and SUR are the object of this research. Their morphometric, geochemical, and petrographic characteristics, and the crystal size distribution of the hosted plagioclase is used to infer geological conditions when the pillows were formed. Therefore, this study aims to understand the formation conditions of pillow lava and their implication for geologic setting in the Karangsambung area.

D. Gunarsih (✉)
Syiah Kuala University, Banda Aceh, Indonesia
e-mail: dinagunarsih@usk.ac.id

M. Abdurrachman · I. A. Kurniawan
Bandung Institute of Technology, Bandung, Indonesia

F. Lucci et al. (eds.), *Selected Studies in Geomorphology, Sedimentology, and Geochemistry*, Advances in Science, Technology & Innovation, https://doi.org/10.1007/978-3-031-43744-1_5

2 Methods

In this study, some methods were applied to determine the factors that influence the formation of pillow lavas in Karangsambung. Measurements on 222 pillow lavas used the Walker Method (1992), which measures the pillow lava length in parallel (H) and vertical (V) directions with respect to the assumed deposition surface. The statistics were made based on the measurement results. Summary statistics of the measurement results were useful for quantitatively comparing the difference in the size of pillow lavas at the three research locations, besides being based on qualitative field observation. Twenty samples were analyzed using a polarizing microscope and captured for digitizing plagioclase minerals using ImageJ software. The plagioclase that had been digitized in 2D was converted into 3D using Morgan and Jerram's (2006) worksheets and corrected using CSD Correction software (Higgins, 2000). The crystal size distribution of plagioclase was used in calculating the cooling time by assuming a growth rate of 10^{-8} mm/s (Brugger & Hammer, 2010). The calculation of the cooling rate used the parameters of the total crystal density and SiO_2 concentration and water content in the melt (Toramaru et al., 2008). The relative viscosity used the Klein et al. (2018) worksheet which calculated viscosity based on crystal size. Besides, six rock samples were analyzed for major and trace elements by XRF to determine the tectonic setting.

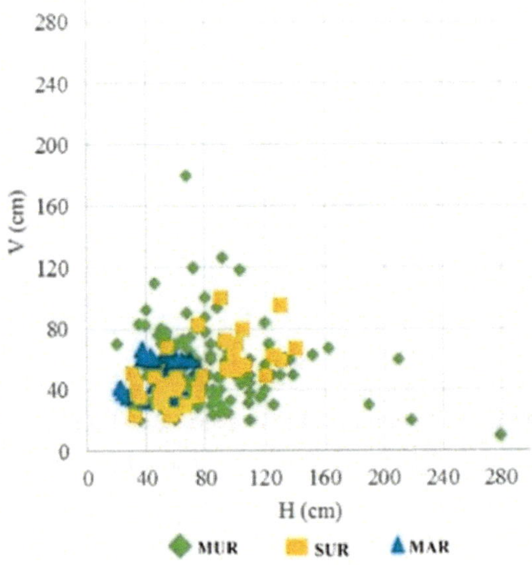

Fig. 1 Graph of the MUR, SUR, and MAR pillow lava size

3 Results

3.1 Morphometry

The pillow lava sizes in MUR, SUR, and MAR are significantly different (Fig. 1). The MAR pillow lavas vary in size, from the small one to the largest one. They are also perfectly shaped and closely packed pillows, so that the matrix is less abundant (Fig. 2a). The matrix is the fragments that are between two or more pillow lavas, in which the form is angular or non-angular fragments. The pillow lavas at SUR are relatively medium, the shape is less perfect, and the matrix is more abundant than the matrix of the pillow at MUR (Fig. 2b). On the other hand, the pillow lava at MAR is small, has an imperfect shape, and is less-packed, so the matrix is very abundant (Fig. 2c). The statistics of pillow lava measurement results are presented in Table 1.

3.2 Petrography and Crystal Size Distribution (CSD)

The mineralogy of basaltic rocks exposed at MUR, MAR, and SUR is similar, which consists of phenocryst and microphenocryst of euhedral–subhedral plagioclase, pyroxene, and olivine, with groundmass dominated by glass and pyroxene (Fig. 3). The textures are generally intersertal and intergranular. However, from the rock samples in MUR, there are glomerocryst and sieve textures, which distinguish them from other samples. This represents a difference in the magmatism process experienced by magma (Renjith, 2014).

Fig. 2 Pillow lavas outcrops located in **a** MUR, **b** SUR, and **c** MAR

Table 1 Statistic of pillow lava measurement results

		Maximum	Minimum	Mean	Samples (N)
Muncar river	H (cms)	280	20	77.89	163
	V (cms)	180	10	52.25	
	H/V ratio	28	0.29	1.89	
Susu river	H (cms)	140	30	75.97	38
	V (cms)	100	24	51.89	
	H/V ratio	2.40	0.59	1.51	
Mandala river	H (cms)	76	22	50.45	21
	V (cms)	67	33	51.10	
	H/V ratio	1.71	0.52	1.02	

Fig. 3 Thin section of **a** MUR **b** MAR **c** SUR basaltic rocks samples

Moreover, differences in the magmatic process can also be seen by the CSD curve (Fig. 4). This had been suggested by Vinet and Higgins (2010) in some examples of processes that influence the shape of the CSD curve. The curved lines show that magma mixing occurs in pillow lava at MUR, so the crystallization is slow. On the other hand, igneous rocks at SUR and MAR exhibit a straight line in CSD, which is related to an increase in the cooling rate. However, the pillow lava at MAR crystallizes more slowly than the pillow lava at SUR (Fig. 5). This can be explained due to differences in magma temperature. Vinet and Higgins (2010) explained that during coarsening texture, small crystals could experience dissolution, then large crystals grew. This process occurs when the temperature is high and remains stable at the liquidus temperature. In petrography, olivine, the mineral that is formed at high temperatures, is more abundant in basaltic rocks at MAR. The relative viscosity of MAR lava is also significantly different from that of the others (Fig. 6). This was due to Klein's calculation of relative viscosity, in which the parameter used is the crystal size.

3.3 Geochemistry

The Ti-V-Sc discrimination diagram (Vermeesch, 2006) places rock samples in a mid-ocean ridge tectonic setting (Fig. 7). In 2012, Puswanto et al. concluded that the igneous rocks at MAR and SUR had a tholeiitic affinity. Suparka (1988) stated that the massive rocks in northern Karangsambung were related to the mid-ocean ridge. This is related to tectonic evolution in Western Indonesia and Java Island that was started with the oceanic spreading floor 127 million years ago (Asikin, 1974; Liu et al., 1983) and followed by the collision of Argoland and Eurasian Plate on the Late Cretaceous (Abdurrahman et al., 2019) to expose pillow lavas in Karangsambung area as pre-tertiary rocks.

4 Discussion

Pillow lavas in this study are associated with the mid-ocean ridge tectonic setting. The different magmatism processes that magma underwent affected the time and rate of

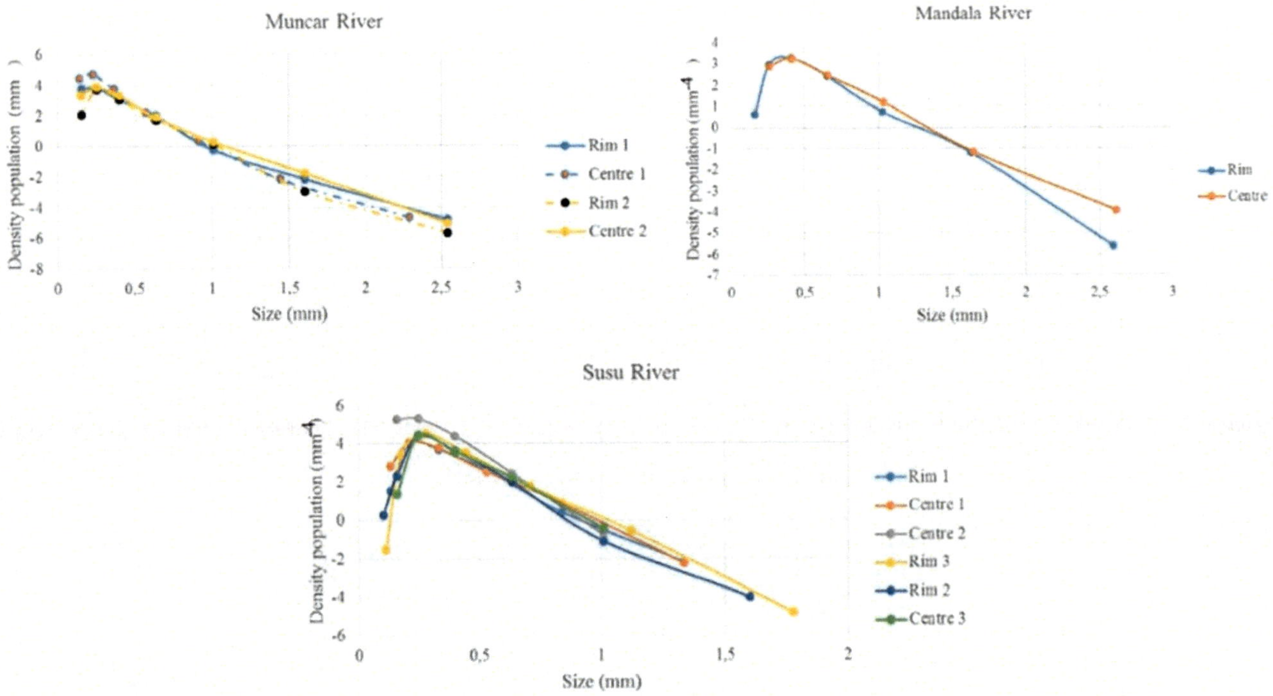

Fig. 4 Crystal size distribution curve

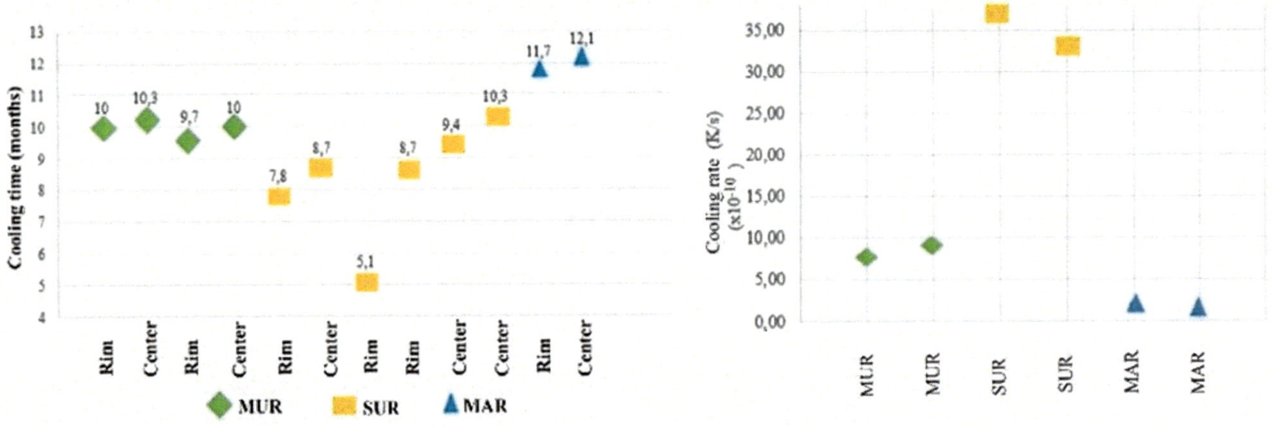

Fig. 5 Graph of cooling time and cooling rate

crystallization of the lava. The fast cooling rate resulted in an abundant matrix among pillow lava and less-packed pillow lava. Thereby, pillow lava at SUR has a more abundant matrix than pillow lava at MUR which tends to be closely packed. Conversely, pillow lava at MAR has an abundant matrix, but its cooling rate and time were slow. Matrix abundance can also be controlled by the slope and depth of the water column.

In this case, the pillow lava at MAR was highly controlled by slope and depth. The shallow water column resulted in the pillow lava being imperfectly formed with an abundant matrix, while the steep slope resulted in an abundant matrix and a less-packed pillow. This slope control also produced a small pillow of the pillow lava at MAR, so that the high viscosity control to produce a large pillow did not have any effect.

In pillow lava at MUR, the deep water column provided greater pressure, thus, giving the pillow and matrix a little perfect form. As they are closely packed and large pillow sizes show that pillow lava in MUR was formed on the most-gentle slope. In pillow lava at SUR, a steeper slope and shallower depth gave medium size and more matrix than that in pillow lava at MUR.

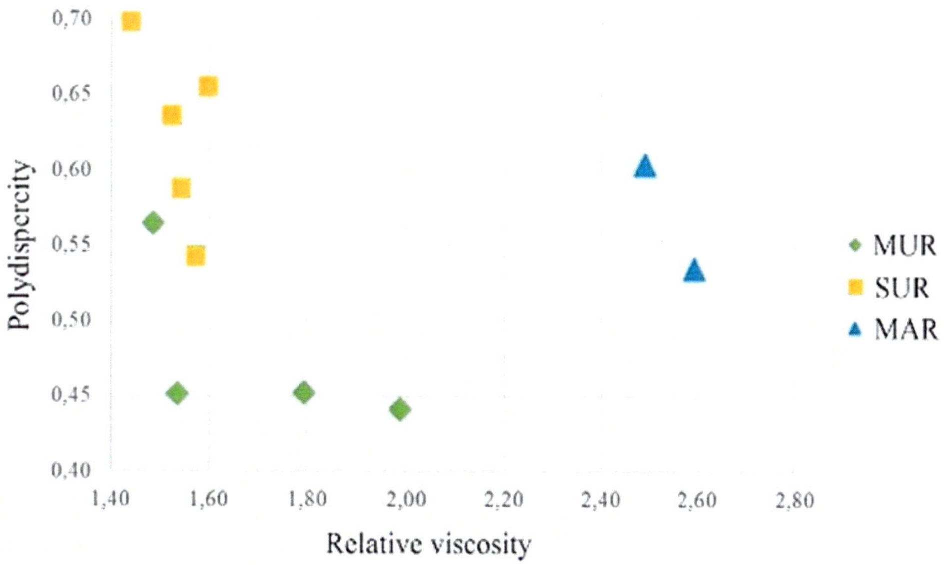

Fig. 6 Graph of relative viscosity

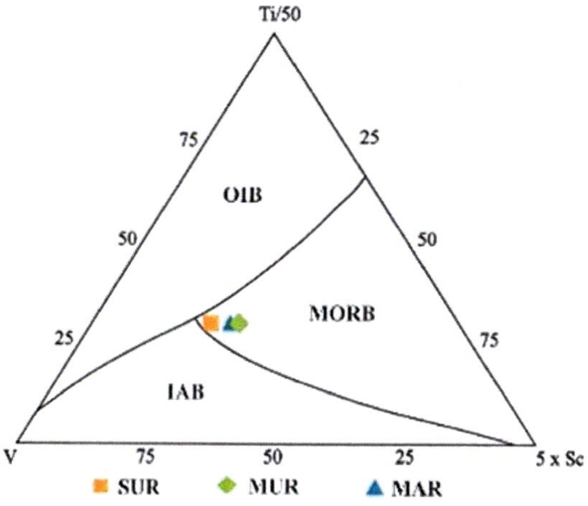

Fig. 7 Ti-V-Sc discrimination diagram shows the pillow lavas in this study are related to MOR

5 Conclusions

Based on the geochemical whole-rock analysis, pillow lavas at MUR, MAR, and SUR were formed in the mid-oceanic ridge tectonic setting. The process of forming these pillows was greatly controlled by the slope and depth of the water column. The results of the study conclude that there were three geological conditions when oceanic spreading floor occurred: (1) the most-gentle slope and very deep water column depth, (2) steep slope and shallower water column depth, (3) very steep slope and very shallow water column depth.

References

Abdurrachman, M., Widiyantoro, S., Saepuloh, A., and Kurniawan, I. A. (2019). *Tracing the missing Argoland beneath Java: Evidence from geochemical signature and seismic tomogram*. Springer Nature.

Asikin, S. (1974). *Geological evolution of Central Java* (pp. 71–92). Doctoral Dissertation, Institut Teknologi.

Brugger, C. R., & Hammer, J. E. (2010). Crystal size distribution analysis of plagioclase in experimentally decompressed hydrous rhyodacite magma. *Earth and Planetary Science Letter, 300*, 246–254.

Higgins, M. D. (2000). Measurement of crystal size distributions. *American Mineralogist, 85*, 1105–1116.

Klein, J., Mueller, S. P., Helo, C., Schweitzer, S., Gurioli, L., & Castro, J. M. (2018). An expanded model and application of the combined effect of crystal size distribution and cystal shape on the relative viscosity of magmas. *Journal of Volcanology and Geothermal Research, 357*, 128–133.

Liu, C., Curray, J. R., & McDonald, J. M. (1983). New constraint on the tectonic evolution of the eastern Indian Ocean. *Earth and Planetary Science Letter, 65*, 331–342.

Morgan, D. J., & Jerram, D. A. (2006). On estimating crystal shape for crystal size distribution analysis. *Journal of Volcanology and Geothermal Research, 154*, 1–7.

Puswanto, E., Hastria, D., & Ansori, C. (2012). Petrogenesis of andesit amygdaloid Jatibungkus in relation to mafic igneous rock of the Ophiolite Complex Karangsambung. In: *LIPI Proceeding*, Indonesia.

Renjith, M. L. (2014). Micro-textures in plagioclase from 1994–1995 eruption, Barren Island Volcano: Evidence of dynamic magma plumbing system in the Andaman subduction zone. *Geoscience Frontiers, 5*, 113–126.

Schnur, S. R. (2007). *An analysis of the morphology and physical properties of pillow lavas of the Nicasio Reservoir Terrane*. Marin County, California: Implications for seamount formation and structure, Theses Magister Program, Carleton College.

Suparka, M. E. (1988). *Petrology and geochemistry of northern Karangsambung Ophiolite Complex, Luh Ulo, Central Java* (pp. 87–90). Doctoral Dissertation, Institut Teknologi.

Toramaru, A., Noguchi, S., Oyoshihara, S., & Tsune, A. (2008). MND (microlite numbe r density) water exsolution rate meter. *Journal of Volcanology and Geothermal Research, 175*, 156–167.

Vermeesch, P. (2006). Tectonic discrimination diagram revisited. *Geochemistry Geophysics Geosystems, 7*.

Vinet, N., & Higgins, M. D. (2010). Magma solidification processes beneath Kilauea Volcano, Hawaii: A quantitative textural and geochemical study of the 1969–1974 Mauna Ulu Lavas. *Journal of Petrology, 51*, 1297–1332.

Walker, G. P. L. (1992). Morphometric study of pillow – size spectrum among pillow lavas. *Bulletin of Volcanology, 54*, 459–474.

Evaluation of Coal Quality: A Case Study of Ankpa Coal, Mamu Formation Anambra Basin, South-Eastern Nigeria

Yusuf Ayoola Jimoh, Ajadi Jimoh and Adewole Adetoro Ajala

Abstract

Coal, being an economic deposit, can be used for various purposes depending on their properties and composition. Hence, the need to carry out a detailed investigation on Ankpa coals. Ankpa coals fall within the Mamu Formation of Anambra Basin in south-eastern Nigeria. The coals were investigated to deduce the rank and the coal quality with respect to power generation, cokability, handling, fueling, combustion, agglomeration, and slagging potential. Nine coal samples were collected from two boreholes (BH-2) and (BH-4) having depths of 16.5 and 27.4 m, respectively, and were subjected to proximate, ultimate, and elemental analyses. Moisture content ranges from 2.10 to 8.75 wt%, ash content, 4.45–26.8 wt%, volatile matter, 40.10–56.9 wt%, and the fixed carbon 20.45–39.24 wt%. The moderate moisture, low to high ash content, and high volatiles matter in the coals suggest high volatile sub-bituminous, non-coking coal. The carbon content range from 23.3 to 52.4%, hydrogen 1.73–6.88%, oxygen 9.57–18.9%, nitrogen 1.42–2.93%, and sulphur 0.73–4.15%. The sulphur content (>0.8%) and nitrogen content (>1.5%) indicate non-coking coal and it would pollute the atmosphere during combustion by releasing flue gases. The agglomeration and slagging were deduced by the ratio of C/H, FC/V, SiO_2/Al_2O_3, Fe_2O_3/CaO, silica ratio (G), and base/acid ratio (B/A). The agglomeration level and tendency of the coals to slag varies from weak to strong. The coals are not suitable for coking, but can be used for power generation, combustion applications, and fuelling due to its high combustibility.

Y. A. Jimoh (✉) · A. Jimoh
Department of Geology and Mineral Science, Kwara State University, Malete, P.M.B 1530, Nigeria
e-mail: yusuf.jimoh@kwasu.edu.ng

A. A. Ajala
Research Divisions, Stalwart Ventures LLC, Newark, USA

Keywords

Agglomeration · Coal · Cokability · Formation · Proximate

1 Introduction

Coals exist in many parts of the world such as China, Australia, USA, Russia, and Germany. In Nigeria, coal deposits are found mainly in Benue Trough and Anambra Basin. Minor occurrences have been reported in Bida and Sokoto Basins. The Cretaceous Anambra basin (Fig. 1) is one of the inland sedimentary basins in Nigeria formed during the break up and separation of the African and South American plates (Burke et al., 1971). The current energy crisis facing the country has made Nigeria embrace power generation from coal as a major source of energy, in addition to conventional hydro and thermal sources. Within the Anambra Basin, the Campanian–Maastrichtian Mamu Formation has coal deposits in localities such as Owukpa, Okaba, and Ogboyaga. The Awo Akpali coal mine in Ankpa is recently discovered and little or no information on the coal has been provided with respect to its quality and possible utilization. The studies were principally carried out to determine the rank, cokability, power generation potential, combustion, agglomeration, and slagging potential of the coal in order to ascertain their potential relevance to possible utilization and resource potential.

2 Materials and Methods

Nine coal samples from two boreholes were collected from Awo Akpali coal mine in Ankpa. The coordinates of the sample area fall in Anambra basin within latitudes N 7° 23′ and N 7° 26″ and longitudes E 7° 36′ and E 7° 39″. The studied boreholes are BH-2 and BH-4 having depths of 16.5 m and 27.4 m, respectively. Four coal samples were

F. Lucci et al. (eds.), *Selected Studies in Geomorphology, Sedimentology, and Geochemistry*,
Advances in Science, Technology & Innovation, https://doi.org/10.1007/978-3-031-43744-1_6

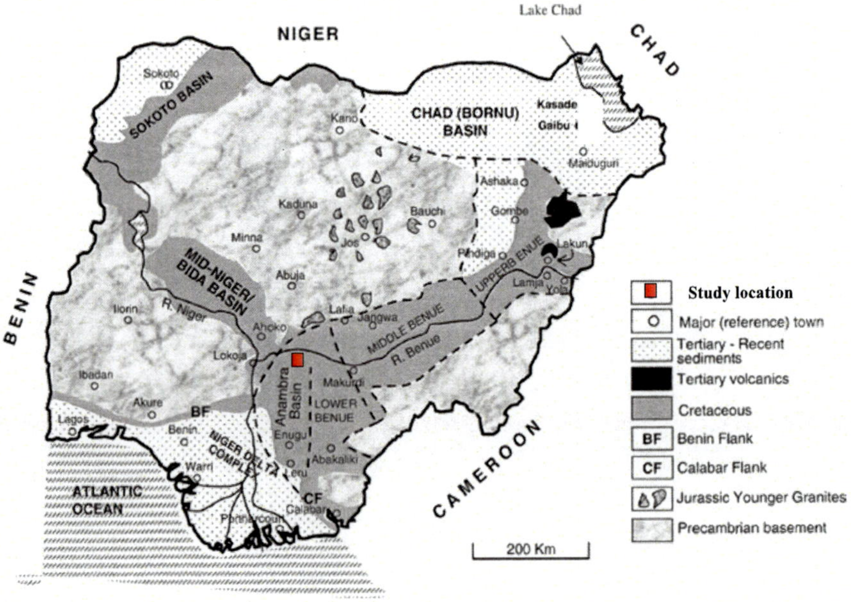

Fig. 1 Geological map of Nigeria showing the location of study (Anambra basin) modified after (Obaje, 2009)

obtained from BH-2, while five coal samples were collected from BH-4. Seam 1 and seam 2 of BH-2 are 1.4 and 0.5 m thick, respectively, while Seam 1 and 2 of BH-4 have thicknesses of 0.6 m and 1 m, respectively. The proximate analysis of the coals was carried out by Proximate Analyser (VG0STBR model), ultimate by Ultimate Analyser (Advant'X model), and elemental analysis was performed with XRF machine (Thermo scientific advant'X 1200 model). All the analyses were carried out according to American Society for Testing and Materials (ASTM) D388 (1992) at National Steel Raw Material Exploration Agency (NSRMEA) Kaduna Nigeria. Nine coal samples were subjected to proximate analysis to determine the moisture content, volatile matter, ash, fixed carbon, and calorific value, while five samples were analysed for ultimate analyses to determine the carbon, hydrogen, oxygen, nitrogen, and sulphur content. The inorganic geochemical analyses were carried out mainly for the major oxide elements.

3 Results

The results of the proximate, ultimate, and major oxide geochemistry are presented in Tables 1, 2, and 3.

4 Discussion

4.1 Coal Quality, Rank, and Agglomeration

Based on the proximate data (Table 1), the moisture content ranges from 2.10 to 8.75 wt%, ash content varies between 4.45 and 26.8 wt%, volatile matter ranges from 40.10 to 56.9 wt%, while the fixed carbon is between 20.45 and 39.24 wt%. The coal has moderate moisture, low to high ash content, and high volatile matter; thus classified as high volatile sub-bituminous coal (Guo et al., 2018). The volatile

Table 1 Proximate data and Calorific values of Ankpa coal samples (air dried)

S/N	Samples code	Sample type	Lithology	M (%)	Ash (%)	VM (%)	FC (%)	CV (MJ/Kg)	FC/VM
1	AK1A	Borehole	Coal	4.11	20.04	51.90	23.95	26.24	0.46
2	AK1B	Borehole	Coal	6.69	21.30	43.70	28.31	23.94	0.65
3	AK1C	Borehole	Coal	5.41	10.05	45.82	38.72	25.01	0.85
4	AK1D	Borehole	Coal	7.61	4.45	48.70	39.24	26.70	0.81
5	AK3A	Borehole	Coal	8.75	13.90	56.90	20.45	28.85	0.36
6	AK3B	Borehole	Coal	8.52	7.70	52.30	31.48	27.71	0.60
7	AK3C	Borehole	Coal	2.70	19.60	40.10	37.60	21.21	0.94
8	AK3D	Borehole	Coal	2.10	26.80	52.90	28.20	26.99	0.53
9	AK3E	Borehole	Coal	3.95	23.90	43.70	28.45	22.51	0.65

Note M: moisture content; VM: volatile matter; FC: fixed carbon; CV: calorific value

Table 2 Ultimate data of Ankpa coal samples

S/N	Sample code	Carbon (%)	Hydrogen (%)	Oxygen (%)	Nitrogen (%)	Sulphur (%)	C/H
1	AK3A	72.42	6.14	9.57	1.83	1.82	11.79
2	AK3B	73.61	6.88	11.94	2.90	1.59	10.69
3	AK3C	23.28	1.73	18.96	1.42	4.15	13.46
4	AK3D	58.78	4.68	14.43	2.41	0.73	12.56
5	AK3E	70.22	5.97	12.35	2.51	1.51	11.76

Note C/H: carbon/hydrogen

Table 3 Major elemental composition of Ankpa coal samples analysed

S/N	Sample code	Lithology	Sample Type	SiO_2	Al_2O_3	Fe_2O_3	TiO_2	MgO	P_2O_5	K_2O	CaO	Na_2O
1	AK1A	Coal	Borehole	24.24	7.99	2.37	2.65	0.91	1.26	0.61	0.48	0.38
2	AK1B	Coal	Borehole	16.26	6.59	4.55	2.07	0.45	0.89	0.46	0.71	0.40
3	AK1C	Coal	Borehole	11.02	3.25	2.53	1.38	0.81	1.44	0.24	0.56	0.42
4	AK1D	Coal	Borehole	25.56	6.12	3.39	-	0.61	0.87	-	0.36	0.20
5	AK3A	Coal	Borehole	11.51	3.80	4.62	0.93	0.51	1.47	0.60	1.34	0.51
6	AK3B	Coal	Borehole	6.52	3.93	2.69	1.05	0.23	1.58	-	0.46	0.63
7	AK3C	Coal	Borehole	58.04	19.29	8.69	2.02	1.13	1.86	2.96	1.22	0.57
8	AK3D	Coal	Borehole	36.82	5.80	1.04	-	0.70	1.21	1.01	0.31	0.44
9	AK3E	Coal	Borehole	11.92	4.59	3.89	1.08	0.56	0.71	0.31	0.70	0.59

matter contents and ultimate analysis data (Table 2) are in accordance with rank. The results of ultimate analyses (Table 2) and rank determinations classify these coals as sub-bituminous B/C (Guo et al., 2018). These types of coals are characterized by low nitrogen but high oxygen and relatively moderate sulphur contents. In the evaluation of coal quality, the most important relationship is that between volatile matter and ash content (Dıez et al., 2002), and in the studied samples, this relationship was found to be negative.

With respect to cokability, the moisture content (up to 8%), the high volatile matter (>36%), and high ash (>10%) indicate non-coking coals (Dıez et al., 2002; Zhang et al., 2014). The high volatiles in the coals are favourable for combustion applications and fuelling (Guo et al., 2018). The ratio of FC/V was also used to deduce the agglomeration and slagging tendency.

The percentage of carbon from the ultimate analysis of coal sample from BH-4 is not consistent with the fixed carbon obtained from proximate analysis. The carbon content ranges from 23.3 to 72.4%, hydrogen 1.73 to 6.88%, oxygen 9.57 to 18.9%, nitrogen 1.42 to 2.93%, and sulphur, 0.73 to 4.15%. Coal quality with respect to cokability, agglomeration tendency, and rank was deduced. The coal has high amount of sulphur (> 0.8%) which suggests it is not suitable for coking (Mochizuki et al., 2013). The high nitrogen content (>1.5%) in the coal would pollute the atmosphere during combustion by releasing flue gases (Carpenter et al., 2007). The agglomeration and slagging tendency were deduced by the ratio of C/H and indices

Table 4 Limits of the discriminant index for slagging (Guo et al., 2018)

Tendency of slagging			
Index	Weakly	Medium	Strongly
B/A	<0.206	0.206–0.4	>0.4
G	>78.8	78.8–66.1	<66.1
SiO_2/Al_2O_3	<1.87	1.87–2.65	>2.65
Fe_2O_3/CaO	Out of 0.3–3	0.3–3	Near 1

generated from the major elemental composition (Tables 4 and 5). The agglomeration level of the coals varies between weak, medium, and strongly (Table 5).

5 Conclusion

Proximate analysis and calorific value revealed that Ankpa coal has moderate moisture content, high volatile matter and ash, suggesting it can be used in coal-fired power plant generation and fuelling due to its high combustibility. The high volatile matter and sulphur in the coal classify it as non-coking. The variation in the agglomeration level is due to combination of different agglomeration indices which ranges between weak, medium, and strong. Coal beneficiation and blending are required. It is been utilized in coal power plants to minimize environmental pollution that would arise during combustion due to high sulphur content.

Table 5 Indicators and agglomeration level of Ankpacoal samples analysed

Sample code	C/H	FC/V	B/A	G	S/A	I/C	Agglomeration degree
AK1A	–	0.46	0.14 (Weakly)	86.6 (Weakly)	3.03 (Strongly)	4.94 (Weakly)	Weakly
AK1B	–	0.65	0.26 (Medium)	74.0 (Medium)	2.47 (Medium)	6.41(Weakly)	Medium
AK1C	–	0.85	0.29 (Medium)	73.9 (Medium)	3.39 (Strongly)	4.52 (Weakly)	Medium
AK1D	–	0.81	–	85.4 (Weakly)	4.18 (Strongly)	9.42 (Weakly)	Weakly-Strongly
AK3A	11.79	0.36	0.47 (Strongly)	64.0 (Strongly)	3.03 (Strongly)	3.45 (Weakly)	Strongly
AK3B	10.69	0.60	–	65.9 (Strongly)	1.66 (Weakly)	5.85 (Weakly)	Weakly-Strongly
AK3C	13.46	0.94	0.18 (Weakly)	84.0 (Weakly)	3.01 (Strongly)	7.12(Weakly)	Weakly
AK3D	12.56	0.53	–	94.7 (Weakly)	6.35 (Strongly)	3.35(Weakly)	Weakly
AK3E	11.76	0.65	0.34 (Medium)	69.8 (Medium)	2.60 (Medium)	5.56 (Weakly)	Medium

Note C/H = carbon/hydrogen; FC/V = fixed carbon/volatile matter; B/A, base/acid ratio = $(Fe_2O_3 + CaO + MgO + Na_2O + K_2O)/(SiO_2 + Al_2O_3 + TiO_2)$; G, silica ratio = $100 \cdot SiO_2/(SiO_2 + Fe_2O_3 + CaO + MgO)$; I/C = Fe_2O_3/CaO, S/A = SiO_2/Al_2O_3

The rank of the coal is high volatile sub-bituminous based on the calorific value. Desulphurization of the coals should be carried out before burning to mitigate the effect of sulphur dioxide emission.

References

American Society for Testing and Materials (ASTM) D388. (1992). Standard classification of coals by rank. In *Annual Book of ASTM Standards, section 5, Petroleum products, Lubricants and fossil fuels, Gaseous Fuels: Coal and coke* (vol. 0505, pp. 192–195). ASTM.

Burke, K. C., Dessauvagie, T. F. J., & Whiteman, A. J. (1971). The opening of the Gulf of Guinea and the geological history of the Benue depression and Niger delta. *Nature Physical Science, 233*, 51–55.

Carpenter, A. M., Niksa, S. R. I., Scott, D. H., & Wu, Z. (2007). *Fundamentals of coal combustion*. IEA Clean Coal Center.

Dıez, M. A., Alvarez, R., & Barriocanal, C. (2002). Coal for metallurgical coke production: Predictions of coke quality and future requirements for cokemaking. *International Journal of Coal Geology, 50*(1), 389–412.

Guo, L., Zhai, M., Wang, Z., Zhang, Y., & Dong, P. (2018). Comprehensive coal quality index for evaluation of coal agglomeration characteristics. *Fuel, 23*, 379–386.

Mochizuki, Y., Ono, Y., Uebo, K., et al. (2013). The fate of sulfur in coal during carbonization and its effect on coal fluidity. *International Journal of Coal Geology, 120*, 50–56.

Obaje, N. G. (2009). *Geology and mineral resources of Nigeria* (p. 221). Lecture Notes in Earth Sciences. Springer-Verlag.

Zhang, L., Liu, W., & Men, D. (2014). Preparation and coking properties of coal maceral concentrates. *International Journal of Mining Science and Technology, 24*(1), 93–98.

3D Modeling of the Hydrothermal Alteration Zoning and Correlation to Ore Formation, the Perama Hill Epithermal Au Deposit, Thrace, NE Greece

Stavros Triantafyllidis and Filippos Drakoulakos

Abstract

A total number of 65 samples from 13 drill holes from the Perama Hill epithermal Au deposit were analyzed by X-ray diffraction, Scanning Electron Microscopy, and Transmitted and Reflected Light Microscopy. The major lithologies and the four identified hydrothermal alteration zones, including an Argillic to Advanced Argillic alteration zone, an Argillic alteration zone, a Silicification zone, and an Intermediate Silicic-Argillic alteration zone, were imported in the Geovia SURPAC™ software and 3D models of host rock lithology and hydrothermal alteration zoning of the deposit were modeled. The models show that the most intense hydrothermal alteration phenomena occurred at the central and southern part of the Perama Hill deposit. At the central part, a large depression is present indicating the pathways of the main feeder zones which are closely associated to the Advance Argillic to Argillic hydrothermal alteration and Silicification zones with evidence of overprinting between the two. The 3D modeling also showed a sudden spatial dispersion of the hydrothermal alteration zones (Argillic alteration) at the transition between the lower cohesive andesitic breccia and the overlying unconsolidated and highly porous felsic volcanic sandstone. The precious and critical metal content of the deposit is rapidly increased at the intermediate conglomerate, indicating that the sudden change in host rock hydraulic conductivity played a major role in the formation of the Perama Hill Au deposit.

Keywords

Epithermal Au system · 3D modeling · Hydrothermal alteration · Precious and critical metals

1 Introduction

The Perama Hill epithermal Au deposit is located approximately 30 km W–NW of Alexandroupolis city, Thrace, NE Greece (Fig. 1a). The deposit is mushroom shaped, and fine grained free Au mineralization is hosted in the upper felsic sandstone (Fig. 1b). The calculated reserves are 9.4 Mt with average grade of 3.20 gr/tn Au and 3.75 gr/tn Ag. Within the sandstone, Au grades range between 2 and 10 g/tn, whereas in the underlying sector (refractory ore), the Au grades are below 1 g/tn (Juras et al., 2010).

2 Analytical Methods-Methodology

A total number of 65 samples from 13 boreholes from the Perama Hill epithermal Au deposit were used in this study (Fig. 1a). The host rock lithology and hydrothermal alteration parageneses were determined by combination of transmitted and reflected light optical microscopy, X-ray Diffractometry and Scanning Electron Microscopy.

The Geovia SURPAC™ software was employed in designing the 3D models of the lithology and hydrothermal alteration zoning of the deposit. Two separate models were developed using the triangulation solids method and the block model method, with the latter providing the best results, namely due to the relatively limited number of drill holes and samples. The block model grid dimensions were set to 5 m × 5 m × 5 m, based on drill hole spacing, with sub-blocks of the same size. The model was spatially estimated using the nearest neighbor method, which assigns a value from the nearest sample. For the lithology model, the maximum search radius was set to 110 m and the maximum

S. Triantafyllidis (✉) · F. Drakoulakos
School of Mining and Metallurgical Engineering, National Technical University of Athens, Iroon Polytechneiou 9, 157 80, Zografou, Athens, Greece
e-mail: striantafyllidis@metal.ntua.gr

F. Lucci et al. (eds.), *Selected Studies in Geomorphology, Sedimentology, and Geochemistry*,
Advances in Science, Technology & Innovation, https://doi.org/10.1007/978-3-031-43744-1_7

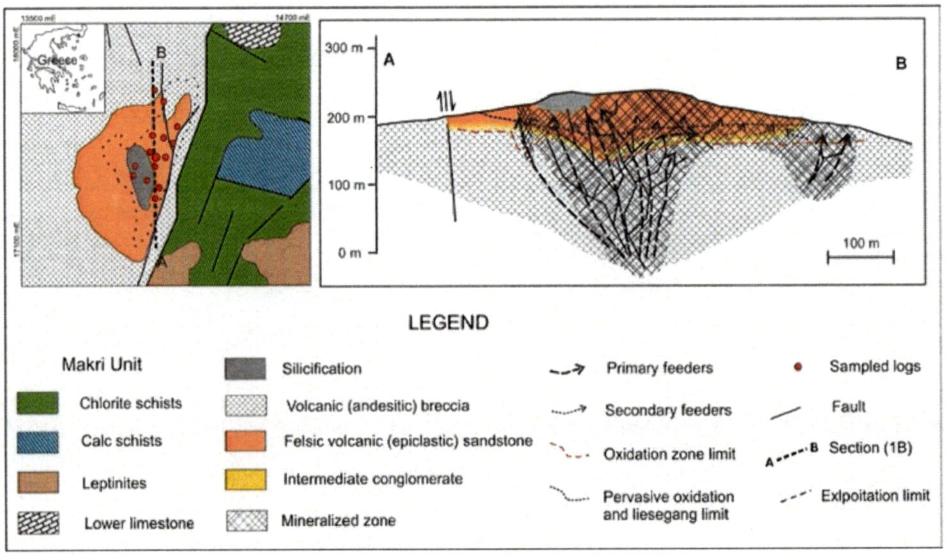

Fig. 1 Simplified geological map (**a**) and cross section of the Perama Hill epithermal Au deposit, with modifications after (Triantafyllidis & Skarpelis, 2015)

vertical search distance was set to 20 m, depending on drill hole available data, whereas for the hydrothermal alteration zoning, the maximum search radius was set to 70 m and the maximum vertical search distance was set to 20 m.

3 Results

Three main lithologies were distinguished, including the lower volcanic (andesitic) breccia (VB), the upper felsic volcanic (epiclastic) sandstone (FES), and the intermediate conglomerate (IC) marking the transition from the lower andesitic breccia to the overlying sandstone. Four hydrothermal alteration zones were defined based on gangue mineralogy and texture including an Argillic to Advanced Argillic alteration (AAZ) with quartz, kaolinite and subordinate Aluminum Phosphates-Sulfates (APS-svanbergite, woodhouseite, gorceixite and crandallite), an Argillic alteration (AZ) with quartz, kaolinite and minor illite, a Silicification (SZ) zone with predominant quartz and an Intermediate Silicic-Argillic alteration zone (ISAZ) with predominant quartz and minor kaolinite, marking the transition between the Argillic and Silicification zones. Major precious metal phases identified include free Au grains, Ag and Au–Ag tellurides (Table 1), Au–Ag ± Cu alloys, whereas major critical metal phases include bismuthinite (Bi_2S_3), kawazulite [$Bi_2(Te,Se,S)_3$], native Bi, and unidentified Bi-Se and Bi-Se ± Cu phases.

4 Discussion

Based on the 3D modeling, most intense hydrothermal alteration phenomena occurred at the central and southern parts of the Perama Hill deposit, with the development of the SZ and the AAZ hydrothermal alterations. Evidence of overprinting is present, particularly at the central part where the AAZ hydrothermal alteration alternates with the SZ and the ISAZ. The outcome of the geology model demonstrates the collapse (depression) at the central part of the deposit, indicating the pathways of the main feeder zones which are closely associated to the AAZ and SZ hydrothermal alterations. Precious metal phases, including Au–Ag ± Cu alloys, free Au–Ag grains, Ag- and Au–Ag tellurides (Table 1), and Bi-Te-Se phases are closely related to the AAZ and SZ alterations. The increased precious and critical metal content at the intermediate conglomerate (IC) is related to the sudden change in host rock hydraulic conductivity, separating the cohesive lower andesitic breccia–VB (refractory ore) and the unconsolidated, highly porous and syngenetic

Table 1 Representative SEM–EDS analyses (in wt%) of Au and Ag phases and host rock lithology from the Perama Hill epithermal Au deposit

Ore phase	Lithology	Au	Ag	Cu	Te	Total
Petzite	IC	25.61	42.08	n.d	33.11	100.79
Hessite	IC	n.d	63.57	n.d	37.30	100.87
Au–Ag–Cu alloy	IC	90.38	8.17	0.71	n.d	99.26
Au–Ag–Cu alloy	IC	91.57	7.52	0.76	n.d	99.85
Free Au	FES	100.69	n.d	n.d	n.d	100.69

n.d. Not detected

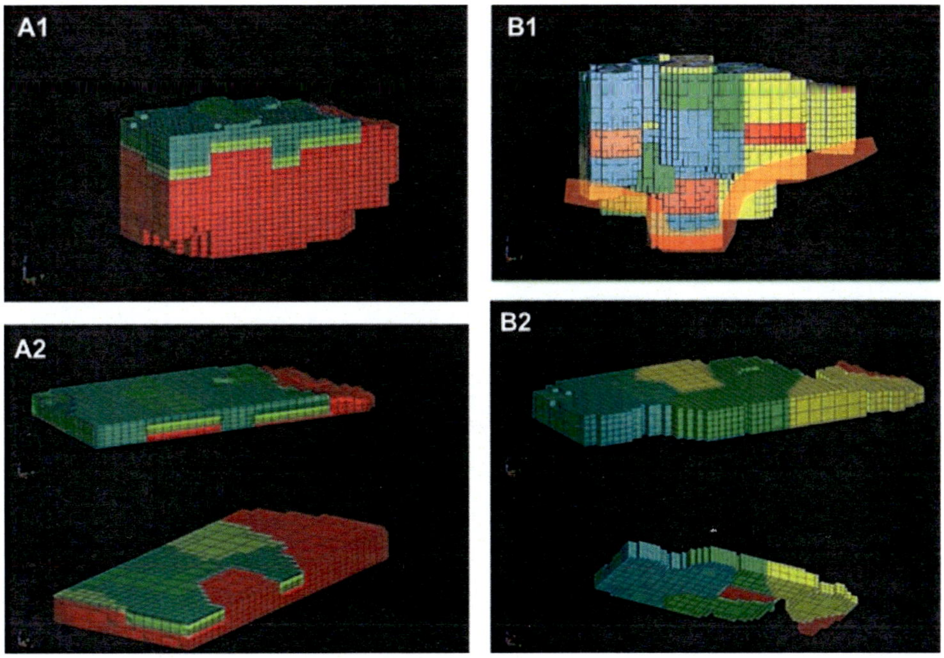

Fig. 2 3D models of the lithology and hydrothermal alteration zoning of the Perama Hill epithermal system based on the Geovia SURPAK™ block model (Drakoulakos, 2019). A1. Lithology block model. A2. Lithology sections at 200 m (up) and 150 m (low) a.s.l. (above sea level). VB-red, IC-yellow, FES-green. B1. Hydrothermal alteration zoning block model, with the IC depicted (shaded orange). B2. Sections at 200 m a.s.l. AAZ-red, SZ-cyan, ISAZ-green, AZ-yellow

to the ore overlying felsic sandstone (FES). This observation is also depicted in the 3D modeling of the hydrothermal alteration zoning (AZ), with sudden spatial dispersion of the Argillic alteration zone (AZ) toward the flanks of the deposits in both IC and the overlying FES relative to the lower VB (Fig. 2).

5 Conclusions

The 3D modeling of the Perama Hill epithermal Au deposit showed that differences in hydraulic conductivity between the lower cohesive VB and the overlying unconsolidated IC and FES played a key role in the shape, texture, hydrothermal alteration zoning and precious metal content of the system. Precious and critical metal phases are largely concentrated in the IC and the lower part of the FES and is strongly correlated to the AAZ and SZ hydrothermal alterations.

References

Drakoulakos, F. (2019). *Three dimensional modeling of the Perama Hill epithermal Au system based on ore mineralogy and hydrothermal alteration zoning, Thrace, NE Greece*. Unpubl. MSc Thesis, School of Mining and Metallurgical Engineering, National Technical University of Athens.

Juras, S., Miller, R., Perkins, P. (2010). *Technical report, of the Perama Hill project, Thrace, Greece*. Unpublished Report, Eldorado Gold Corporation.

Triantafyllidis, S., & Skarpelis, N. (2015). Supergene alteration of a refractory gold mineralization: The Perama hill deposit, NE Greece. *Bulletin Geological Society Greece, 49*, 59–75

Minerals with Economic Potential, Present in the Pegmatites from Conțu, Cindrel Mountains, (Romania)

Nicolae Călin, Ciprian Constantina, Oana-Claudia Ciobotea-Barbu, Aurora-Măruța Iancu, Adriana-Mariana Ion, Delia-Georgeta Dumitraș, Diana Perșa, Ștefan Marincea, Oana-Gabriela Rădoi and Maria-Angela Anason

Abstract

In this study, the characterization of the minerals from Conțu pegmatites, which are of economic interest, will be presented. The investigation was carried out using Polarized Light Microscopy (PLM), X-Ray Powder Diffraction (XRD), and Electron Probe Micro-Analyzer (EPMA). The main minerals are represented by albite (cleavelandite variety), spodumene, muscovite, and quartz. Accessory minerals are sphene, beryl, tourmaline (schorl variety), sillimanite, cassiterite, microcline, uraninite, garnets (almandine and spessartine variety), columbite-tantalite, vivianite, triphylite, ferrisicklerite, heterosite, amblygonite–montebrasite, monazite, fluorapatite, hydroxylapatite, wolfeite, maricite, and "ferrogatehouseite". The Conțu pegmatitic field's names (Conțu Inferior and Conțu Superior) are located in the area of the pegmatitic body, between the Lotru valley and the Sebeș valley close to the kyanite ferrous rock deposit from Negovanu (Lotru mountains). Conțu pegmatites can be classified into Li-rich pegmatites and feldspathic pegmatites. The two types of pegmatites have different evolutionary stages. The first stage is represented by plagioclase (albite cleavelandite and oligoclase variety), which is joined by biotite–phyllosilicate and quartz. The second stage is characterized by the presence of the potassium feldspar (microcline) developed by the transformation of plagioclase feldspar, and muscovite formed by differential biotite or by substitution of plagioclase feldspar. Li-rich pegmatites constitute an evolved stage that corresponds to the accumulation of minerals rich in rare elements. The presence of lithium in silicates (spodumene) and phosphates (triphylite, ferrisicklerite, amblygonite, montebrasite) can certainly highlight a green energy source, which can be recycled indefinitely.

Keywords

EPMA · XRD · Triphylite · Spodumene · "Ferrogatehouseite"

1 Introduction

The Conțu fields pegmatites are located in gneiss, mica schists, and amphibolites formation correspond to the Carpathian geodynamic evolution. Li-rich pegmatites from two fields Conțu Inferior and Conțu Superior are the only formations with rare metal present in Romania.

2 Analytical Methods Used

2.1 Polarized Light Microscopy (PLM)

To identify minerals from Conțu pegmatites with polarizing microscope, we utilized thin sections. The used optical microscopes were Leica type optical microscope within the "Laboratoire de Minéralogie et Cristallochimie Département de Géologie, Ličge, Belgium". Figure 1 is indicated "ferrogatehouseite" with spodumene; ferrisicklerite, heterosite, muscovite, albite association; "ferrogatehouseite" in spodumene with spodumene and quartz; ferrisicklerite with albite.

N. Călin (✉) · C. Constantina · O.-C. Ciobotea-Barbu · A.-M. Iancu · A.-M. Ion · D.-G. Dumitraș · D. Perșa · Ș. Marincea · O.-G. Rădoi · M.-A. Anason
Geological Institute of Romania, Bucharest, Romania
e-mail: nicolae_cln@yahoo.com

O.-C. Ciobotea-Barbu
Faculty of Applied Chemistry and Materials Science, University POLITEHNICA of Bucharest, Bucharest, Romania

F. Lucci et al. (eds.), *Selected Studies in Geomorphology, Sedimentology, and Geochemistry*, Advances in Science, Technology & Innovation, https://doi.org/10.1007/978-3-031-43744-1_8

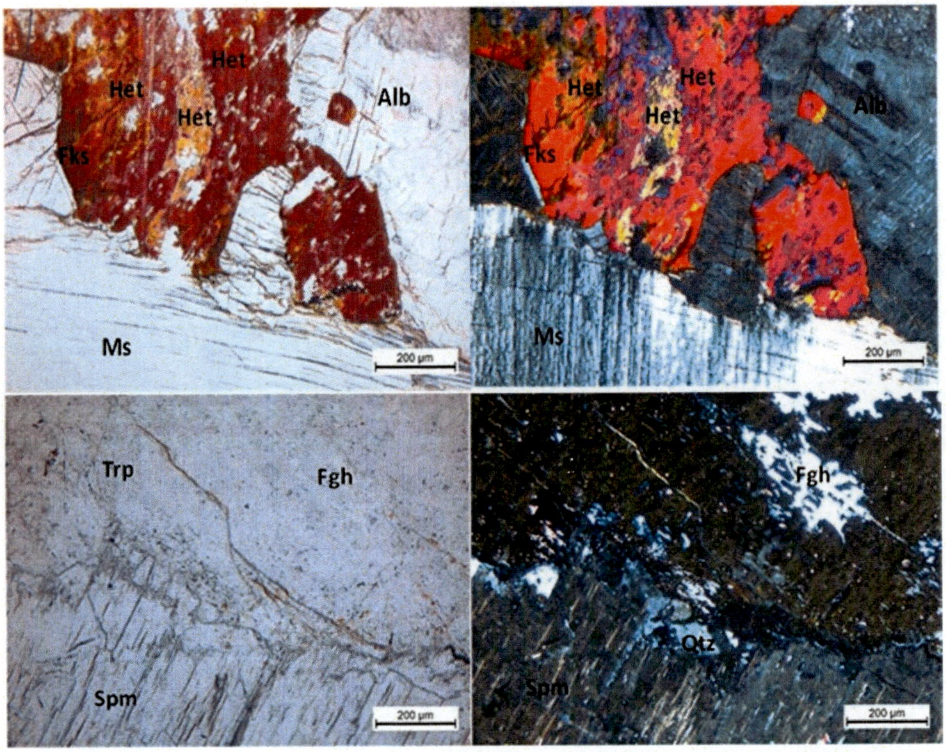

Fig. 1 On the left side–plane-polarized light; on the right side–crossed Nicols. "Ferrogatehouseite" (Fgh) appears on a cleavage plane of Triphylite (Trp), Heterosite (Het) appear in yellow and violet, and Ferrisicklerite (Fks) in red tones. Albite (Alb); Spodumene (Spm); Quartz (Qtz), Muscovite (Ms)

2.2 X-Ray Powder Diffraction (XRD)

To achieve the structure of minerals, we used X-ray powder diffraction measurements, diffractograms were acquired with BRUKER D8 ADVANCE (CuKα radiation, $\lambda = 1.5060$ Å), with an accelerating voltage of 40 kV and a beam current of 40 mA, with database ICDD, in the Laboratory Radiometry and RX, Geological Institute of Romania.

2.3 Electron Probe Micro-Analyzer (EPMA)

Quantitative chemical analyses were conducted with a CAMECA SX-50 electron microprobe mineralogy in the Laboratory of the Ruhr University, Bochum, Germany, operating in the wavelength-dispersion mode with a beam current of 20 mA and an accelerating voltage of 15 kV. The calibration was made with standard minerals like: titanite $CaTiSiO_5$; graftonite $(Fe^{2+},Mn,Ca)_3(PO_4)_2$; pyrope $Mg_3Al_2(SiO_4)_3$; andradite $Ca_3Fe_2(SiO_4)_3$; spessartine $Mn_3Al_2(SiO_4)_3$; jadeite $NaAlSi_2O_6$; and topaz $Al_2SiO_4(F,OH)_2$.

3 Results

In Li-rich pegmatite, the mineralogical association shows that we have granitic composition represented by the main minerals albite, spodumene, quartz, and muscovite. Accessory minerals are beryl, uraninite, petalite, almandine, spessartine, cassiterite, schorl, zircon, topaz, titanite, pyrite, fluorite, sillimanite, epidote, zoizite, biotite, microcline, microclineperthite, oligoclase, rutile, magnetite, hematite, columbite-tantalite, ilmenite, cookeite, kaolinite, illite, sericite, vivianite, triphylite, ferrisicklerite, heterosite, amblygonite, montebrasite, monazite, fluorapatite, hydroxylapatite, tavorite, variscite, hureaulite, alluaudite, sphalerite, maricite, wolfeite, and "ferrogatehouseite" $Ca_{0,0018}(Fe_{2,9631}Mn_{1,9575}Mg_{0,0355})_{4,9562}(PO_4)_2(OH)_4$. PLM indicate the isomorphic series triphylite–ferrisicklerite–heterosite; the inclusion "ferrogatehouseite" in primary phosphate triphylite, and in final the replacement of triphylite with wolfeite. XRD indicates the structure of minerals sample CS 2B1-spodumene a = 9.454 (Å), b = 8.406 (Å), c = 5.229 (Å), β = 110.08 (°), V = 391.20 (A^3); petalite a = 11.758 (Å), b = 5.146 (Å), c = 7.617 (Å), β = 113.06 (°), V = 424.05 (A^3); beryl a = 9.258 (Å), c = 9.204 (Å), V = 683.22 (A^3); triphylite a = 4.718 (Å), b = 10.378 (Å), c = 6.033 (Å). EPMA confirm minerals in sample 11/c2 triphylite $Li_{0,998}(Fe_{0.657}Mn_{0.332}Mg_{0.011})_{1,000}PO_4$; sample 42/c1 ferrisicklerite $(Li_{0,970}Ca_{0,014})_{0,985}(Fe_{0,663}Mn_{0,421}Mg_{0,003})_{1,088}PO_4$, sample 152/c1 maricite $(Na_{0,905}Ca_{0,006})_{0,911}(Fe_{0,624}Mn_{0,380}Mg_{0,004})_{1,008}PO_4$, sample 114/c6 wolfeite $Ca_{0,001}(Fe_{1,232}Mn_{0,718}Mg_{0,016}Al_{0,003})_{1,979}(OH)_{1,00}PO_4$.

4 Discussion

Fundamental research shows us a wealth of rare minerals. Li-rich pegmatites constitute an evolved stage in the pegmatitic process, corresponding to accumulation of minerals taking in their structure chemical elements that we find in the rare elements class. In spodumene mineral, we try to obtain lithium carbonate through an industrial process, and then extract lithium and use it. On the contrary, phosphate from isomorphic series triphylite–ferrisicklerite–heterosite presents an interest in green energy, since it can be used as electrodes in batteries (Xu et al., 2004; Zhou et al., 2004), particularly triphylite which can be used for recharging lithium batteries as an active electrode (Aldon et al., 2010; Arumungam et al., 2009; Dodd et al., 2007; Fehr et al., 2007).

5 Conclusions

The pegmatites of the Conţu field show practically no net contact with a magmatic body, but based on this mineralogical study, we can admit through the relations between minerals (graphic structures given by quartz and spodumene, quartz and potassium feldspars), as well as the transformations observed in the isomorphic series triphylite–ferrisicklerite-heterosite, that we are probably dealing with a magmatic intrusion located at considerable distances from pegmatitic bodies, which shows that the pegmatites are very fractionated and influenced by adjacent rocks in which they are emplaced.

References

Aldon, L., Perea, A., Womes, M., Ionica-Bousquet, C. M., & Jumas, J. -C. (2010). Determination of lamb—Mössbauer factors of $LiFePO_4$ and $FePO_4$ for electrochemical in situ and operando measurements in Li–ion batteries. *Journal of Solid State Chemistry, 183*, 218–222.

Arumungam, D., Kalignan, G. P., & Manisankar, P. (2009). Synthesis and electrochemical characterizations of nano-crystalline $LiFePO_4$ and Mg-doped $LiFePO_4$ cathode materials for rechargeable lithium-ion batteries. *Journal of Solid State Electrochemistry, 13*, 301–307. Author, F. (2016).

Author, F. (2016). Article title. *Journal* 2(5), 99–110.

Dodd, J. L., Halevy, I., & Fultz, B. (2007). Valence fluctuation of Fe-57 in disordered $Li_{0.6}FePO_4$. *Journal of Physical Chemistry C, 111*, 1563–1566.

Fehr, K. T., Hochleiter, R., Schmidbauer, E., & Schneider, J. (2007). Mineralogy Mössbauer spectra and electrical conductivity of triphylite $Li(Fe^{2+}, M^{n2}+)P_04$. *Physics and Chemistry of Minerals, 34*, 485–494.

Xu, Y. N., Ching, W. Y., & Chiang, Y. M. (2004). Comparative studies of the electronic structure of $LiFePO_4$, $FePO_4$, Li_3PO_4, $LiMnPO_4$, $LiCoPO_4$, and $LiNiPO_4$. *Journal of Applied Physics, 95*, 6583–6585.

Zhou, F., Kang, K. S., Maxisch, T., Ceder, G., & Morgan, D. (2004). The electronic structure and band gap of $LiFePO_4$ and $LiMnPO_4$. *Solid State Comm., 132*, 181–186.

The Genesis and Spatial Distribution of Ore Mineral Species of the Tschudi Copper Deposit, Namibia; Insights from Sequential Essay Modeling and Petrographic Studies

Josua Kalunde and Absai Vatuva

Abstract

Tschudi copper mine is situated within a platform sequence of the Neoproterozoic Damara Belt in northern Namibia. The Tschudi orebody is stratiform, dipping ~ 35° north, with a strike length of ~ 2.5 km, extending below 300 m. The deposit is hydrometallurgically complicated due to the presence of a variety of sulfide and oxide copper ores. The sulfide ores include bornite, chalcocite, covellite, and chalcopyrite, while the oxide ores consist of malachite, azurite, and cuprite. The copper is therefore extracted from the ores through selective leaching. Thus, understanding the spatial distribution of different mineral species and genesis is crucial for the economic exploitation of the Tschudi deposit. For this study, sequential assay data obtained from the mine were grouped in different oxidation percentages; < 5% is chalcopyrite, 5 to < 10% is bornite-covellite, 10 to < 50% is chalcocite, and > 50% are oxides. Individual mineral wireframes were created using SURPAC software and polished section petrographic studies, which enabled to model the distribution of ore minerals. The models were coupled with geological observations to discuss the genesis of the deposit. The mineralization is disseminated through sandstone and the intermittent basal conglomerate unit of the Tschudi Formation. The Tschudi Formation lies unconformably on the dolomitic carbonate sediments of the Huttenberg Formation. The model shows that the more acid-soluble copper oxides and chalcocite are proximal to the dolomite footwall. The less acid-soluble bornite, covellite, and chalcopyrite are closer to the hanging wall. The study shows that the Tschudi deposit is a sediment-hosted copper deposit that has undergone in situ supergene copper sulfide oxidation with limited copper supergene enrichment.

Keywords

Supergene oxidation · Mineral wire frame · Material type classification · Sedimentary hosted copper deposit · Fifth keyword

1 Introduction

The Tschudi deposit is mineralogically and texturally complex due to various sulfides (bornite, chalcocite, covellite, and chalcopyrite) and oxides (malachite, azurite, and cuprite) from which the copper is extracted through the hydrometallurgical process. As outlined by Bogdanovic et al. (2016), malachite, chalcocite, and azurite are leached with an acid solution, while covellite and bornite are leached using cyanide solutions. However, copper is leached from chalcopyrite at a slower rate than chalcocite, covellite, and bornite (Hiroyoshi et al., 2000). There is thus a need for selective leaching to optimize the mining operation. The weathering classification system employed at the mine does not account for the mineral type and metallurgical processes. Mineral type classification can be achieved by calculating the oxidation ratio from sequential assays for ore intervals sampled (e.g., Beaumont, 2016; Beaumont & Musinguini, 2019; Jacobs, 2016). Furthermore, the genesis of the deposit is still not resolved, Viviers (1992) postulated a copperbelt model (epigenetic) and a hydrothermal model (diagenetic).

In this study, the spatial distribution of mineral species was determined using oxidation percentages that necessitated modeling individual wireframes of mineral species using SURPAC software. Furthermore, the mineral

J. Kalunde · A. Vatuva (✉)
University of Namibia Southern Campus, Keetmanshoop, Namibiae-mail: avatuva@gmail.com

F. Lucci et al. (eds.), *Selected Studies in Geomorphology, Sedimentology, and Geochemistry*,
Advances in Science, Technology & Innovation, https://doi.org/10.1007/978-3-031-43744-1_9

wireframes are coupled with the polished section's petrographic description and geology of the deposit to discuss the Tschudi deposit's genesis.

2 Methods

A total of 189 collars, 291 surveys, and 3078 geochemical assay data was validated before being imported into the database for pit5a3. A sequential data geometallurgical mineral type classification modeling was employed by calculating the oxidation percentage using Equation: Oxidation (%) = AsCu/TCu * 100, whereby AsCu is acid-soluble copper and TCu is total copper analyzed.

The calculation of the oxidation percentage was done on assay data, whereby if the oxidation ratio percentage is equal to or more than 50% color-coded green, this represents oxides (mainly malachite and azurite), 10–50% red, chalcocite; 5–10%, bornite-covellite; and < 5%, chalcopyrite (Fig. 1). Cross sections were generated across pit5a3 using the generated database, and different mineral species were domained/modeled using SURPAC software. The polished section's petrographic description of minerals was done using a reflected light microscope.

3 Results

The Tschudi ore body is hosted within the Tschudi Formation's basal arenite (hanging wall), unconformably overlying the dolomites of the Huttenberg Formation (Hanging wall). The Tschudi deposit is located on the southern limb of the Tschudi-Uris Syncline as a planar stratiform deposit, dipping at approximately 25°–35° to the NNW. The dolomite is in some parts brecciated, probably due to small isolated solution collapse zones.

Three major sets of fault are identified in the deposit. The major set trends at 045°–070° and is sub-parallel to bedding and dips steeply toward the south. Other sets trend at 100°–120° and 140°–160° and are both steeply dipping.

Two zones of mineralization have been identified in pit5a3, upper and lower ore zones; they have similar lateral mineral zonation, Fig. 2. In the lower ore zone, oxidation percentages range between 10 and 100, which includes chalcocite and oxides, which are proximal to the dolomite footwall across the pit, while in the upper ore zone, these minerals are concentrated from the center to the east of the pit. Petrographic studies have confirmed the modeled spatial distribution of mineral species.

The data also shows that assays with oxidation percentages ranging from 0.1 to < 10 which includes bornite-covellite and chalcopyrite mineralization in the lower zone are toward the hanging wall across pit5a3, while in the upper zone, these minerals are constrained to the east of the pit.

4 Discussion

Sequential leaching is an important procedure used as a diagnostic tool to establish the exact location of copper minerals (Parkison, 1995). The use of sequential data assays to classify mineral types was proved successful by Jacobs (2016), Beaumont (2016), and Beaumont and Musinguini (2019) at the Kansanshi Mine by using oxidation ratios. Our study identified the spatial distribution of mineral species in pit PIT5A3 of the Tschudi deposit, recognizing two zones of mineralization. The lower ore zone forms a continuous layer lying on the dolomite, while occasional subparallel lenses/pods of mineralization occurring several meters above constitute the upper ore zone. The ore zones show similar lateral mineral zonation enabling four major

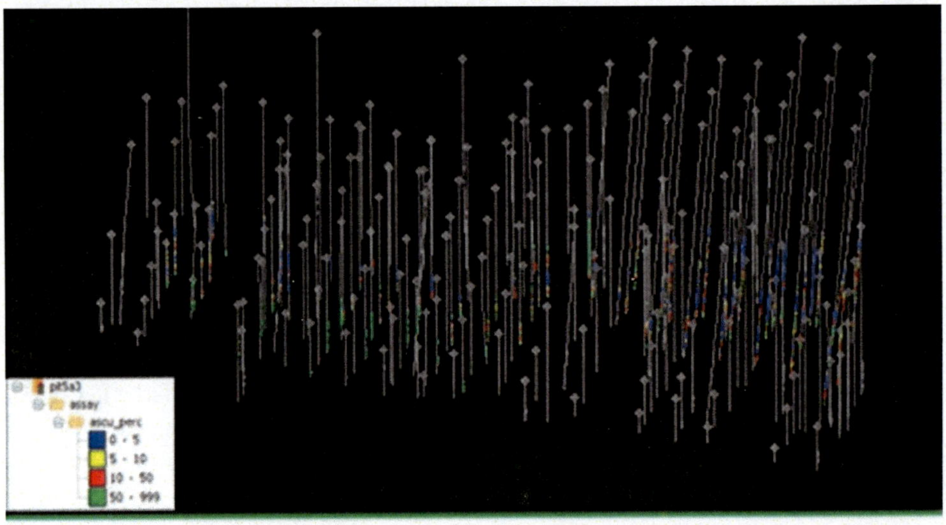

Fig. 1 Pit5a3 database showing acid-soluble drill hole assay results (green = oxides, yellow = chalcocite, blue = bornite-covellite and red = chalcocite)

Fig. 2 Pit5a3 plan view of mineral zonation at 1178 to 1172 elevation, note the oxide (green) and chalcocite (yellow) on footwall, bornite-covellite (blue), and chalcopyrite (red) on the hanging wall. Oxide and chalcocite were observed in the center of the pit due to in situ sulfide oxidation facilitated by faults

mineral wireframes to be modeled, namely oxides, chalcocite, bornite-covellite, and chalcopyrite.

Microscopy studies and pit wall mapping coupled with selected drill holes logging data within the pit shell confirmed the mineral groupings' spatial distribution within the ranges of the specified oxidation percentage. The modeled wireframes are to be used hand in hand with the mining composites such that in each ore composite, there are segments of mineral types that will allow selective mining, stockpiling, and crushing.

The geology and spatial distribution of ore minerals support a hydrothermal (diagenetic) genesis of the deposit postulated by Viviers (1992) instead of the epigenetic model. The oxide zone correlates well with major fault sets within the deposit. These features must have facilitated the weathering/oxidation of sulfide minerals at depth as alluded by many researchers (e.g., Lovering, 1948; Sillitoe, 2005; Spencer, 1917). The study shows that this sedimentary copper deposit has undergone an in situ sulfide supergene oxidation. The distribution of mineral species is controlled by their solubility, acidity, and faulting.

5 Conclusions

The study delineated the spatial distribution of different mineral species within the pit; hence, the modeled wireframes can be used for selective mining and stockpiling. The upper and lower orebody has similar mineral zonation; thus, sequential analysis material classification can be used even when mining progresses. The Run of Mine can be processed or stockpiled separately depending on the minerals' species. The majority of the readily leachable copper minerals (chalcocite and oxides) are proximal to the dolomite footwall while the less soluble (bornite-covellite) and insoluble copper minerals (chalcopyrite) are distributed toward the arenites in the hanging wall. Deep-seated thrust faults provide much of the permeability required to descent solution for supergene oxidation at greater depth. The effects of supergene oxidation are generally beneficial because they may generate copper ores amenable to heap leaching and SX-EW processing.

References

Beaumont, C. (2016). *A geo-metallurgical strategy for improving ore quality and mineral processing efficiency at Kansanshi mine in Zambia*. MSc Research Report. University of the Witwatersrand.

Beaumont, C., & Musinguini, C. (2019). Application of geometallurgical modelling to mine planning in a copper-gold mining operation for improving ore quality and mineral processing efficiency. *The Journal of African Institute of Mining and Metallurgy., 119*, 243–252.

Bogdanovic, G., Stankovic, V., Antic, D., Trumic, M., & Milicevic, D. (2016). *Leaching of low grade copper ores: A case study for "Kraku Bugaresku-Cementacija" deposits (Eastern Serbia)*. University of Belgrade.

Hiroyoshi, N., Miki, H., Hirajima, T., & Tsunekawa, M. (2000). A model for ferrous-promoted chalcopyrite leaching. *Hydrometallurgy, 57*, 31–38.

Jacobs, T. T. (2016). *Process mineralogical characterisation of the Kansanshi copper Ore, NW Zambia*. MSc thesis, University of Cape Town. https://open.uct.ac.za/handle/11427/20490

Lovering, T. S. (1948). Geothermal gradients, recent climatic changes, and rate of sulfide oxidation in the San Manuel district. *Arizona: Economic Geology, 43*, 1–20.

Parkison, G. A. (1995). The sequential copper analysis method-geological, mineralogical, and metallurgical implications. *Organization: Society for Mining, Metallurgy and Exploration*, 10.

Sillitoe, R. H. (2005). Supergene oxidized and enriched porphyry copper and related deposits. In *Society of economic geologists, Inc. Economic geology 100th anniversary* (pp. 723–768).
Spencer, A. C. (1917). *The geology and ore deposits of Ely* (p. 189). U.S. Geological Survey Professional Paper 96.
Viviers, G. J. (1992). *Geological report on the Tschudi copper/silver deposit*. Unpublished TCL company report.

Geology of Digum Vermiculite Deposit, Tigray, Northern Ethiopia

Bheemalingeswara Konka, Amaha Alemayohu Tesfay and Atakilt Araya

Abstract

Vermiculite deposit, about 65 mt, reported by Ezana Mining Company in northern Ethiopia is studied for petrography and geochemistry. The deposit, located near Digum village, is confined to a small elongated intrusive complex, about 0.7 Km^2. The complex is comprised of ultramafic, syenite, and alkali granitic rocks. The intrusive rocks have been emplaced into the Neoproterozoic basement rocks comprised of metavolcanic and phyllite rocks of Tsaliet Group. Vermiculite occurs up to a depth of ~ 25 m, formed by weathering of phlogopite present in ultramafic rocks. Petrography indicates the presence of phlogopite, and XRD data also suggest presence of vermiculite. Intrusive rocks geochemical data plot in fields of Ijolite/Ijolite-gabbro, syenite, and alkali granite. Data also suggest volcanic-arc tectonic setting. These rocks are comparable, spatially and temporally, with the nearby post-tectonic Negash igneous complex which varies in composition from ultramafic to felsic though dominantly intermediate. Digum complex is conspicuous in rock composition and facilitate vermiculite and provides an opportunity to study the petrogenesis of the intrusive complexes.

Keywords

Vermiculite · Phlogopite · Alkali pyroxenite/Ijolite · Digum · Ethiopia

B. Konka (✉)
Department of Mining and Geological Engineering, FET, BIUST, Palapye, Botswana
e-mail: Kbheema2006@gmail.com

A. A. Tesfay
Consultant Geologist, Mekelle, Tigray, Ethiopia

A. Araya
Ezana Mining Development PLC, Mekelle, Tigray, Ethiopia

1 Introduction

Vermiculite is a mica, a hydrated, magnesium–iron–aluminum trioctahedral sheet silicate of varied composition. It is known from 1930, sand now expanded its way into the modern life needs. It occurs as flakes, appears like mica, and belongs to mica group with light brown color. It is an altered product from biotite or phlogopite micas and is crushed and sorted into different sizes, large, medium, and fine. Higher lattice distance (11A$^{\circ}$) and the presence of water in the structure of vermiculite allow exfoliation about 30 times of its original size when heated. This allows trapping of air which reduces the bulk density from 640 to 960 kg/m^3 to 56–19 kg/m^3, becomes light weight. It is fire resistant, good absorbent, odorless, and has a high cation exchange capacity (CEC). It is effectively used in packaging, construction, automotive, and horticulture.

Vermiculite occurs as an alteration product of biotite, phlogopite, diopside, hornblende, and serpentine by weathering (about 100 m) or by low temperature (<350 °C) hydrothermal alteration of pyroxenites, peridotites, dunites, carbonatites, and amphibole schists cut by alkaline granitoids or pegmatites or mix of the two processes. Commonly associated mineral phases are corundum, apatite, serpentine, and talc.

Reserves of vermiculite, reported globally, are about 50 million tones and have an annual production of about 0.5 million tones (USGS, 2020). Major vermiculite producing countries are USA, South Africa, China, Russia, Brazil, and others. These deposits are reported to occur in Precambrian (1500–3000 Ma) rocks (e.g., Parabola region, South Africa; NW corners of China; Virginia and South Carolina, eastern Appalachian range, USA) and also in Triassic age (225 Ma) from Libby, Montana deposits in U.S.A.

There is a considerable gap between the demand and supply of vermiculite, globally. At the same time, many countries are reporting the presence of minable vermiculite

F. Lucci et al. (eds.), *Selected Studies in Geomorphology, Sedimentology, and Geochemistry*, Advances in Science, Technology & Innovation, https://doi.org/10.1007/978-3-031-43744-1_10

deposits in India, presence of about 50 million tones Namekara deposit in Uganda. Similarly, in Ethiopia, a small 0.65 million tones deposit is reported for the first time by Ezana Mining Development PLC in 2003 (EMD, 2003) near village Digum in the eastern zone of Tigray, northern Ethiopia. Digum is about 3 km from Abreha Atsbeha, 10 km from Hawzein, 10 km from Negash, and about 10 km from Wukro (Fig. 1). The deposit covers < 1 km^2 area and is bounded by the following UTM co-ordinates 542000E–5520000E and 1533000N–1543000N. The objective of the present study was to investigate different aspects of the deposit in terms of petrography, geochemistry, and petrogenesis.

2 Methodology

A detailed geological, structural map is prepared. 26 chip rocks from percussion drills and surface exposures were chosen for thin section preparation. 20 rock powders were chosen for major, minor, and trace elements using X-ray Fluorescence Spectrometer (XRF) and 3 rock powders were analyzed for mineralogy using X-ray Diffractometer.

3 Results

3.1 Geology of the Area

Basement rocks of Ethiopia form the southern-most part of Arabian Nubian Shield (ANS) and are characterized by low grade green schist facies island-arc assemblages, East dipping sutures, and N–NW trending shear zones and products of "Pan African Orogeny" (Stern, 1994; Abdelselam et al., 1996, and others). Basement of northern Ethiopia is comprised of metavolcanic-metasedimentary assemblage referred as Tsaliet Group, the younger metasedimentary as Tambien Group (Beyth et al., 1971), followed by intrusive plutons, Phanerozoic sediments and Cenozoic volcanics. Syn-tectonic (~800–735 Ma), confined to Tsaliet. Post-tectonic (~620–~520 Ma) (Beyth, 1972; Alene et al., 2006).

Metavolcanic and phyllite represent the basement and show sharp contact (Fig. 1) Phyllite shows N–S and NE–SW foliation, with vertical to sub-vertical dips and at places shearing. Quartz veins suggest hydrothermal activity in the area, like in Abrha Atsbeha, Negash, and Hawzein. Quartz veins are relatively more in phyllite, and they show different orientations (N10° E, N–S, E–W, and N70° E). A variety

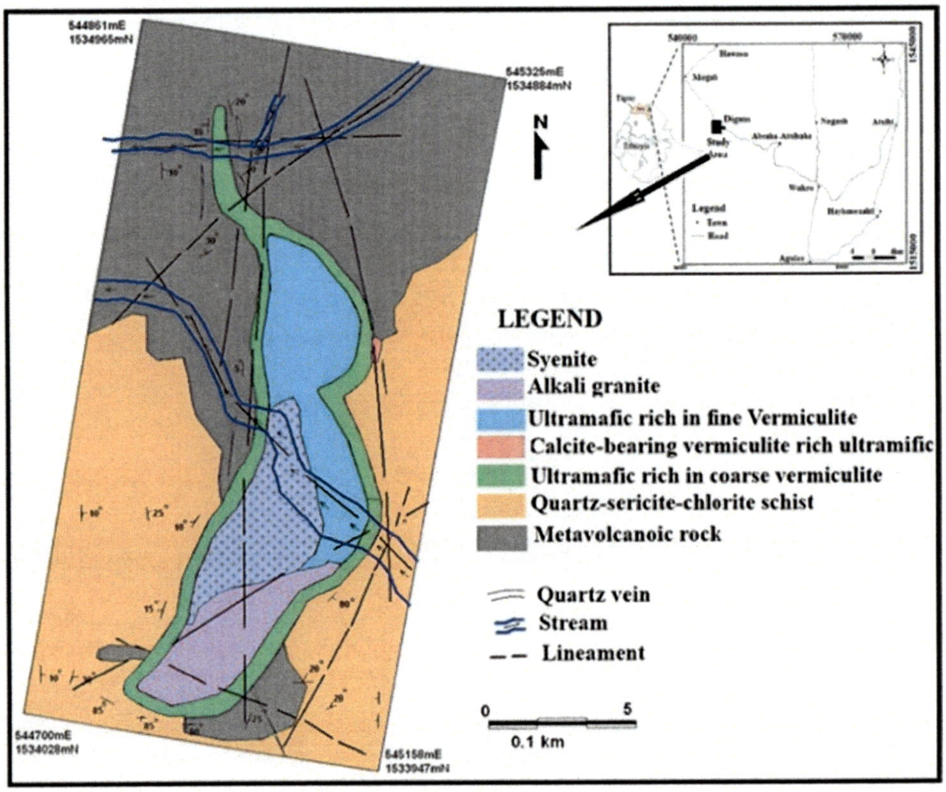

Fig. 1 Geological map of study area with location map, Ethiopia (after EMD, 2003)

of intrusive rocks intruding basement rocks are forming an elongated igneous complex. The complex is structurally controlled and post-tectonic. The intrusive rocks' effects are evident in phyllite, development of schistocity (quartz-sericite-chlorite schist), crenulations, and change in dip directions. Chloritization, epidotization, silicification, and kaolinization are common.

3.2 Mineralized Intrusive Complex

The intrusive complex hosts vermiculite deposit and is comprised of a series of igneous rocks, ultramafic (alkali pyroxenite/ Ijolite and Ijolite-gabbro), syenite, syeno-diorite, and alkali granite and cut by quartz, calcite veins. The composition, position, and orientation of the rocks of the complex suggest that each intrusive igneous rock belongs to different episode of intrusion, the younger in the inner part, the older is the periphery. The whole complex attained elongated shape following regional trend of NNE-SSW though bulging at the center indicating structural control.

The company has conducted percussion drilling and established presence of vermiculite deposit, varying in thickness from 2 to 21 m with an average of 8.23 m for coarse vermiculite along the boundary and from 4 to 18 m with an average of 13.8 m for fine.

3.3 Petrography and XRD

The petrographic study revealed presence of pyroxene, phlogopite, orthoclase, plagioclase, quartz, and calcite representing alkali granite, syenite, and ultramafic rock compositions. There has been considerable confusion between biotite and phlogopite, optically both have similar appearances. X-ray data (Fig. 2) compared with reported and standard charts and found that the samples contain significant amounts of vermiculite.

3.4 Geochemistry

The XRF data indicate that the vermiculite composition is comparable with the reported elsewhere. Depth-wise analysis clearly demonstrates an increase of Fe_2O_3, Al_2O_3, and MgO in the zone of vermiculite. In TAS diagram, the data plot in the fields of Ijolite, Ijolite-gabbro, syenite, alkali feldspars. Harker diagrams indicate positive relation

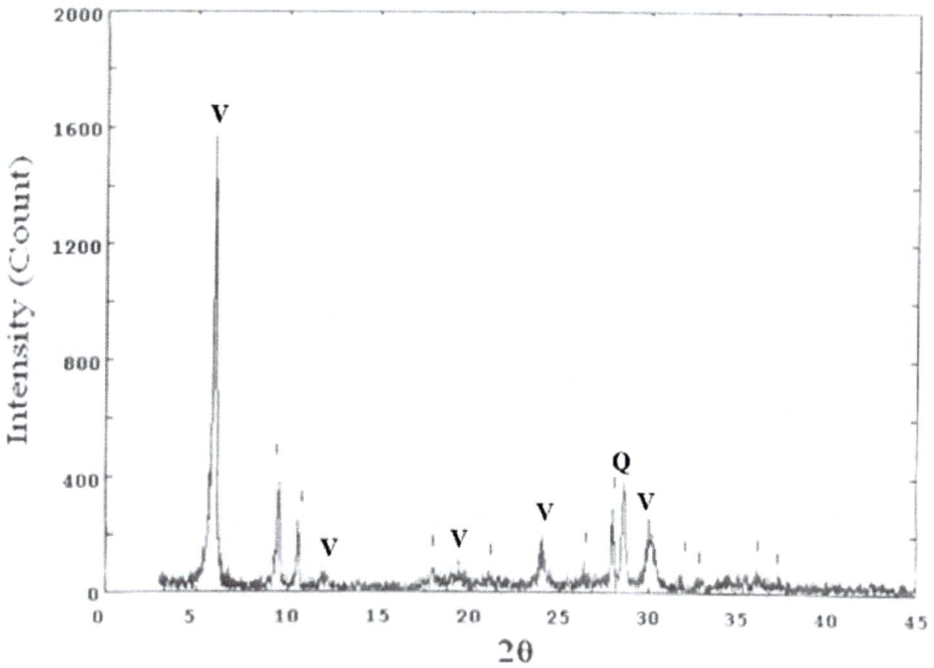

Fig. 2 Vermiculite diffractogram for samples from Digum, Tigray, Ethiopia

of SiO_2 with Na_2O, K_2O and negative with Fe_2O_3, MgO, and TiO_2. In discrimination diagram (Zr-Ti), the data plot mainly in island-arc tectonic setting.

4 Discussion

At Digum, an intrusive complex comprised of ultramafic to alkali granite trending NE-SW has intruded the Neoproterozoic basement. The intrusive hosted vermiculite deposit is in association with ultramafic rocks. The rock contains a considerable amount of phlogopite and pyroxenes. Geochemical data of the ultramafic plot in Ijolite/alkali pyroxenite fields. The paragenetic sequence suggests the oldest being ultramafic with and without calcite, and the youngest being alkali granitic rock. The intrusive is unique in the area and is spatially related to the nearby Negash Igneous Complex dominantly of diorite/granodiorite composition (Asrat et al., 2001) and compared its chemistry with the Cenozoic volcanics of northern Ethiopia (Le Bas and Mohr, 1968).

5 Conclusions

Vermiculite formed by weathering is ~ 0.38 mt coarse and ~ 0.27 mt fine. The intrusive rocks ultramafic (Ijolite/Ijolite-gabbro), syenite, syeno-diorite, and alkali granite form a complex, elongated and structurally controlled. Digum complex compares well with similar deposits reported elsewhere. The complex, post-tectonic, indicates spatial relation with the nearby, Negash Igneous Complex, intermediate in composition and post-tectonic. Digum may be an off shoot of the same magma chamber.

References

Abdelsalam, M. G., & Stern, R. J. (1996). Sutures and shear zones in the Arabian-Nubian Shield. *Journal of African Earth Sciences, 23*, 289–310.

Alene, M., Jenkin, G. R.T., Leng, M. J., Darbyshire, F. D. P. (2006). The Tambien group, Ethiopia: An early Cryogenian (ca. 800–735 Ma) Neoproterozoic sequence in the Arabian-Nubian shield. *Precambrian Research, 147*, 79–99.

Asrat, A., Barbey, P., & Gleizes, G. (2001). The Precambrian geology of Ethiopia: A review. *Africa Geoscience Review, 8*, 271–288.

Le Bas, M, J., Mohr, P. A. (1968). *Feldspathoidal rocks from the Cenozoic volcanic province of Ethiopia. Geologische Rundschau, Bd., 58* (1968).

Beyth, M. (1972). *The geology of Central-Western Tigre*. Ph.D. Thesis, Rheinische Friedrich-Wilhems Universität, Bonn, Germany, 159.

Beyth, M., Dow, D. B., Levitte, D., Temesgen, H., & Tsegaye, H. (1971). *Geological mapof Mekelle sheet area ND 37–11*. TigreProvince. Geol. Survey of EthiopiaPublication.

Ezana Mining Development PLC. (2003). *Geological map of vermiculite deposit (1:5000)*. Mekelle, Ethiopia (unpubl.).

Stern, R. J. (1994). Neoproterozoic (900–550 Ma) arc assembly and continental collision in the East African Orogen: Implications for consolidation of Gondwanaland. *Annual Review of Earthplanetary Sciences, 22*, 319–351.

U.S. Geological Survey (USGS). (2020). *Mineral commodity summaries* (p. 183).

The Gunung Gupit High- and Low-Sulfidation Epithermal Gold Mineralization in Menoreh, Central Java, Indonesia

Arifudin Idrus, Iswahyudi Agus Sinugroho and I Wayan Warmada

Abstract

Gunung Gupit in the Menoreh range, situated in Magelang regency, Central Java, is one of several Neogene volcanoes in Java Island, Indonesia. Hornblende basaltic andesite, hornblende dacitic andesite porphyry, and microdiorite are exposed in a half-circular depression, where hydrothermal alteration and ore mineralization occurred in northern Kulon Progo mountain. The presence of diagnostic minerals such as enargite-luzonite as well as chalcopyrite—pyrite—goethite and hematite, with acidic hydrothermal minerals including alunite, kaolinite, pyrophyllite, dickite, and halloysite, along a NE-trending silicified alteration zone, together with a vuggy-quartz texture, indicating a high-sulfidation epithermal deposit type. The highest grade of ore obtained from in situ channel sampling is 1.37 g/t Au, 595 g/t Ag, and 1.1% Cu. This mineralization system is overprinted by NE- and NW-trending banded quartz-carbonate ± pyrite veins containing 3 g/t Au, a typical LS epithermal deposit type. This deposit type is also supported by a moderate temperature (266–389 °C) and salinity (2.7–8.2 wt% NaCl eq.) of fluid inclusions in sulfide-quartz veins. These findings indicate that the Neogene intermediate magmatism in Java Island could have formed high-sulfidation epithermal-type deposits in Menorehas previously found in Tumpangpitu (East Java) and Cijulang (West Java), etc., which are situated along the West–East trending Sunda-Banda magmatic arc. This implies that, in the future, new LS and HS epithermal prospects might still be discovered along the Neogene magmatic belt.

Keywords

Epithermal · Gold · Gunung Gupit · Central Java · Indonesia

1 Introduction

Java is one of the islands located along the West–East trending Sunda-Banda magmatic arc (Fig. 1). This island is an emerging part of the region for gold and copper endowment (Maryono et al., 2018; Prihatmoko & Idrus, 2020). A number of occurrences of epithermal-type mineralization are known in Java Island, particularly in the southern Mountains of West Java, such as Pongkor (Basuki et al., 1994) and Cibaliung (Rosana et al., 2002). In Central Java, several deposits have been discovered, such as Sangon in Kulon Progo (Pramumijoyo et al., 2017) and Paningkaban in Banyumas (Idrus et al., 2015). Several major HS epithermal deposits are found in East Java including Kumbokarno in Trenggalek (Aldan et al., 2020) and Tumpangpitu in Banyuwangi (Harrison et al., 2018).

The study area, Gunung Gupit, is situated in the northern part of Menoreh range, which is administratively included in the Magelang district. This region is occupied by Old Andesite Formation (van Bemmelen, 1949). Gunung Gupit has been described as a high-sulfidation (HS) epithermal deposit which is overprinted by low-sulfidation (LS) epithermal mineralization in the distal portion (Sinugroho, 2014). This paper deals with the geological framework and ore characteristics of the deposits and their implication for gold exploration in the region.

A. Idrus (✉) · I. A. Sinugroho · I. W. Warmada
Department of Geological Engineering, Universitas Gadjah Mada, Yogyakarta, Indonesia
e-mail: arifidrus@ugm.ac.id

F. Lucci et al. (eds.), *Selected Studies in Geomorphology, Sedimentology, and Geochemistry*,
Advances in Science, Technology & Innovation, https://doi.org/10.1007/978-3-031-43744-1_11

Fig. 1 Location map of Gunung Gupit (Gn Gupit) situated in Central Java as a part of Neogene Sunda-Banda magmatic arc. The distribution of major epithermal-porphyry as well as other deposit types and coal is also indicated (modified from Maryono et al., 2018)

2 Materials and Methods

Representative samples of rocks and veins taken during geological and alteration mappings were analyzed in terms of mineralogy, ore chemistry, and fluid inclusions. The mineralogical analysis consists of petrography, ore microscopy, and XRD, conducted at Universitas Gadjah Mada. FA-AAS and ICP-OES analyses for ore chemistry were undertaken at Intertek Laboratory, Jakarta. Micro-thermometric analyses of fluid inclusions in quartz veins were conducted by a Linkam THMS-600 heating and freezing stage at Geotechnology of LIPI in Bandung, Indonesia.

3 Results

The study area shows a half-circular depression of approximately 10 km in diameter, which is thought to be an ancient volcanic crater (Fig. 2). Rock types exposed in the circular depression comprise andesitic breccia (Bxa), hornblende basaltic andesite (Vba), and hornblende dacitic andesite porphyry (Vpa). Ore mineralization is localized by hornblende dacitic andesite porphyry (Vpa), and hornblende basaltic andesite (Vba). The oldest rock, namely sandstone (Sst) identified as a member of Nanggulan Formation, is observed as xenoliths within Vpa. Microdiorite has been identified on the top of Gunung Gupit, forming a small conical body, about 15 m in diameter (not mappable). The youngest rock unit, namely limestone (Slm), crops out in the southern part of study area (Fig. 3).

Mineralization of both HS and LS epithermal gold types was identified in the study area. Four hydrothermal alteration zones related to HS epithermal type consist of chlorite-epidote-carbonate ± pyrite (propyllitic), illite–smectite mixed layer-kaolinite ± pyrite-montmorillonite-halloysite (argillic), quartz-alunite-luzonite-dickite-pyrophyllite ± zunyite ± diaspore (advanced argillic), and quartz ± pyrite-paragonite (silicified) (Fig. 4). APS-zunyite-diaspore assemblage may indicate a temperature of up to 375 °C. Advanced argillic and silicified altered rocks commonly show vuggy-quartz texture. HS epithermal mineralization is also indicated by NE-trending sulfide-quartz veins. Stockworks are also present. Banded LS epithermal-type quartz veins are observed in the southern portion. Ore mineralization is mainly controlled by ENE and NNW trending structures (Fig. 4). Sulfide-quartz veins accompanied by enargite-luzonite, pyrite, and oxide-clay contain the highest grades of 1.37 ppm Au, 595 ppm Ag, and 11% Cu. A

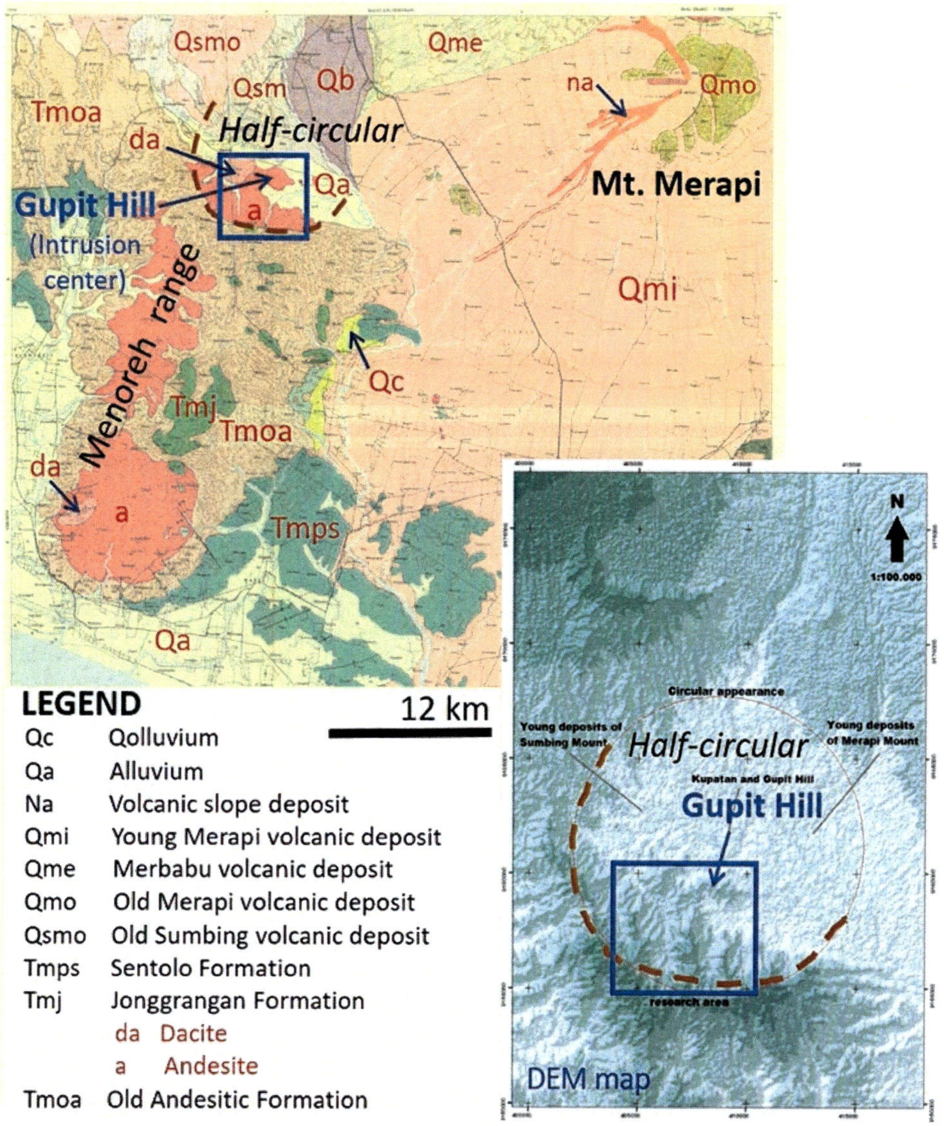

Fig. 2 Gunung Gupit prospect situated in half-circular depression of an ancient volcanic crater plotted on regional geological map of Yogyakarta sheet (Rahardjo et al., 1995) and DEM image. *Map source* https://www.opendem.info/link_dem.html

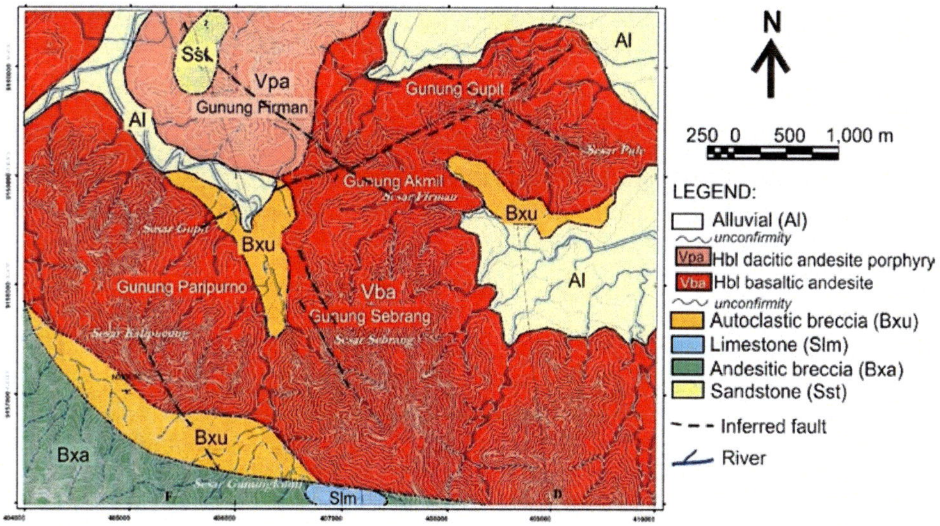

Fig. 3 Geological map of Gunung Gupit and its vicinity (modified from Sinugroho, 2014)

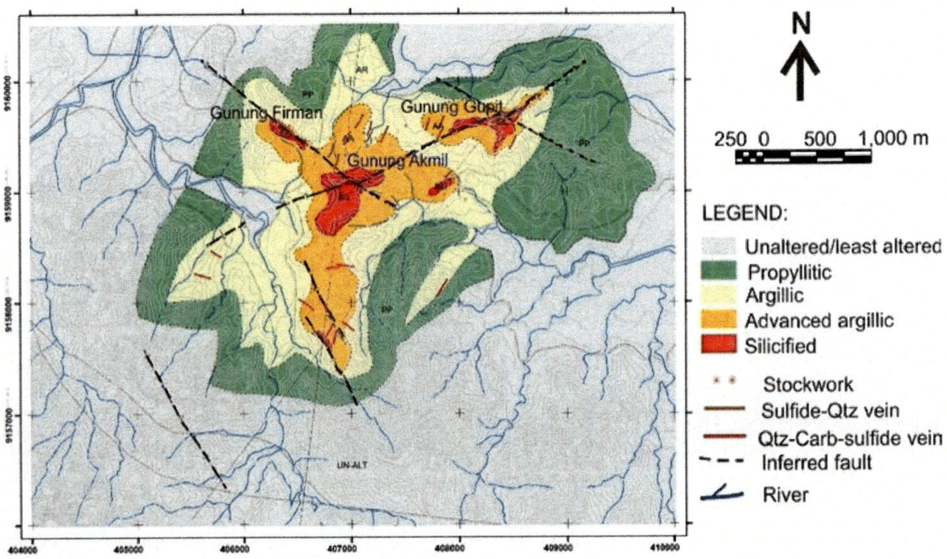

Fig. 4 Alteration and veining map of Gunung Gupit area (modified from Sinugroho, 2014)

quartz-carbonate vein accompanied by pyrite-clay contains elevated grades of 3 ppm Au, 4.6 ppm Ag, and 236 ppm Cu.

Fluid inclusions in sulfide-quartz veins related to HS epithermal mineralization are dominated by Type II vapor-rich (V) rather than Type I-biphase (L-V) inclusions. The abundance of vapor-rich inclusions may indicate the boiling condition. Type III-polyphase (L-V-S) inclusions consisting of vapor, liquid, and platy crystals of probably hematite are also observed. Polyphase inclusions show temperature of homogenization (Th) ranging from 266 to 389 °C (mostly > 350 °C) with a salinity of ~ 8.2 wt.% NaCl eq. The temperature of homogenization (Th) displays a very large range, which may represent a large range of the formation temperature of alteration stages. In contrast, fluid inclusions in LS epithermal quartz-carbonate veins exhibit lower temperatures varying between 215 and 350 °C with a salinity of ~ 2.7 wt% NaCl eq.

4 Discussion: Implication for Exploration

The occurrences of advanced argillic alteration composed of dickite, pyropyllite, and variable proportion of alunite, vuggy-quartz texture, and presence of gold in association with enargite-luzonite suggest that the hydrothermal deposit is of a HS epithermal type (cf. Hedenquist et al., 1996). Alunite is relatively rare, which may imply the deposit has been deeply eroded. Banded quartz-carbonate ± pyrite veins associated with near neutral wallrock alteration suggest a LS epithermal deposit type. At least two major HS epithermal gold deposits are confirmed in Java Island, namely Cijulang in Garut, West Java (Rosana et al., 2013) and Tumpangpitu in Banyuwangi, East Java (Harrison et al., 2018). The Eastern Sunda arc covering Java, Lombok, and Sumbawa Islands evidently hosts several world-class porphyry deposits (Maryono et al., 2018). In Gunung Gupit, concealed porphyry might be present. Ground magnetic and IP surveys also indicate the presence of a possible causative intrusion anomaly beneath Gunung Gupit (SEG UGM-SC, 2015). Moderate temperature fluid inclusions and the presence of hematite daughter inclusions may indicate high magma fertility beneath (e.g., Idrus et al., 2009). This implies that HS and LS epithermal gold, as well as porphyry copper–gold deposit types, have the potential to be discovered in the Menoreh and Kulon Progo mountain ranges. Therefore, systematic exploration is certainly needed in the region.

5 Conclusions

Gunung Gupit is a transition complex of porphyry-HSE telescoping by LSE mineralization. A concealed porphyry deposit might also be present. Hydrothermal alteration and ore mineralization of HS epithermal type is mainly controlled by ENE and NNW trending structures. Both HS and LS epithermal deposit types are hosted by hornblende basaltic andesite and hornblende dacitic andesite porphyry. HS epithermal mineralization is characterized by the presence of diagnostic minerals such as enargite-luzonite and acidic alteration such as alunite, dickite, and pyrophyllite. Vuggy-quartz is observed, which is thought to be the center of HSE mineralization system. The highest grade of ore obtained from in situ channel sampling is 1.37 g/t Au, 595 g/t Ag, and 1.1% Cu. This deposit type is also supported by moderate temperature (266–389 °C) and salinity (2.7–8.2 wt% NaCl eq.) of fluid inclusions. This system is overprinted by NE- and NW-trending banded quartz-carbonate ± pyrite veins containing high grades of 3 g/tAu and 4.6 ppm Ag, as well as low Cu grade of 236 ppm, typical of a LS epithermal-type deposit. LS epithermal quartz-carbonate ± pyrite veins exhibit relatively low temperature fluids in the range of 215 and 350 °C with salinity of ~ 2.7 wt% NaCl eq. The discovery of HS and LS epithermal, as well as possible concealed porphyry in Gunung Gupit, suggests that similar deposit types are still open to be discovered in the region, particularly in the Menoreh and Kulon Progo mountain ranges, as a part of the eastern Sunda magmatic arc.

Acknowledgements The authors express many thanks to the managements of PT. Oksindo Eksplorasi Indonesia and PT. Antam Tbk for their supports during this study. We also thank to Department of Geological Engineering, Universitas Gadjah Mada for supporting this work.

References

Aldan, F. A., Idrus, A., Takahashi, R., Kaneko, K. (2020). Spatial and temporal constraints of leached Cu–Au porphyry shoulder high-sulfidation epithermal deposit: Insight from new discovered Kumbokarno Prospect, Trenggalek District, East Java. *Journal of Physics: Conference Series 1367, 012037*, 1–15.

Basuki, A., Sumanagara, A. D., & Sinambela, D. (1994). The Gunung Pongkor gold-silver deposit, West Java, Indonesia. *Journal of Geochemical Exploration, 50*, 371–391.

Harrison, R. L., Maryono, A., Norris, M., Rohrlach, D., Cooke, D. R., Thompson, J. M., Creaser, J. A., & Thiede, D. A. (2018). Geochronology of the Tumpangpitu porphyry Au–Cu–Mo and high-sulfidation epithermal Au–Ag–Cu deposit: Evidence for pre- and post-mineralization diatremes in the Tujuh Bukit district, Southeast Java, Indonesia. *Economic Geology, 113*, 163–192.

Hedenquist, J. W., Izawa, E., Arribas, A., White, N. C. (1996). Epithermal gold deposits: Styles, characteristics and exploration. *Resource Geology*, 165–182.

Idrus, A., Hakim, F., Warmada, I. W., Aziz, M., Kolb, J., & Meyer, F. M. (2015). Geology and ore mineralisation of Neogene sedimentary rock-hosted LS epithermal gold deposit at Paningkaban, Banyumas regency, Central Java, Indonesia. *Journal of Southeast Asian Applied Geology, 7*, 75–81.

Idrus, A., Kolb, J., & Meyer, F. M. (2009). Mineralogy, lithogeochemistry and elemental mass balance of the hydrothermal alteration associated with the gold-rich Batu Hijau porphyry copper deposit, Sumbawa Island, Indonesia. *Resource Geology, 59*, 215–230.

Maryono, A., Harrison, R. L., Cooke, D. R., Rompo, I., & Hoschke, T. G. (2018). Tectonics and geology of porphyry Cu–Au deposits along the eastern Sunda magmatic arc, Indonesia. *Economic Geology, 113*, 7–38.

Pramumijoyo, P., Idrus, A., Warmada, I. W., & Yonezu, K. (2017). Geology, geochemistry and hydrothermal fluid characteristics of low sulfidation epithermal deposit in the Sangon area, Kokap, Special Region of Yogyakarta. *Journal of Applied Geology, 2*, 48–58.

Prihatmoko, S., & Idrus, A. (2020). Low-sulfidation epithermal gold deposits in Java, Indonesia: Characteristics and linkage to the volcano-tectonic setting. *Ore Geology Reviews, 121*, 103490.

Rahardjo, W., Sukandarrumidi, Rosidi, H. M. D. (1995). *Geological map of Yogyakarta sheet, scale 1:100,000*. Pusat Survei Geologi, Bandung.

Rosana, M. F., Sunarie, C. Y., Siregar, E. S. A., Verdiansyah, O., Yuniardi, Y., Syafri, I., Watanabe, K. (2013). Characteristics of epithermal high sulfidation of Cijulang prospect, Garut, West Java, Indonesia. In *Proceedings of the 3rd Asia Africa mineral resources conference, Ulan Bator—Mongolia* (pp. 1–4).

Rosana, M. F., & Matsueda, H. (2002). Cikidang hydrothermal gold deposit in western Java, Indonesia. *Resource Geology, 52*, 341–352.

Sinugroho, I. A. (2014). Geologi dan karakteristik alterasi—Mineralisasi endapan epitermal sulfidasi rendah (LS) dan sulfidasi tinggi (HS) di Pegunungan Menoreh, Kabupaten Magelang, Jawa Tengah, Indonesia, Universitas Gadjah Mada, unpublished Master Thesis, p. 223.

SEG UGM-SC. (2015). *Geophysical prospecting and hazard assessment in mineral prospect area (Gunung Gupit and its Vicinity)*, Borobudur, Fieldcamp Proceedings, p. 76.

van Bemmelen, R. W. (1949). The geology of Indonesia (vol. IA, 9. 732). Government Printing Office, Martinus Nijhoff, The Hague.

Polymetallic Mineralization in the Western Part of the Numidic Chain (North-East Algeria)

Oumeima Benmebarek and Azzedine Bouzenoune

Abstract

Iron and polymetallic mineralization host-rocks are located in the Tellian domain of the Maghrebides range. These host-rocks crop out as E-W oriented Jurassic complex anticlines. They are essentially composed of Liassic dolomite and limestones characterized by two types of ore deposits. Cluster iron oxides and hydroxides are associated with fissural polymetallic mineralization (Cu-Ba, Pb-Zn). Moreover, three distinctive carbonate microfacies were recognized, and they are characterized by particular components and microstructures that revealed an intertidal to infratidal depositional environment. Stable isotopes study of O and C performed on calcite gangue and micritic limestone host, which show that $\delta^{13}C_{micrite}$ %/PDB values are between +1.02 and +2.09, while they are from −1.98 to +1.02 for $\delta^{13}C_{calcite}$ %/PDB. This indicates an inorganic carbon source. The $\delta^{18}O_{micrite}$‰/PDB results are from −6.98 to −5.84 and are $\delta^{18}O_{calcite}$‰/PDB from −7.53 to −6.29. The $\delta^{18}O$ data under −4 ‰ suggest that ore deposition could have been produced by mineralizing fluid with depletion in ^{18}O and/or with implication of hotter fluid than the host-rocks.

Keywords

Liassic limestones · Iron ore deposits · Tellian domain · Stable isotopes

1 Introduction

Two types of ore deposits are commonly hosted within the Tellian limestones of Lesser Kabylia. They are partly located in Kef Dardja, Kef Sidi Marouf, Kef Boulahmem, and Kef M'cid Aicha. These uplifted carbonate rocks cropping out in the southeastern part of Jijel province (Fig. 1) which are similar to ore-bearing carbonate anticlines known in the western part of the Tellian domain, and particularly to the Baborian zone which extends through the western limit of Jijel (Glaçon, 1967). Indeed, these similarities can be useful to clarify some metallogenic characteristics of the studied ore deposits.

The iron mineralization has been classified as vein ore bodies and hydrothermal-metasomatic carbonates replacement deposits (Popov, 1976) associated with barite, tetrahedrite, galena, and sphalerite. A supergene enrichment followed the mineralizing process and thus fully or partially altered the ore minerals.

The current understanding of the ore deposits genesis in western part of Numidic chain consists of few studies that reported merely petrographic aspects of the ore-bearing carbonates, but some have conducted mineralogical analyses (DRX, MEB-EDS) on iron ore (Manchar, 2007; Ouaar, 2009). Although the geological aspects of the host's setting have been widely described (Durand-Delga, 1955; Vila, 1980), only scarce geochemical studies are available in the literature (Bouzenoune et al., 2016). Hence, the objective of this article is to provide a presentation of some geological features, petrographic observations, and preliminary stable isotopes' data characterizing the Liassic carbonates rocks of the Tellian domain and their related ore deposits.

O. Benmebarek (✉) · A. Bouzenoune
Geological Engineering Laboratory, University Mohamed Seddik Benyahia, Jijel, Algeria
e-mail: o.benmebarek@univ-jijel.dz

F. Lucci et al. (eds.), *Selected Studies in Geomorphology, Sedimentology, and Geochemistry*, Advances in Science, Technology & Innovation, https://doi.org/10.1007/978-3-031-43744-1_12

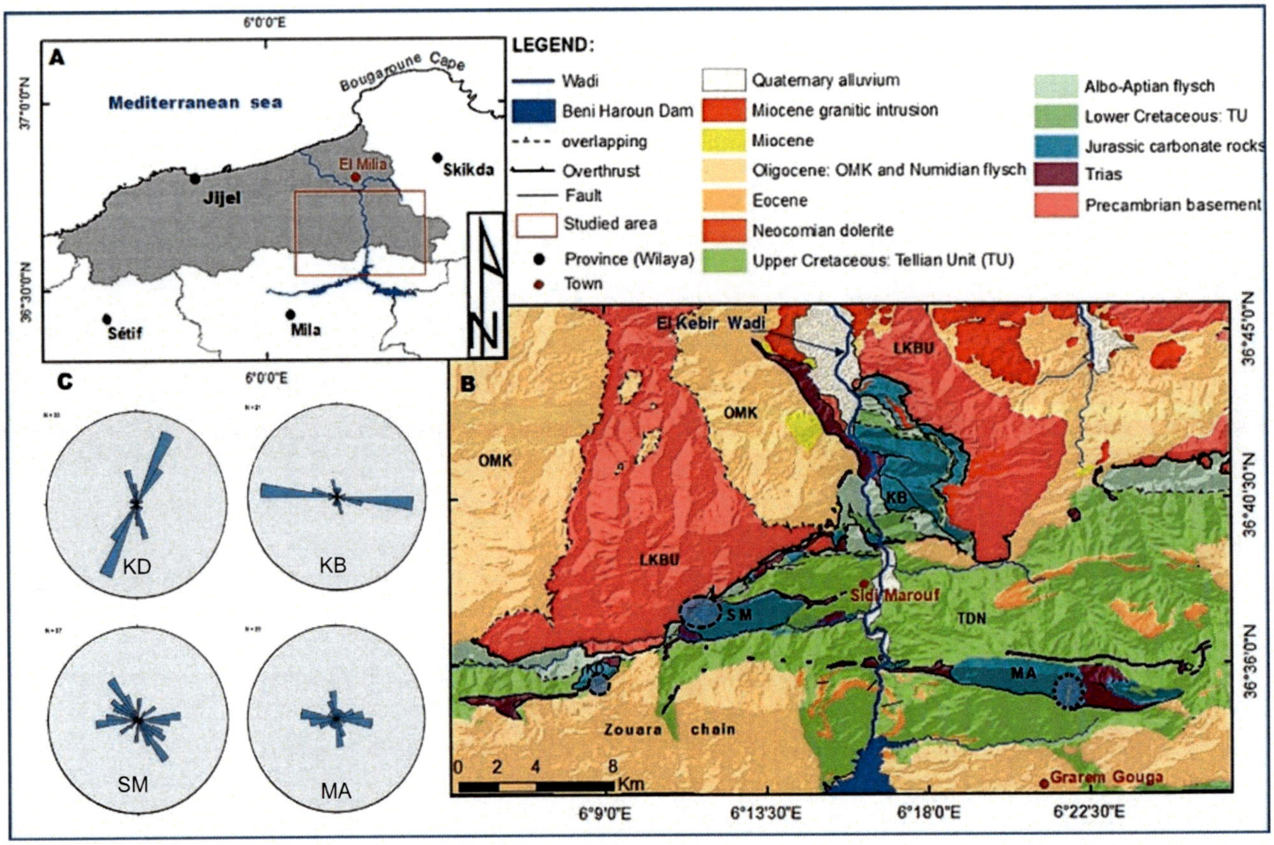

Fig. 1 **a** The studied area location (in red square) in NE Algeria, **b** geological map of the West Numidic zone (after Durand-Delga, 1955, modified) (in black circle: host-rock and mineralization sampled area). **c** rose diagrams for orientations of the ore bodies. Abbreviations: LKBU: Lesser Kabylia Basement Units, TDN: Tellian Djemila's nappe, OMK: Oligo-Miocene Kabyle, KD: Kef Dardja, SM: Sidi Marouf, KB: Kef Boulahmem, MA: M'cid Aicha

2 Geological Setting

The Liassic dolomites and limestones investigated in this work belong to the northern outcrops of the Tellian Djemila's nappe (Vila, 1980) of the Maghrebides chain. They are linked to the large tectonic slices of carbonate rocks in the Baborian domain and appear as folds elongated in an E-W direction within Cretaceous formation (Fig. 1b).

The Tellian zone is represented by successive thrust sheets that consist of the Jurassic to Cretaceous sedimentary formations. It is separated to the North by Albo-Aptian flysch and the overthrust internal units of the Precambrian Kabylian basement. The south drift of this internal basement during the Paleogene over the flysch zone and external Tellian domain caused several tectonic deformations and deep faults oriented generally N-S or E-W. The most important tectonic fault zone is the one that runs E-W on the southern front of the Kabylian basement. This faulting system could have acted as mineralizing fluid pathways, and thus assigned to the mineralization controlling event (Fig. 1c).

The basement is mainly formed by orthogneiss and mica schists with a thick cipolin intercalation mass (20 m) affected by multiple iron oxides filled veins. The Triassic formation is exposed toward the flanks of the Jurassic anticlines. It consists of chaotic evaporitic formation with marls, gypsum, brecciated dolomite, sandstone, and sometimes dolerite intrusions with oligist (Durand-Delga, 1955) which could be brought to the surface by the Triassic extrusion. Eruptive mafic rocks are also observable within Jurassic or Cretaceous formations (Fig. 1b) to the northern and eastern limits of Kef Boulahmem.

The Cretaceous formations are characterized by clays, black marls, and marl-limestone alternations in the Tellian allochthonous zone. On the other hand, the Albo-Aptian massylian flysch is composed of epimetamorphic shales and sandstones situated in narrow-band oriented NE-SW along the south side of the Kabylian basement. The Cenozoic (Eocene to Miocene) sedimentary deposits mainly consist of transgressive formations composed of sandstones, clayey sediments, and conglomerate rocks. However, the Oligocene in Zouara chain (Fig. 1b) is composed of interbedded deposits of shale and highly quartzose coarse sandstones. It forms the Numidian nappe which is the highest structural unit of the western Numidic chain. Furthermore, a post-tectonic calc-alkaline magmatism cross-cut during Miocene various rocks of both internal and external zones of the Maghrebides. It outcrops in El Miliaregion and Bougaroune Cape (Fig. 1).

3 Results and Discussion

3.1 Petrography of the Host-Rock

Macroscopically, the carbonate rocks are mostly fine-grained of brownish, grayish (Fig. 2a–c), and a pinkish color (Fig. 2d) due to their dolomitization (Fig. 3c, f), with stratified form. However, the stratification may locally be indistinct due to the intense scaling of outcrop units. The limestones bedding, oriented E-W, is 30 cm to 1 m in thickness and intercalated with clay and flint beds to the top of the Jurassic series.

Microscopic analysis has revealed dominance of three main microfacies (Fig. 3): massive dolomite mud, wackstone with peloids (Pel.), sponge spicules, and grainstone with ooids (Oo.) and peloids. Intraclasts (lit.), bipyramidal quartz (Qz bi.), rhombohedral dolomite (Dol.) including some bioclasts identified as foraminifera (for.), echinoderm, lamellibranch (Lamll.), and algae were recognized. Microstructures such as calcite and/or iron oxides veins (fis.), stylolite (St.), and bird's-eye were also observed in the sections. Accordingly, these microstructure, textural, and paleontological criteria indicate a shallow moderate to high-energy depositional environment; it consists of intertidal to infratidal zones.

3.2 Petrography of the Mineralization

Mineralization is composed of hematite-limonite/goethite, calcite, and barite. These deposits are clustered (Fig. 2c) and have vein morphology. On the other hand, tetrahedrite, galena, sphalerite, calcite, and barite are present within narrow veins cross-cutting the host-rock. As a matter of fact, sulfides like galena (Fig. 2f) and sphalerite are only observable in Kef M'cid Aicha.

The iron carbonate protore is not identified yet as a particular carbonate mineral. It is generally associated with tetrahedrite which is frequently altered to malachite-azurite (Fig. 2b–e). According to the literature (Bouzenoune et al., 2016), this tetrahedrite contains more antimony than arsenic which is comparable to other mineralization related to carbonate rocks in northern Algeria's geological domains.

The ore samples revealed the following paragenetic sequences: (i) Formation of barite; (ii) main ore deposition: polymetallic sulfides, then iron mineralization; (iii) white coarse calcite formation; and (iv) supergene ore enrichment. Regarding the mineralization textures (Fig. 2), we mostly

Fig. 2 Photographs showing outcrop and macroscopic textures of the studied ore deposits. Our structural compilation indicates that iron mineralization is related to fractures' zones affecting all carbonate outcrops, the main orientations are N150°–170° and N80°–100°. However, tetrahedrite, malachite, azurite, and coarse calcite or barite occur within millimetric to centimetric fissure veins oriented N 25°E and N 170°E (Fig. 1c, Fig. 2a)

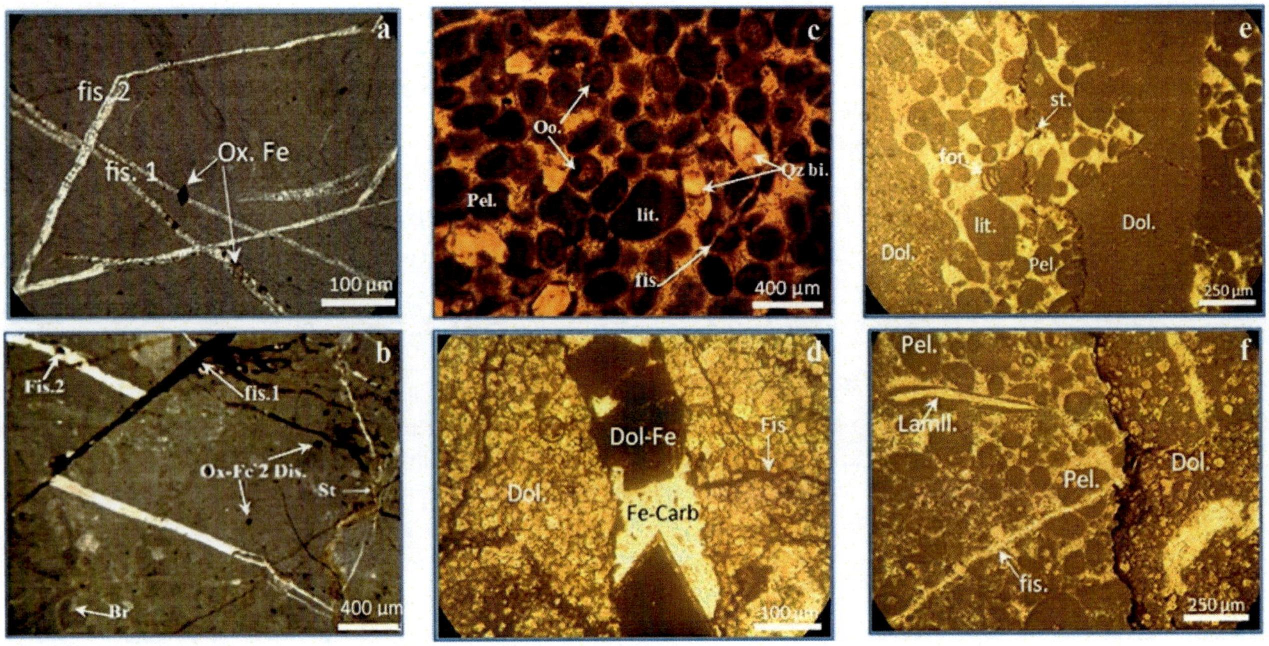

Fig. 3 Photographs showing microscopic characteristics of the ore-bearing carbonates

observed massive, brecciated, botryoidal, and collomorphic habits. Also, calcite and barite may show geode texture.

3.3 Stable Isotopes Data

Micritic limestone samples of Kef Dardja (Fig. 1b) show $\delta^{13}C_{micrite}$ values about +1.02 to +2.09‰/PDB and $\delta^{18}O_{micrite}$ values that vary from −6.98 to −5.84‰/PDB. Besides stable isotope for calcite gangue (Fig. 2a) shows $\delta^{13}C_{calcite}$ values ranging between −1.98 and +1.02, whereas $\delta^{18}O_{calcite}$ data vary from −7.53 to −6.29.

Thus, the low $\delta^{13}C$ values of micrite and calcite (average of +1.6 $\delta^{13}C_{micrite}$ and −2.16 of $\delta^{13}C_{calcite}$) are related to marine limestones indicating an inorganic carbon origin. The excursion from the norm of $\delta^{18}O$ micrite and $\delta^{18}O$ calcite values, about a mean of −6.43 and −6.90, respectively, is probably produced by a hydrothermal fluid depleted in ^{18}O or hotter than the ambient host-rocks. It could also have an origin by a mix of both. Most accurately, the same hydrothermal fluid seems to be responsible for the recrystallization of the limestone host and the mineralization deposition.

4 Conclusion

Liassic dolomite and limestones that appear in the Tellian zone of Lesser Kabylia are mainly formed by marine carbonates rocks of intertidal to infratidal zones. They are generally massive and show sometimes breccia appearance which could be related to syn-sedimentary distension faults.

These Liassic carbonates host polyphased epigenetic mineralization, it comprises iron oxides/hydroxides and polymetallic sulfides. The iron oxides and hydroxides result from the replacement of the iron carbonate protore during a supergene enrichment. The ore bodies display broadly as clusters, veins and lens-like veins deposits that clearly follow the same trends (N-S and E-W) as the post-thrust faulting system of West Numidic chain. We can conclude that ore formation had strong tectonic control.

The stable isotopes' signatures of the limestone and calcite gangue show an interaction between the host-rock and a hydrothermal fluid, perhaps epithermal, poor in ^{18}O. This fluid replaced locally the host-rock and is responsible for ore deposition.

References

Bouzenoune, A., Boufaa K., & Remoum K. (2016). Lithostratigraphy of the carbonated Liassic in the Sidi Marouf region and description of the associated mineralization (Fe, Cu-Ba) (Jijel, North-eastern Algeria). *Geological Service Member of Algeria, 19*, 27–37.

Durand-Delga, M. (1955). *Geological study of the western Numidic chain (Algeria)*. Thesis Sciences.

Glaçon, J. (1967). *Investigations into the geology and metalliferous deposits of the Setifien Tell (Algeria)*. Thesis, University of Montpellier.

Manchar, N. (2007). *Comparative study between two iron ore deposits in distinctive geological settings*. The Sidi Marouf deposits

(Lesser Kabylia, Jijel) and the Boukhadra deposit (Mellegue Mountains, Tebessa). North-Eastern Algeria. Magister, University of Constantine.

Ouaar, B. (2009). *Geological and gitological study of the Sidi Marouf iron ore deposits (Jijel-North-eastern Algeria).* Magister, University of Badji Mokhtar Annaba.

Popov, A. (1976). Iron ore deposits in Algeria. In: H. W. Walther & A. Zitzmann (Eds.), *The iron ore deposits of Europe and adjacent areas* (Vol. 1, pp. 83–89). Publ. Federal Institute of Geosciences and Natural Resources.

Vila, J. M. (1980). *The alpine chain of eastern Algeria and the Algero-Tunisian frontier* (Vol. I–II). Thesis Es-sciences. University of Paris VI.

Sedimentological and Geochemical Aspects of the Late Paleocene-Early Eocene Phosphorites Cropping Out in North Western Tunisia (Sra Ouertane Basin)

Naima Ahmadi, Anna Travé, Elisabeth Playa, Jaime Frigola, Christopher J. Hollis, Noura Kotti, Besma Mardassi, Mongi Felhi and Ali Tlili

Abstract

The present work aims to reconstruct the paleoenvironments using a multidisciplinary approach including sedimentology, petrography, and geochemistry. These studies were undertaken on the late Paleocene–early Eocene succession of the Sra Ouertane area, Northwestern Tunisia. These deposits are mainly composed of phosphorites interbedded with limestones and marls. Phosphatised coprolites and indeterminate pellets contain a poorly preserved radiolarian assemblage, which can only be differentiated to the genus level. The presence of *Calocyclas*, *Phormocyrtis*, *Lychnocanium*, and *Theocorys*spp indicates a late Paleocene–early Eocene age. Geochemical study of major (Fe, Si, P, Ti, K) and trace (Zn, Y, Zr, Sr) elements reveals that the phosphorites at the base and top of the studied section, are characterized by high Sr concentration in addition to high P. In contrast, Zr, Ti, Si, and K are more abundant in samples from the middle part of the section. This enrichment of Zr, Ti, and K was favored by an efficient weathering of neighboring hinterlands during humid episodes, which favored the detrital inputs. This humid period was contemporaneous with the warmest global climate event.

Keywords

Late Paleocene-early Eocene · Phosphorites · Sra Ouertane Basin · Radiolarians · Detrital inputs · Humid episode

1 Introduction

Geological studies of the late Paleocene–early Eocene sedimentary strata deposited around Kasserine paleo-island in Tunisia have shown that these strata are commonly constituted of phosphatic layers interbedded with limestones, porcelanites, and dolomitic marls (Ahmed et al., 2014; Murray et al., 1990; Soudry et al., 2002; Zaïer, 1999; Zaïer et al., 1998). Phosphorites of the Southern basin have been widely studied due to their economic potential (Soudry et al., 2002). Contrarily, much less attention has been paid to the Sra Ouertane phosphatic strata located in the north of Kasserine paleo-island. During the last two decades, several authors have reported that the phosphatic strata outcropping in the Sra Ouertane area is deposited in a shallow open-marine platform (Garnit et al., 2012). Phosphorites of open-marine environments are enriched in chemical elements compared to other marine facies (Tlili et al., 2012) due to the depositional environment conditions and the terrigenous supply (Baioumy, 2011). This latter can be assessed using Ti, K, Zr, and Al elements (Sassi, 1974). The aim of this work is to examine geochemical features and to identify radiolarians in the late Paleocene-early Eocene strata, in order to determine the depositional environment and paleoclimatic context.

N. Ahmadi (✉) · M. Felhi · A. Tlili
GEOGLOB Laboratory, Science Faculty, University of Sfax, 3038 Sfax, Tunisia
e-mail: naimaahmadi38@gmail.com

A. Travé · E. Playa
Departament de Mineralogia, Petrologia i Geologia Aplicada, Facultat de Ciències de la Terra, Universitat de Barcelona (UB), 08028 Barcelona, Spain

J. Frigola
GRC Geociències Marines, Department De Dinàmica da Terra i de l'Oceà, Facultat de Ciències de la Terra, Universitat de Barcelona, 08028 Barcelona, Spain

C. J. Hollis
GNS Science, Lower Hutt, New Zealand
N. Kotti · B. Mardassi Water Energy and Environment Laboratory (3E), National Engineering School of Sfax, 3038 Sfax, Tunisia

F. Lucci et al. (eds.), *Selected Studies in Geomorphology, Sedimentology, and Geochemistry*, Advances in Science, Technology & Innovation, https://doi.org/10.1007/978-3-031-43744-1_13

2 Geological Setting

In Tunisia, the late Paleocene-early Eocene age corresponds to the main period of phosphate genesis. These sedimentary marine successions are deposited in three basins around the Kasserine paleo-island (Burollet & Oudin, 1980). The Sra Ouertane basin is located in the Northwestern part of Tunisia, between 35° and 36° N latitude and 8° and 9° E longitude. This basin is considered as a part of the central Tunisian Atlas (Ayed, 1980). The studied TS14 trench, which is 50 m thick and which corresponds to the Metlaoui Formation, is composed of a succession of phosphorites interbedded with limestones and marls, frequently burrowed, belongs to the phosphatic sequence of the North flank of Jebel Ayata. This latter corresponds to a large asymmetric syncline basin, trending NW–SE. It is characterized by a N130 to N150 oriented accident, the layers are of low dip, limited to the south by an accident N130, 85° NE, reflecting an abnormal contact between the clays of El Haria formation (Paleocene) and the limestones of El Garia formation (Lutetian) (Fig. 1).

3 Materials and Methods

Systematic sampling was performed depending on facies variations. Samples were investigated as thin sections under a petrographic microscope in order to identify the allochems. Pellets and coprolites were isolated from 3 samples to examine their endogangue based on SEM observations. The broken pellets and coprolites were mounted on scanning electron microscope stub using FEI Quanta 200 device (FEI Europe B.V., Eindhoven the Netherlands) with EDS spectrometer (EDAX-Genesis). The radiolarians are observed and identified using SEM observations of five samples. The elemental composition was measured directly in thin sections of 23 samples by means of an Avaatech XRF core scanner.

4 Results

4.1 Optical Microscopy and SEM Observations

Phosphorites are mainly composed of phosphatic pellets with rounded to ovoid shapes (Fig. 2a), pelecypods, benthic foraminifera (Fig. 2b), fish remains, coprolites, pellets, oolites, and glauconites. Most pellets were completely opaque, which may indicate richness in organic matter (Fig. 2c). The size of the pellets ranges from 50 to 100 μm, and have a proto-oolitic feature showing an obvious cortex (Fig. 2a). The pellets are often contiguous as a cluster, due to the presence of organic matter in the edges of the pellets. Pellets and bioclastic fragments, with random orientations, are bathing in dolomitic cement (Fig. 2d).

SEM observations on five samples show the presence of radiolarians. Poor preservation means that the specimens can only be identified to the genus level. The association of *Calocyclas* (E), *Phormocyrtis* (F), *Lychnocanium* (G), and *Theocorys* (H) spp. indicates a Paleocene–early Eocene age (Hollis, 2006). Other specimens remain undetermined due to poor preservation. Previous studies of phosphatic strata of southwestern and central-eastern Tunisia (Murray et al., 1990) found that radiolarian skeletons were pseudomorphed by carbonate fluorapatite and somewhat lost their ornamentations in the process, as often detected in the phosphatic series of southwestern and central-eastern, Tunisia. However, radiolarians can sometimes be exceptionally well-preserved in phosphatised sediments (Paleocene) (Jaballi et al., 2019).

4.2 Geochemical Analysis

The phosphatic samples analyzed from the TS14 trench are rich in Ca and P contents and present relatively lower contents of Si, Fe, Ti, and K. These major elements, as well as the trace elements (Zn, Y, Zr, and Sr), indicate distinct geochemical fluctuations. The records of Zr, Ti, Fe, Si, Zn, and K exhibit approximately similar behaviors. They are more abundant in the middle of the section and less abundant in the lower and the upper parts of the section. Opposite trends are observed for P, Y, and Sr elements. These geochemical behaviors provided above are clearly depicted by using principal components analysis (PCA). The results of PCA clearly display two main groups separated along the axis1. Group 1, plotted near the lower values of axis1, includes P, Y, and Sr and groups the samples from the lower and the upper parts of the section. Group 2, plotted in the higher values of axis 1, includes Ti, Si, Zn, Fe, and K and corresponds to the samples from the middle part of the section (Fig. 3).

5 Discussion

The content of P, Y, and Sr are higher at the base and the top parts of the late Paleocene-early Eocene succession of the Sra Ouertane Basin. Deposits show clear variations in terms of composition, fauna content, and weathering impact. From bottom to top, phosphate layers become thinner at the expense of marls and limestones. Phosphate genesis and the vertical distribution of the associated chemical elements seem under the control of sea-level changes where diagenetic processes allow the dissolution of fauna remains (bones and

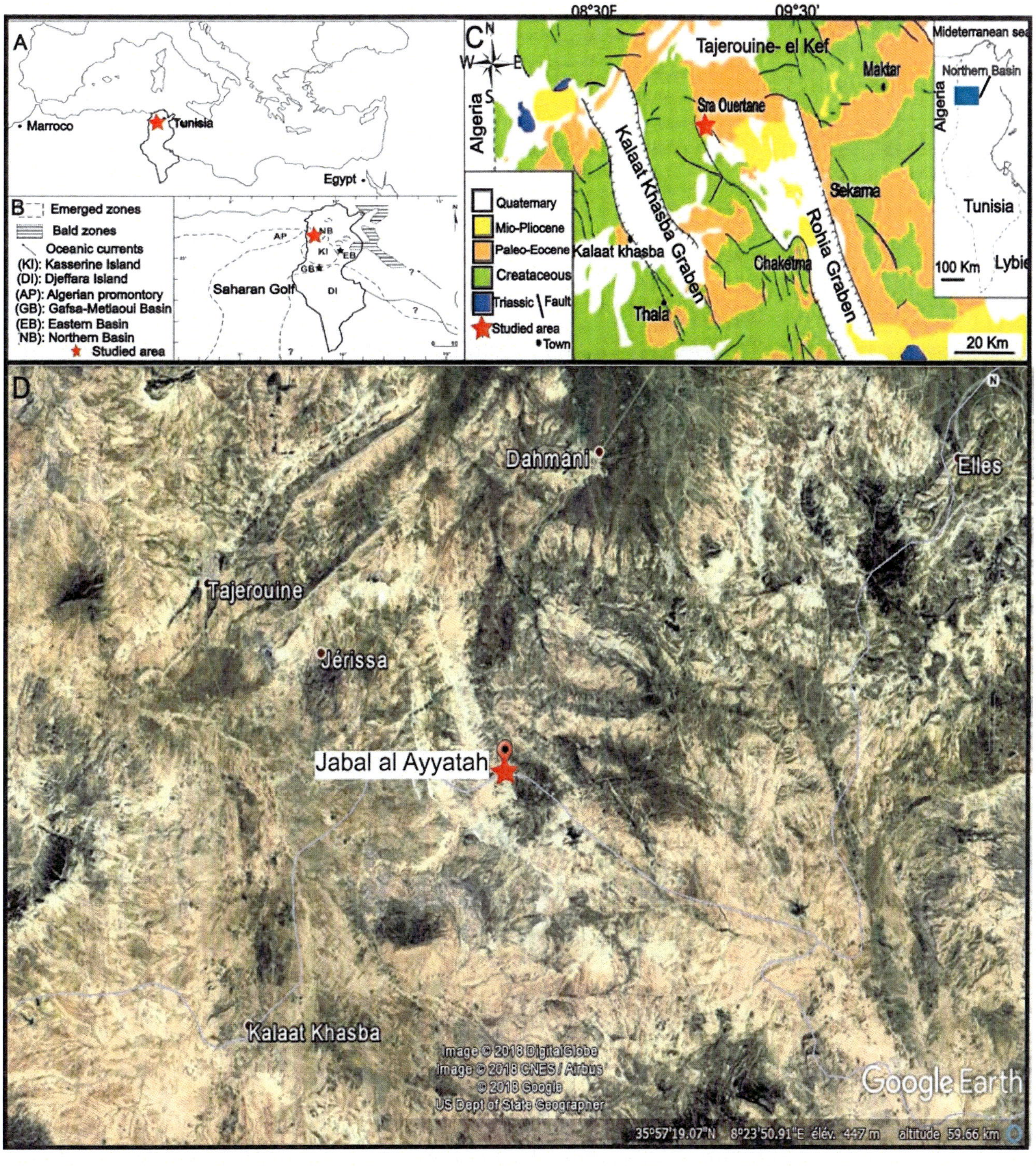

Fig. 1 **a** General location of the central Tunisia study area in North Africa. **b** early Eocene palaeogeography of Tunisia (Zaïer et al., 1998), **c** geological map of Northern Basins showing the locations of the study sites (Garnit et al., 2012), **d** photo from Google map showing Jebel Ayata

squale teeth). The phosphate samples are mainly composed of phosphatic pellets and coprolites which are associated with late Paleocene-early Eocene radiolarians (Fig. 1). Compared to the southern basins, phosphorites deposited in several marine contexts at the vicinity of Kasserine paleo-island, are characterized by the different geochemical composition of the bulk rocks, depending on the depositional environment properties. The enrichment of Ti, K, Zr, and Al contents for the middle part is owed to detrital inputs (Sassi, 1974), whose supply was likely prevailing during a humid episode.

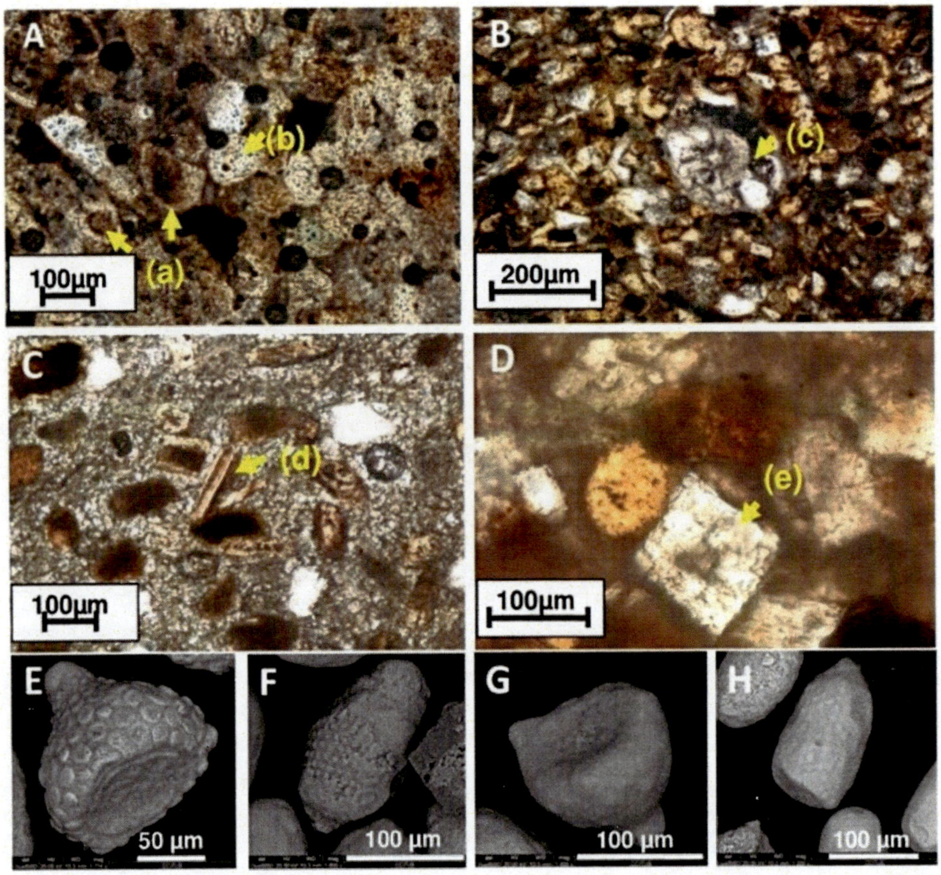

Fig. 2 Thin section photomicrographs showing the main components: **A** sample C2-3 illustrating rounded to ovoid phosphatic pellets (a) and oolithe (b). **B** sample C2-8 showing benthic foraminifera (c), buried in contiguous phosphatic pellets. **C** sample C2-10 showing phosphatic pellets and bone fragments (d) with a predominance of dolomitic cements (e) **D** sample C2-23 showing micron-sized dolomite. SEM micrographs showing the radiolarian tests, which constitute the inner part of the pellets recovered from three samples (**E** *Calocyclas*; **F** *Phormocyrtis*; **G** *Lychnocanium*; and **H** *Theocorys*)

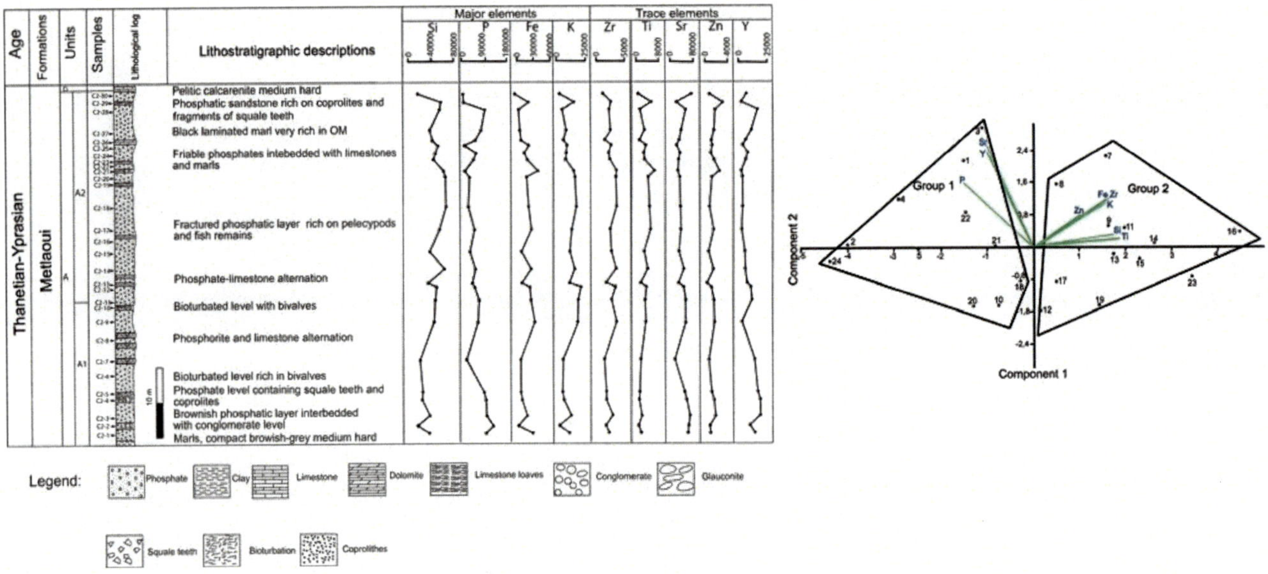

Fig. 3 Lithostratigraphic section of TS14 showing the variation of major and trace elements along the studied section. Ordination biplot of samples according to a principal component analysis (PCA) based on the correlation matrix of the geochemical data of specimens sampled from TS14 trench with late Paleocene-early Eocene age

6 Conclusion

This study shows that pellets and coprolites of the Sra Ouertane Basin contain radiolarians, corresponding to *Calocyclas, Phormocyrtis, Lychnocanium,* and *Theocorysspp*, which indicate a late Paleocene-early Eocene age. In addition, these rocks display a chemo-stratigraphic zonation along the studied section. Thus, P, Ca, and Sr contents are more abundant at the lower and the upper parts of the section. Whereas, the middle part is characterized by relatively higher contents of Ti, K, Zr, and Al. This occurrence should be related to the increasing of detrital input during the sedimentation of the middle part of the section enhanced by more humid conditions. The stratigraphy and variation of the sedimentary paleo-environment were under the control of sea-level changes.

References

Ahmed, A. H., Tlili, A., Zalat, A. A., & Jeddoui, Y. (2014). Fossil diatoms from endogangue of the Ypresian phosphatic pellets of the Gafsa-Metlaoui basin: Implication on the origin of biogenic silica and depositional environment. *Arabian Journal of Geoscience, 8*(2), 1077–1087.

Baioumy, H. (2011). Rare earth elements and sulfur and strontium isotopes of Upper Cretaceous phosphorites in Egypt. *Cretaceous Research, 32*, 368–377.

Ben Ayed, N. (1980). The role of E-W dropout in the structural evolution of the Tunisian Atlas. *C. R. Soman Geological Society of France Fasc, 1*, 29–32.

Burollet, P. F., Oudin, J.-P. (1980). *Paléocène et Eocène en Tunisie—Pétrole et phosphate, Géologie Comparée des Gisements de Phosphate et de Pétrole* (pp. 205–216). BRGM.

Garnit, H., Bouhlel, S., Barca, D., & Chtara, C. (2012). Application of LA-ICP-MS to sedimentary phosphatic particles from Tunisian phosphorite deposits: Insights from trace elements and REE into paleo-depositional environments. *ChemErde-Geochemistry, 72*, 127–139.

Hollis, C. J. (2006). Radiolarian faunal change across the Paleocene-Eocene boundary at Mead Stream, New Zealand. *Eclogae Geologicae Helvetiae, 99*, S79–S99. Hollis, C. J., Stickley, C. E., Bijl, P. K., Schiøler, P., Clowes, C. D., Li, X., & Campbell, H. (2017). The age of the Takatika Grit, Chatham Islands, New Zealand: Alcheringa. *An Australasian Journal of Palaeontology*, 1–14.

Jaballi, F., et al. (2019). Mineralogical and geochemical behavior of heated natural carbonate-apatite of the Ypresian series, Maknassy-Mezzouna basin, central Tunisia. *Carbonates and Evaporites Journal, 34*, 1689–1702.

Murray, R. W., Brink, M. R. B., Gerlach, D. C., & Gerlach Russ, G. P. (1990). Rare earth elements as indicators of different marine depositional environments in Chert and Shale. *Geology, 18*(3), 268–271.

Sassi, S. (1974). *La sédimentation phosphatée au Paléocène dans le Sud et le Centre Ouest de la Tunisie* (p. 300). Thèse de Doctorat ès Sciences, Université de Paris Sud Orsay.

Soudry, D., Ehrlich, S., Yoffe, O., & Nathan, Y. (2002). Uranium oxidation state and related variations in geochemistry of phosphorites from the Negev (Southern Israel). *Chemical Geology, 189*, 213–230.

Tlili, A., Saidi, R., Fourati, A., Ammar, N., & Jamoussi, F. (2012). Mineralogical study and properties of natural and flux calcined porcelanite from Gafsa-Metlaoui basin compared to diatomaceous filtration aids. *Applied Clay Science, 55*, 47–57.

Zaïer, A. (1999). *Evolution tectono-sédimentaire du bassin phosphaté du centre-ouest de la Tunisie minéralogie, pétrographie, géochimie et genèse des phosphorites*. Université de Tunis II.

Zaïer, A., Béji-Sassi, A., Sassi, S., & Moody, R. T. J. (1998). Basin evolution and deposition during the early Paleocene in Tunisia. In D. S. Macgregor, R. T. J. Moody, & D. D. Clark-Lowes (Eds.), *Petroleum geology of North Africa* (Vol. 132, pp. 375–393). Geological Society of London Special Publication.

Mineral Characterizations of Langer Heinrich Uranium Rocks, Republic of Namibia

Sarunya Promkotra, Thidarat Cotanont,
Neeranuch Mahayossanan and Tawiwan Kangsadan

Abstract

Mineral characterizations of uranium rocks in Langer Heinrich Uranium, Republic of Namibia, Southern Africa, are intended to determine the mineral compositions of uranium rocks related to the uranium mineral distribution. The research studies consist of lithological descriptions of rocks, petrological analysis, and mineral analysis. Rock-forming minerals of uranium deposits occur in sedimentary rocks (Arkose). Polished section analysis appears carnotite included in quartz–calcite matrix, margaritasite incorporated in calcrete matrix, and orthobrannerite inserted between fractured rocks. According to thin-section analysis, rock samples consist of quartz, feldspar, and lithic fragment. Physically, granular rock materials are made of coarse to fine sands whose packing is of moderately sorted grains including calcareous cement, angular to sub-angular, with high sphericity. Chemical compositions of uranium rocks are analyzed by X-ray fluorescence spectroscopy presenting Al, Ca, Fe, K, Mg, Mn, Na, Ni, Pb, Si, Ti, Zn, and U. These elements of Ca, K, Fe, and Ti are mostly found more than 2 wt%. However, uranium (U) element exhibits on average a concentration of 0.5 wt%. Mineral compositions of rocks by X-ray diffractometry indicate calcite and quartz as main constituents. Uranium minerals of carnotite, orthobrannerite, margaritasite, uraninite, and davidite are less than 4 wt% of a whole rock.

Keywords

Mineral characterizations · Uranium rock · Namibia · Langer Heinrich uranium

1 Introduction

Uranium rocks in the Republic of Namibia are collected in Langer Heinrich uranium deposits in Erongo Region, southwestern Namibia. Uranium deposits in the central Namib Desert, a surficial calcrete-type uranium deposit (Toens & Hambleton-Jones, 1980), are located around 80 km east of Swakopmund at the foot of the Langer Heinrich Mountains. The Cenozoic sediments, sediments of the Namib groups, are congregation to surface uranium mineralization throughout the Namib Desert, comprising the Langer Heinrich ore mass (Toé et al., 2013). Uranium ores in oxidizing environment with vanadium secondary minerals as carnotite (Hambleton-Jones, 1976) are close to the ground surface counting on the palaeo-channel width, of approximately 1–30 m thickness and 50–1100 m width. Erosion of uranium deposition in host sediments is proceeded by the river flows (Becker & Kärner, 2009). The study of mineral assemblages in uranium rocks is essential to understand the mineral forming of uranium ores in Namibia. The aims of this investigation are to determine mineral compositions of uranium ore related to the uranium mineral distribution in Langer Heinrich Uranium, Republic of Namibia, southern Africa.

2 Materials and Methods

Uranium rocks are collected nearby Langer Heinrich uranium deposits, Republic of Namibia. Hard textures of rocks are prepared for thin and polished section to analyze optical properties and micro-structural characteristics by optical microscopy. Whole-rock samples are grinded for powder

S. Promkotra (✉) · T. Cotanont
Department of Geotechnology, Faculty of Technology, Khon Kaen University, Khon Kaen 40002, Thailand
e-mail: sarunya@kku.ac.th; spromkotra@gmail.com

N. Mahayossanan
Primary Industries and Mines Group, Chonburi Provincial Industry Office, Ministry of Industry, Chonburi 20000, Thailand

T. Kangsadan
Chemical and Process Engineering Program, The Sirindhorn International Thai-German Graduate School of Engineering (TGGS), King Mongkut's University of Technology North Bangkok (KMUTNB), Bang Sue, Bangkok, Thailand

F. Lucci et al. (eds.), *Selected Studies in Geomorphology, Sedimentology, and Geochemistry*,
Advances in Science, Technology & Innovation, https://doi.org/10.1007/978-3-031-43744-1_14

X-ray diffractometer (XRD), D8 Advance (Bruker AXS Co. Ltd., Germany), and also for elemental analysis by X-ray fluorescence spectroscopy (XRF).

Rock samples are reduced in size by grinding and sieving the oversize until all powder passes the sieve on a number of 60 and 100 with an opening size of 250 and 150 microns. Size option is obtained for elemental analysis (XRF) of 250 microns and X-ray diffractometry of 150 microns. Particularly, to separate sand particles, oriented sample is prepared by dispersing sample size of 150 microns in chemical solution. Sodium hexametaphosphate solution (or calgon-commercial name) is prepared for 10% wt/vol and added for clay mineral dispersion. When a container of sample suspensions stands for 10, 20, and 30 min, these suspensions are withdrawn from the upper section (zone 1), middle section (zone 2), and lower section (zone 3) based on time interval. This method is available for separating clay minerals (upper part or zone 1), and heavy minerals (lower part or zone 3).

3 Results and Discussions

3.1 Lithological Descriptions

Rock samples are typically of gray-yellowish color. Their grain sizes are in the range of fine sand to cobbles (0.125–30 mm), angular to sub-angular, and moderately sorted-to-poorly sorted. Abundant crystal grains of feldspar and quartz occur together with merged crystal fragment. Rock texture contains the calcareous cementing material. The summary of lithological uranium rocks is presented in Table 1. Rock-forming minerals of uranium deposits occur in Arkose, a sedimentary rock classified based on Pettijohn (1975).

3.2 Optical Microscopy (Polished and Thin Section)

Ore microscopy involves not only the identification of individual mineral grains, but also the interpretation of ore mineral texture, which are the spatial relationships between grains (Fig. 1). Microscopically, the rock texture is grain-supported. Granular rock materials are made of coarse to fine sands which are moderately sorted grains including calcareous cement, angular to sub-angular, with high sphericity, as shown in Fig. 2.

3.3 Chemical Compositions and Mineral Components

Chemical composition of uranium rocks is estimated by X-ray fluorescence spectroscopy, which indicated the presence of Al, Ca, Fe, K, Mg, Mn, Na, Ni, Pb, Si, Ti, Zn, and U.

Table 1 Lithological descriptions of uranium rock samples

Rock sample	Crystal grains	Grain size (mm)	Rock composition	Rock texture
Uranium rocks	Fine sand to cobbles with angular to sub-angular	0.125–30	Quartz (45–50%) Feldspar 45–50%) Lithic fragment (5%)	Calcareous cementing materials

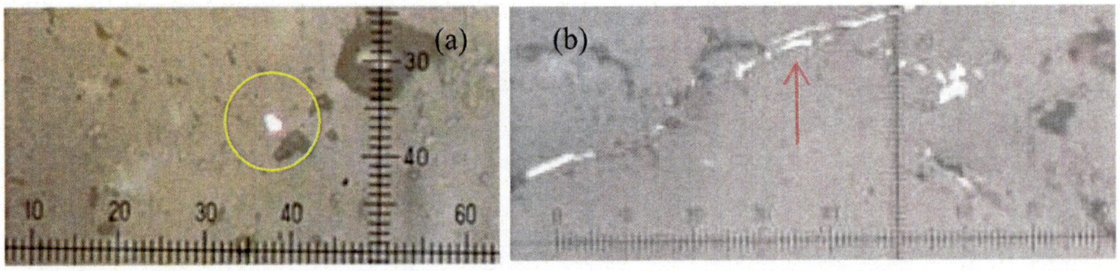

Fig. 1 Polished section **a** carnotite grains (white) and **b** orthobrannerite grain (white) inserted between fractured rocks

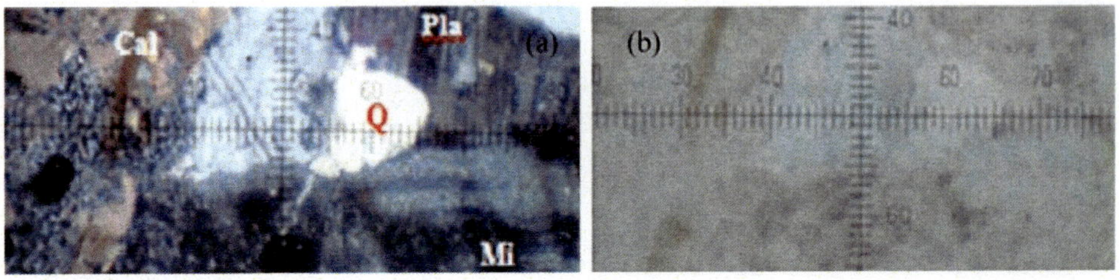

Fig. 2 Thin section of uranium rock sample **a** plane-polarized light (PPL) and **b** cross polarized light (XPL) (Q = quartz, Cal = Calcite, Mi = Microcline, Pla = Plagioclase)

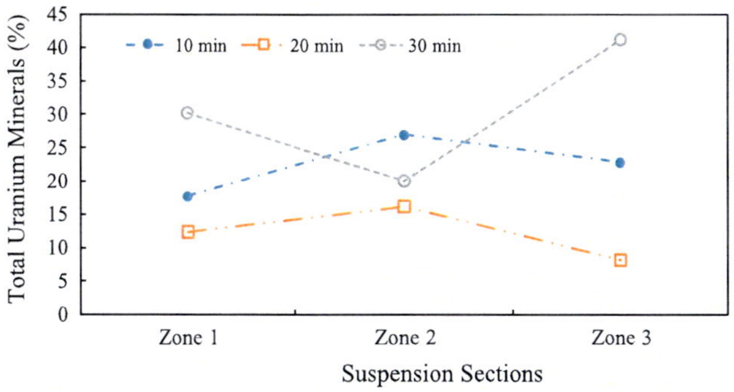

Fig. 3 Total uranium minerals as a function of the suspension section by time in 10, 20, and 30 min

Particularly, major elements of Ca and Si are found more than 10 wt%. Besides, K, Mg, Fe, and Ti are found in the range of 2.0–6.0 wt%. Uranium elements are totally discovered on an average of 0.5 wt%. Mineral compositions of rocks by XRD indicate calcite and quartz as main constituents. Uranium minerals of carnotite, orthobrannerite, margaritasite, uraninite, and davidite are less than 4 wt% of a whole rock. Similarly, XRD analysis of the oriented samples has identified the betafite, carnotite, and uraninite minerals in all three suspended sections (upper, middle, and lower sections) at 10, 20, and 30 min in the amount of 13, 6.3, and 26.7%, respectively (Fig. 3).

4 Conclusions

The mineral composition of Langer Heinrich uranium deposits, obtained by XRD analysis, indicates that these deposits are contained in sedimentary rocks (Arkose). Polished sections indicate that uranium minerals are in carnotite, margaritasite, and orthobrannerite grains. The carnotite grains are included in the quartz–calcite matrix within a calcrete; whereas in the margaritasite, grains are incorporated in the calcrete matrix. Lastly, the orthobrannerite grains are located between fractured rocks. Thin sections are composed of quartz, feldspar, and lithic fragments. In addition, the uranium (U) revealed the concentration in the average of 0.5 wt%. Uranium rocks are composed mainly of carnotite, orthobanerite, margaritasite, uraninite, and davidite in the range of 0.2–1.1 wt%, which totally found uranium concentrations of less than 4 wt% as a whole rock. With the oriented technique, the uranium minerals are found in all 3 suspended sections; however, the highest concentration is in the lower section. Moreover, the total uranium minerals are a function of the suspension section for a longer time. Therefore, this technique is suitable for separating clay minerals from heavy minerals, and beneficial for the mineral processing.

References

Becker, E., & Kärner, K. (2009) *Geological setting of the Langer Heinrich uranium deposit, Namibia.* Proceedings of an International Symposium. IAEA-TECDOC-1739, pp. 52–62. Austria.

Hambleton-Jones, B. B. (1976). *The geology and geochemistry of some epigenetic uranium deposits near the Swakop River, Southwest Africa.* Ph.D. Thesis, University of Pretoria, South Africa, pp. 1–306.

Pettijohn, F. J. (1975). *Sedimentary rocks* (3rd ed.). Harper & Row.

Toé, W., Vanderhaeghe, O., André-Mayer, A. S., Feybesse, J. L., & Milési, J. P. (2013). From migmatites to granites in the Pan-African Damara orogenic belt, Namibia. *Journal of African Earth Sciences, 85*, 62–74.

Toens, P. D., & Hambleton-Jones, B. B. (1980). Uraniferous surficial deposits, international geological congress (pp 1–15). Paris.

Check for updates

Exploring Abandoned Alluvial Tin Mines Using Satellite Image and Mineralogical Data: A Case Study of Belitung Island, Indonesia

Andy Yahya Al Hakim, Mohamad Nur Heriawan, Arie Naftali Hawu Hede, Fakhmi Imanuddin Prakasa, Muhamad El Hakim, Rifqy Adwin Hafizsyah, Vieri Muhammad Naufal, Cahyo Okta Ardian and Satyogroho Dian Amertho

Abstract

Tin in Indonesia is mainly mined from alluvial placer deposits located in the tin belt of SE Asia. The cassiterite concentration is in alluvial placer and was formed by erosion and weathering of cassiterite-bearing granites. Although most resistate economic minerals (such as cassiterite and monazite) have been liberated due to weathering processes, economic trace minerals have been overlooked for many decades. Furthermore, the cessation of alluvial tin mining has generated a large amount of mine tailings, most of which represent nonviable waste materials. These tailings and vegetations occasionally capped the undisturbed areas that still have a potential value of tin, heavy, and rare earth bearing minerals. The current investigations focused on the advancing exploration technology using multi-temporal satellite imagery data to assess land changes due to past tin mining activities from year 1994 to 2017. The algorithms employed to evaluate overburden, tailing deposits, and land change were spectral angle mapper (SAM) and supervised classification. Examination based on X-Ray Diffraction (XRD), grain size analyses, and spectrum data have jointly applied to determine the mineral composition of tailings, overburden, concentrate, alluvial placer allochthonous deposits (*kaksa*), and bedrock (*kong*). The results showed that multi-temporal satellite images and mineralogy can be implemented for identifying a change detection due to mining activities and to delineate the potential mineable heavy and rare earth mineral zones, which are being covered by historic tailings or vegetations.

Keywords

Alluvial tin · Tailings · Rare earth minerals · Satellite image · Mineralogy

1 Introduction

Indonesia was the leading supplier of refined tin with an annual production of 80kt in 2019 (U.S. Geological Survey, 2020), which mainly distributed within Bangka and Belitung Islands. In order to maintain its sustainability, the current target of tin exploration is the residual zones that are thought to contain in situ alluvial tin reserves, which are likely to be covered by overburden or historic tailings. This is in part due to the extensive artisanal tin mining on these islands which has produced dramatic land cover changes.

Effective monitoring and spatial patterns changes in land cover are critical for a sustainable management of mineral resources. The assessment of land cover changes using multi-temporal satellite images in mining areas may reveal important spatial information on different types of coverage including waterbody, tailing, overburden, forest, and grassland. The assessment, monitoring, and analysis of land cover changes are important for several main reasons: (i) there is the lack of available information on overburden and tailing changes at local and regional scales, (ii) assessment of land cover classes and their changes over time, and (iii) criteria for determining the priority areas for conservation and mineral exploration.

This study investigates the abandoned alluvial tin mines using integration of satellite remote sensing and

A. Hakim (✉) · M. N. Heriawan · A. N. H. Hede · F. I. Prakasa · M. E. Hakim · R. A. Hafizsyah · V. M. Naufal
Research Group of Earth Resources Exploration, Faculty of Mining and Petroleum Engineering, Bandung Institute of Technology, Jl. Ganesha 10, Bandung 40132, Indonesia
e-mail: andyyahya@itb.ac.id

C. O. Ardian · S. D. Amertho
PT TimahTbk, Pangkal Pinang, Jl. Jenderal Sudirman 51, Bangka 33121, Indonesia

F. Lucci et al. (eds.), *Selected Studies in Geomorphology, Sedimentology, and Geochemistry*, Advances in Science, Technology & Innovation, https://doi.org/10.1007/978-3-031-43744-1_15

mineralogical study, which is recognized as a well-established technology for providing sufficient information on mineral distribution and land cover mapping within large area (PT LAPI ITB, 2019). Furthermore, the application of mineralogical characterization to mine waste, overburden, and concentrate samples has the potential of improving the long-term design, management, and planning for further exploration strategies.

2 Materials and Methods

Multi-temporal satellite data of Landsat OLI/TIRS and Landsat TM imagery of Belitung Island during 1994–2015 were used (USGS, 2020). Radiometric calibration and atmospheric correction were performed to minimize distortion (sensors, sun, atmosphere, and topography). The Normalized Difference Vegetation Index (NDVI) and Modified Normalized Difference Water Index (MNDWI) methods were performed to quantify the change in vegetation cover and waterbody area, respectively.

Land cover mapping was completed in the mining concession of PT TimahTbk in Belitung Island using supervised classification and maximum likelihood classifier algorithms. The value of a pixel at any point will be categorized into a certain class, according to the similarity of the pixels to the pre-determined region of interest class. The Spectral Angle Mapper (SAM) calculates spectral similarities between an unknown and reference spectrum. The reflectance spectroscopy method was obtained from various samples that were collected during the dry season of 2019 and 2020. Verification and validation of satellite imagery processing was performed by ground truth.

Various types of samples, including tailings, topsoil, concentrate, alluvial placer allochthonous deposits (*kaksa*), and bedrock (*kong*) samples, were collected. A sieve analysis was performed for a dry sample with the opening of +48, +100#, and −100# as those disaggregate properly in dry condition. A mineralogical study was carried out by powder X-ray diffraction (XRD) using a Rigaku Smartlab, polished section was studied using reflected light and scanning electron microscopy. All analytical methods were performed at the Faculty of Mining and Petroleum Engineering, Bandung Institute of Technology, Indonesia.

3 Results

3.1 Mineralogy

A summary of mineral distribution is provided in Table 1. Cassiterite and ilmenite are the predominant mineral phases in concentrates; quartz, tourmaline, and K-feldspar are common gangue minerals in most sample types; clay minerals are predominated by kaolinite. Zircon in tailings samples is concentrated in the fine-grained fraction (−100 mesh) rather than the coarse-grained fraction (+48 and +100 mesh). Rare earth minerals (i.e., monazite and xenotime) are observed and distributed randomly in tailings, *kaksa, kong,* and concentrates. Monazite-(Ce) and xenotime are subhedral to anhedral grains (50–400 μm). Exsolution texture was observed in ilmenite, oscillatory zoning was observed in zircon crystals following crystal faces, radiating outwards from the core.

3.2 Images Classification and Land Cover Changes

The mineralogical data showed in Table 1 were used to map the distribution of tailing and overburden using the SAM method. Using NDVI and MNDWI algorithms, the images of the study area were categorized into different classes, namely vegetation (plantation / garden, fields, forest), settlement, open area, agriculture, barren, built-up, which is simplified as others, and water body (Fig. 1). The result showed that, between 1995 and 2014, the amount of overburden and tailings in East Belitung increased from 3.0 to 19.1% and from 1 to 5% of the total area, respectively (Fig. 1a), in response to decreasing other land areas (i.e., 90.7% in 2005 which decrease down to 69.5% in 2017). In West Belitung, overburden in mining concession in West Belitung increased from 12% in 1994 up to 29% in 2015.

Table 1 Representative mineral compositions of Belitung

	Topsoil	Tailings	*Kaksa*	*Kong*	Concentrate
Major	Quartz, tourmaline	Quartz, K-feldspar	Quartz	Quartz	Cassiterite, ilmenite
Trace	Pyrite, ilmenite, magnetite, microcline, sanidine, sodalite, gibbsite	Zircon, ilmenite, cassiterite, tourmaline, phlogopite, monazite, xenotime	Cassiterite, magnetite, microcline, muscovite, goethite, gibbsite	K-feldspar, pyrite,	Monazite, zircon, rutile, xenotime
Clay	Kaolinite, halloysite	Kaolinite, smectite	Kaolinite, smectite	Kaolinite, smectite	Kaolinite, dickite, illite

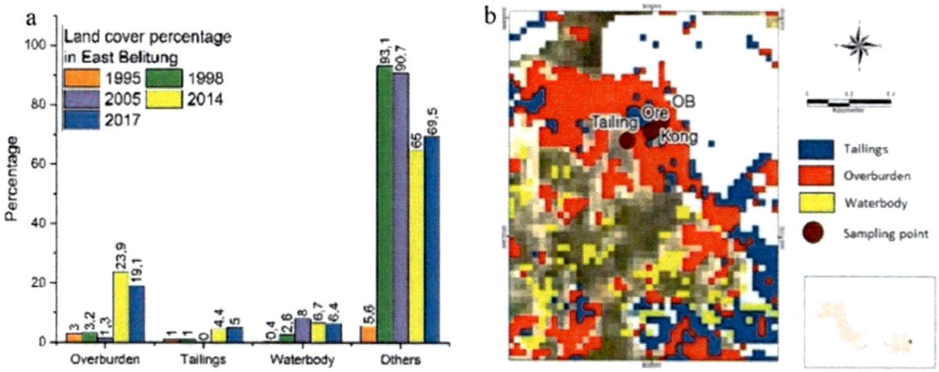

Fig. 1 **a** Land covers' change during the last 22 years in the East Belitung and **b** location of potential areas (overburden) for further exploration

Figure 1b depicts spatial distributional pattern of land cover of the East Belitung that have been verified and validated by the ground truth data.

4 Discussion

During the study period (i.e., 1994–2017), overburden and tailings have been increased due to the expansion of mining activities. The results of this study indicated that heavy and rare earth minerals are still concentrated in tailings. Long-term planning and processing design for Indonesian tin tailings should be designed, as other countries within Malayan tin belt such as Malaysia and Thailand have recovered monazite and xenotime as by-products from the retreatment of processing plant tailings earlier produced from tin placers (van Gosen et al., 2016). Euhedral-subhedral crystals of rare earth minerals may have formed due to in situ weathering and/or the distance of transportation, while anhedral spongy-rounded crystal, usually containing inclusions of clay minerals, may be inferred that those grains may be transported at a specified distance from the weathering location.

5 Conclusions

The development of innovative technologies on remote sensing and mineralogical methods presented in this paper is applicable for the existing business practice in tin mining operations. Tailings should be considered as a potential target for exploration as it contains economic minerals including zircon, monazite, and xenotime. The combination of spectral angle mapper and other algorithms has been successfully applied to detect possible undisturbed areas for exploration target.

References

PT LAPI ITB. (2019). *Multi-temporal land cover mapping using satellite imagery in alluvial tin mine of Bangka Belitung islands*. Unpublished research report.

USGS. (2020). *Mineral commodity summaries 2020*.

USGS. (2020). https://earthexplorer.usgs.gov/, last accessed 2020/07/01.

van Gosen, B. S., & Tulsidas, H. (2016). Thorium as a nuclear fuel. In *Uranium for nuclear power* (pp. 253–296). Woodhead Publishing.

Volcano-Stratigraphy of Anak Krakatau Volcanic Complex (Indonesia) and Comparison Between the 2018 Anak Krakatau Eruption and the 1883 Krakatau Eruption

Muhammad Rahaldi Taher, Mirzam Abdurrachman, Idham Andri Kurniawan, Emmy Suparka, Arif Susanto and Henki Eko Putra

Abstract

In December 2018, Anak Krakatau Volcano, located in the Sunda Strait, Indonesia experienced landslide, eruption, and tsunami that changed the geological setting of the Anak Krakatau Volcanic Complex. We conducted remote sensing analysis and geological mapping to determine the latest geological setting of the Anak Krakatau Volcanic Complex and to compare the eruption processes between violent Krakatau eruption in 1883 and Anak Krakatau eruption in 2018. Based on remote sensing analysis and geological mapping, the research area consists of five geomorphological units, three crowns, five hummocks, and 12 mapped rock/deposit units. The primary found geological structures are sheeting joint, autobreccia, jigsaw fit, vesicular, parallel lamination, cross lamination, and normal and reverse graded bedding; while the secondary geological structures are the Krakatau Fault and normal faults. The 1883 Krakatau eruption deposits show 6 eruption units describing the eruption stage consisting of the Phreatomagmatic, Sub-Plinian, and Plinian Stages. The occurring dominant processes are the interaction between water and magma in the eruption vent, intense erosion in the eruption vent, changes in the eruption intensity, and the collapse of the eruption column from small-large scale which can occur repeatedly. The 2018 Anak Krakatau eruption deposits consist of 5 eruption units describing the eruption stage of the Strombolian and the Phreatomagmatic Stages. The occurring dominant processes are small-scale eruption column collapse, intense erosion in the eruption vent, and intense interaction between water and magma at shallow depths.

Keywords

2018 Anak Krakatau eruption · 1883 Krakatau eruption · Volcano-stratigraphy

1 Introduction

In December 2018, Anak Krakatau Volcano located in the Sunda Strait, Indonesia (Fig. 1) experienced landslide, eruption, and tsunami that changed the geological setting of the Anak Krakatau Volcanic Complex (AK Volcanic Complex). The changes in the geological setting of the AK Volcanic Complex has made the need for a new geological map made through geological mapping to determine the latest geological setting of the AK Volcanic Complex. Furthermore, identification of the 2018 Anak Krakatau eruption deposits is needed to determine the characteristics of the eruption and compare it with the violent 1883 Krakatau eruption to determine the potential eruption of Anak Krakatau in the future.

2 Methods

The applied methods are:

2.1 Remote Sensing

We use satellite imagery from Landsat-8 OLI, Indonesia Digital Elevation Model (DEM), and Google Earth to determine the geomorphology of the AK Volcanic Complex.

M. Taher (✉)
Geological Engineering, Bandung Institute of Technology, Bandung, Indonesia
e-mail: mrahalditaher@gmail.com

M. Abdurrachman · I. A. Kurniawan · E. Suparka · A. Susanto
Petrology, Volcanology and Geochemistry Research Group, Geological Engineering, Bandung Institute of Technology, Bandung, Indonesia
e-mail: mirzam@itb.ac.id

H. E. Putra
Maipark, Jakarta, Indonesia

F. Lucci et al. (eds.), *Selected Studies in Geomorphology, Sedimentology, and Geochemistry*, Advances in Science, Technology & Innovation, https://doi.org/10.1007/978-3-031-43744-1_16

2.2 Geological Mapping

This method is conducted to determine the volcano-stratigraphy and the geological structures of the AK Volcanic Complex. The obtained results are in the form of a Geological Map of the AK Volcanic Complex.

3 Results

3.1 Geology of Anak Krakatau Volcanic Complex

After December 2018, the geological map of AK Volcanic Complex (Abdurrachman et al., 2018) has been changed. Now, the research area (Fig. 2) can be divided into the Old Krakatau Crown, the Young Krakatau Crown (consists of the Rakata Hummock, the Danan Hummock, and the Perbuatan Hummock), the Anak Krakatau Crown (consists of the East Anak Krakatau Hummock and the West Anak Krakatau Hummock), and 12 mapped rock/deposit units.

4 Discussion

4.1 The 1883 Krakatau Eruption

The 1883 Krakatau Eruption (Fig. 3) consisted of three eruption stages, Phreatomagmatic, Sub-Plinian, and Plinian Stages. The Phreatomagmatic Stage produced ash fall, Unit A. In this unit, obsidian fragments were found locally as an indication of the presence of partially solidified magma forming by the interaction between seawater and magma (Self & Rampino, 1981) or due to the degassing process of magma ascending to the surface (Mandeville et al., 1996). The 1883 Krakatau Eruption turned into the Sub-Plinian Stage (Self & Rampino, 1981) which deposited Units B, C, and D. The Sub-Unit B1 is a pumice flow consisting of pumice, obsidian, and lithic fragments. The pumice fragments indicate high gas content in the 1883 Krakatau Eruption which formed the eruption column. The erosion in the eruption vent carried accessory lithic fragments to the surface (Darragh et al., 2006) and widened the size of the eruption vent, which affected the eruption intensity and triggered the collapse of the eruption column (Cas & Wright, 1987). In Sub-Unit B2, pumice flows are found to be alternating with pumice falls describing the instability of the eruption column in a short duration, resulting in a small-scale collapse of the eruption column (Mandeville et al., 1996; Self & Rampino, 1981), changes of the eruption intensity over short timescales (Cas & Wright, 1987), or fluctuations of water-magma ratio in the eruption vent (Darragh et al., 2006). The Sub-Unit B3 is a block and ash flow, the dominant lithic fragments illustrate intense erosion that widened the size of the eruption vent. Unit C is associated with the collapse of the voluminous eruption column due to the increase of eruption intensity. The presence of Unit D as co-ignimbrite breccia indicates that the Sub-Plinian eruption occurred continuously in voluminous numbers with little variation in eruption intensity (Cas & Wright, 1987).The Plinian Stage was preceded by the deposition of Unit E (ashfall), composed of lithic and obsidian fragments from small scale of eruption columns.

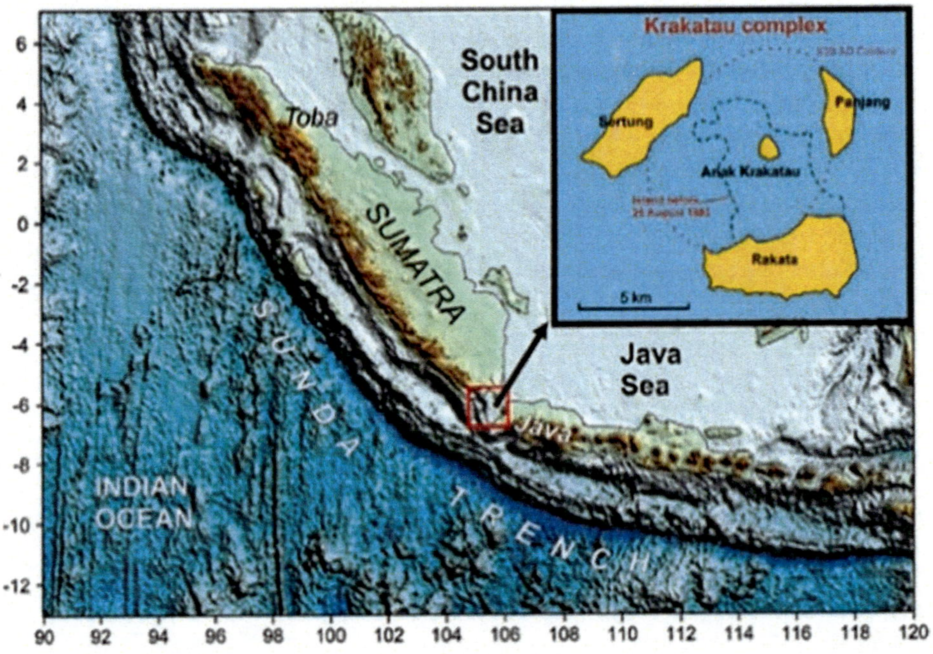

Fig. 1 Anak Krakatau volcanic complex located in the Sunda Strait, Indonesia. Modified from Jaxybulatov et al. (2011)

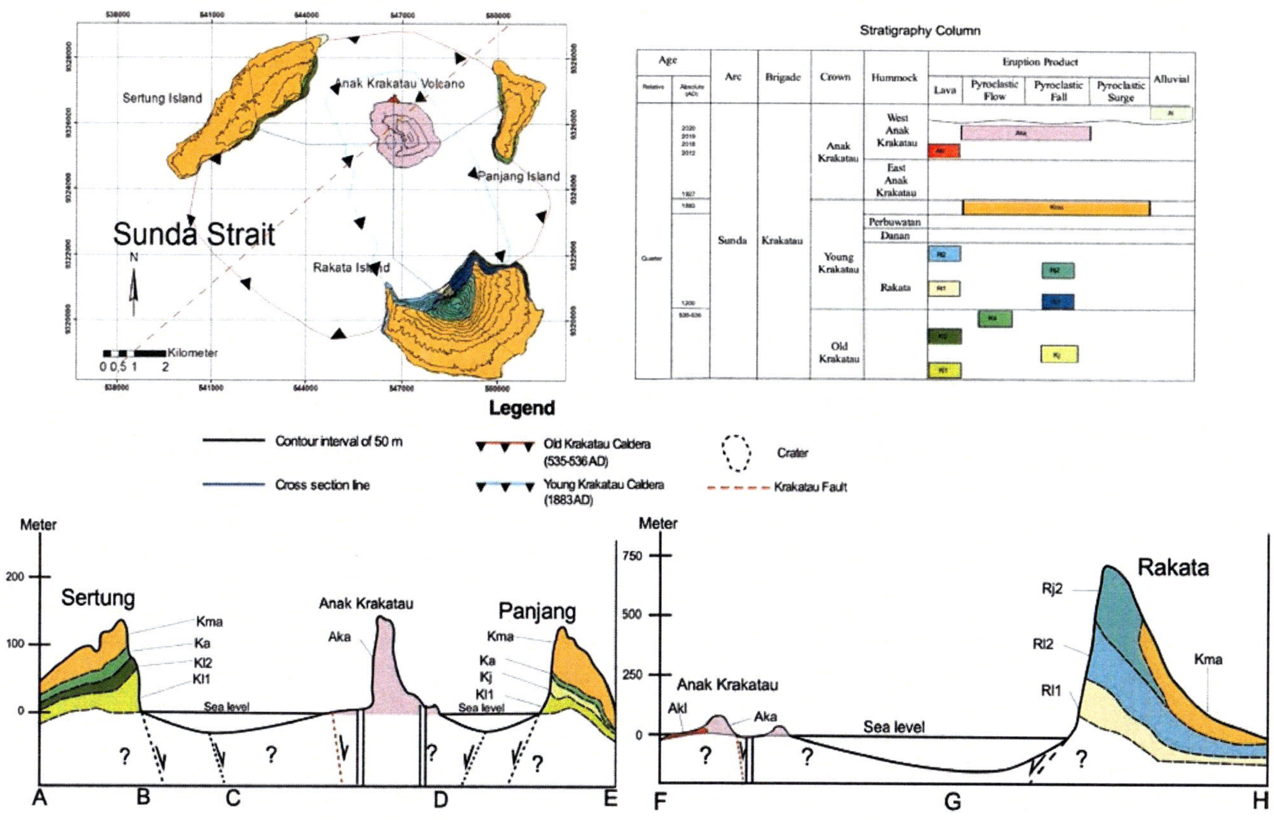

Fig. 2 New geological map of Anak Krakatau volcanic complex

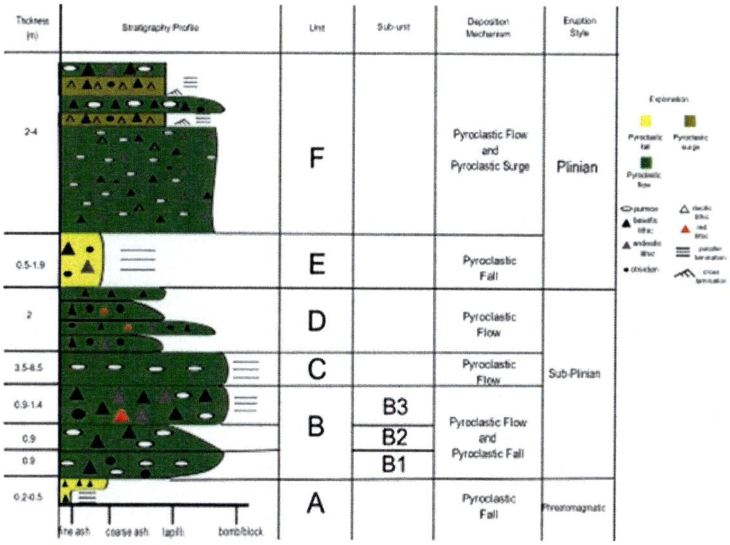

Fig. 3 Eruption unit of the 1883 Krakatau eruption

Above Unit E, Unit F is deposited in the form of alternation between pyroclastic flows and pyroclastic surges, and indicates that there were several small-scale eruption column failures prior to large-scale eruption column collapse (Mandeville et al., 1996).

4.2 The 2018 Anak Krakatau Eruption

The 2018 Anak Krakatau Eruption (Fig. 4) consisted of Strombolian and Phreatomagmatic Stages. The Strombolian stage produced Unit G, which is divided into four sub-units.

Fig. 4 Eruption unit of the 2018 Anak Krakatau eruption

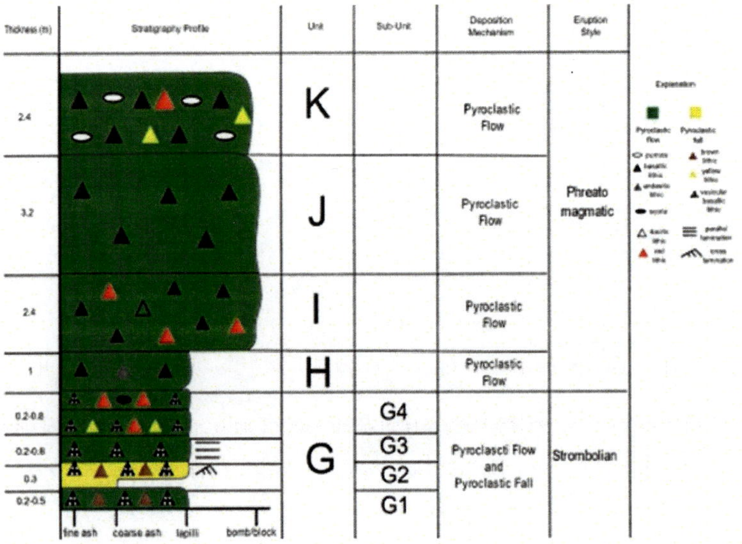

The Sub-Unit G1 is a scoria flow consisting of lapilli vesicular basaltic lithic fragments, solidified from the basaltic-andesitic magma (Cas & Wright, 1987) when the eruption column collapses. The characteristics of Sub-Unit G2, scoria fall, are relatively similar to Sub-Unit G1. The series of Strombolian eruptions continued, producing Sub-Units G3 and G4 as scoria flow. The Phreatomagmatic Stage occurs when seawater interacts with magma below the surface through the exposed eruption vent, and it is estimated that there is a new eruption vent that is directed to the southwest toward the missing part of Anak Krakatau (Williams et al., 2019). The Phreatomagmatic Stage produces Unit H, Unit I, Unit J, and Unit K. Unit H is a block and ash flow that describes the initial stage of the Phreatomagmatic Stage when the pressure increases in a short time due to the interaction between seawater and magma, thus ejecting lapilli lithic fragments. Unit I is a block and ash flow composed of block-lapilli lithic fragments ejected from the eruption vent or lava flow. It is estimated that a new eruption vent opened based on the presence of lapilli-block black lithic fragments in Unit J. The final Phreatomagmatic Stage is thought to occur during the deposition of Unit K based on the presence of whitish pumice fragments associated with the remnants of the gas content that was left when the Phreatomagmatic eruption ended (Manga et al., 2018).

5 Conclusions

1. The research area is divided into three crowns and five hummocks, the Old Krakatau Crown, the Young Krakatau Crown (consist of the Rakata Hummock, the Danan Hummock, and the Perbuatan Hummock), and the Anak Krakatau Crown (consist of the East Anak Krakatau Hummock and the West Anak Krakatau Hummock).
2. The 1883 Krakatau Eruption consists of the Phreatomagmatic, the Sub-Plinian, and the Plinian Stages.
3. The 2018 Anak Krakatau Eruption consists of the Strombolian and the Phreatomagmatic Stages.

Acknowledgements This research was supported by Maipark Research Grant and GL09 Foundation.

References

Abdurrachman, M., Widiyantoro, S., Priadi, B., & Ismail, T. (2018). Geochemistry and structure of Krakatoa volcano in the Sunda Strait. *Indonesia Geosciences, 8*(4), 111.

Cas, R. A. F., & Wright, J. V. (1987). *Volcanic successions: Ancient and modern*. Allen and Unin.

Darragh, M., Cole, J., Nairn, I., & Shane, P. (2006) Pyroclastic stratigraphy and eruption dynamics of the 21.9 ka Okareka and 17.6 ka Rerewhakaaitu eruption episodes from Tarawera Volcano, Okataina Volcanic Centre, New Zealand. *New Zealand Journal of Geology and Geophysics, 49*(3), 309–328.

Jaxybulatov, K., Koulakov, I., Ibs-von Seht, M., Klinge, K., Reichert, C., Dahren, B., & Troll, V. R. (2011). Evidence for high fluid/melt content beneath Krakatau volcano (Indonesia) from local earthquake tomography. *Journal of Volcanology and Geothermal Research, 206*(3–4), 96–105.

Mandeville, C. W., Carey, S., & Sigurdsson, H. (1996). Magma mixing, fractional crystallization and volatile degassing during the 1883 eruption of Krakatau volcano, Indonesia. *Journal of Volcanology and Geothermal Research, 74*(3–4), 243–274.

Manga, M., Mitchell, S. J., Degruyter, W., & Carey, R. J. (2018). Transition of eruptive style: Pumice raft to dome-forming eruption at the Havre submarine volcano, southwest Pacific Ocean. *Geology, 46*(12), 1075–1078.

Self, S., & Rampino, M. R. (1981). The 1883 eruption of Krakatau. *Nature, 294*(5843), 699–704.

Williams, R., Rowley, P., & Garthwaite, M. C. (2019). Reconstructing the Anak Krakatau flank collapse that caused the December 2018 Indonesian tsunami. *Geology, 47*(10), 973–976.

Geomorphology, Geography, Soil Science, Glaciology, Geoarchaeology, Geoheritage

Paleogeographic Background of Eolian Forms on Blagoveshensky Butte Surface (Russia)

Marina Krylenko and Viacheslav Krylenko

Abstract

Anapa bay-bar is a large coastal accumulation form on the Russian Black Sea coast. Blagoveschensky butte is located in the central part of Anapa bay-bar. The dunes located on the surface of the butte are its unique feature. Geological, geomorphological, paleogeographical, archaeological, and other data were used. The eolian deposits on this dune field are identical to the dune deposits of the bay-bar's adjacent area. It is found that the formation of the dunes occurred about 1700 years ago. Several possible versions of eolian sand accumulation and dune field formation are analyzed. To date, the growth of dunes on the Blagoveshchensky butte has stopped, and their transformation is observed.

Keywords

Accumulative form · Dune · Sea-level fluctuations · Coastal zone · Black sea

1 Introduction

Anapa bay-bar is a 47 km-long sandy accumulation form, located in the north-western part of the Russian coast of the Black Sea (Kuban River Delta) (Krylenko, 2015). Blagoveshchensky butte is located in the central part of this bay-bar (see Fig. 1). In this place, the accumulative form is attached to the cliff of the butte. The dune field (about 20 ha) of Blagoveshchensky butte is located on the surface of Neogene loams at an altitude of 15–20 m above sea level. When and how the deposits for the eolian forms were raised from the beach to this height is considered in this paper.

M. Krylenko (✉) · V. Krylenko
Shirshov Institute of Oceanology RAS, Moscow, Russia
e-mail: krylenko@mail.ru

2 Materials and Methods

The results of field studies of the surface and underwater relief, landscape-morphological structure, sediment flows, and the hydrological regime of the Anapa bay-bar were used in the presented work. In addition, archival geological, geomorphological, paleogeographical, archaeological, and other data were analyzed (Izmailov & Arslanov, 2006; Krylenko, 2015; Krylenko et al., 2011; Nevessky, 1957; Porotov et al., 2004) including data on the current dynamics of the dunes (Krylenko & Krylenko, 2019). The methodology is based on analysis of the above materials to create paleogeographic reconstructions of the development of the studied landforms and the geosystem of the Anapa bay-bar.

3 Results

The development of the Anapa bay-bar and Blagoveshchensky butte was determined by sea-level fluctuations and the sediment budget. Paleogeographic reconstructions showed that the formation of the Anapa bay-bar began approximately 5000–5500 years ago during the Holocene transgression. The accumulative body was formed due to abrasion material, shells, and alluvium of the Kuban River. Many regional studies note a significant regression of the Black Sea that occurred between 2500 and 1700 years ago (Izmailov, 2005; Izmailov & Arslanov, 2006; Kaplin et al., 2001). Researchers estimate the decrease of the sea level during the Phanagorian regression at 5.0–5.5 m (see Fig. 2). The age of the studied dune field cannot be reliably determined by the age of the sand that composes it. The composition of eolian deposits on the surface of Blagoveshchensky butte is identical to the composition of the dune deposits of adjacent areas of the bay-bar. The mineral composition of ancient and modern sediments has no significant differences.

F. Lucci et al. (eds.), *Selected Studies in Geomorphology, Sedimentology, and Geochemistry*,
Advances in Science, Technology & Innovation, https://doi.org/10.1007/978-3-031-43744-1_17

Fig. 1 Anapa bay-bar region

The ancient settlement Blagoveshchenskoye-4 (VI century BC–II-III century AD) is located to the North of the studied dune field. Ceramic fragments were found under blowing dunes, on the surface of the underlying loam (Fig. 2). Estimated dates indicate the period of the appearance of ceramics in the IV centuries BC–III centuries AD. Accordingly, eolian accumulative forms on the Blagoveshchensky butte surface did not exist at least until the beginning of the third century AD. Probably, the formation of the dunes began after the Phanagorian regression, already during the beginning of the Nympheum transgression (about 1700 years ago).

4 Discussion

Several models are possible for the accumulation of a great amount of sand, which formed the dunes on the surface of Blagoveshchensky butte:

(1) The Phanagorian regression led to the resumption of solid flow of the Kuban River to the Black Sea. The Nymphaeum transgression caused the redistribution of sediment accumulated during the regression period along the sea coastline. The mineral and mechanical sediment composition (well-sorted alluvial sands) contributed to the growth of dunes.
(2) During the Phanagorian regression, sediment arriving from the North (sand, pebbles, and shells) accumulated in a tombolo or salient. The sediment movement resumed to the South-East after sea-level rise. When this material reached Blagoveshchensky butte, sand entered the beach and cliff.
(3) Earlier, the width of the accumulative terrace near Blagoveshchensky butte was much larger than at present. Dunes were between the beach and cliff. During the development of the bay-bar and the Nymphaeum transgression, the dunes shifted to the cliff and gradually climbed to the butte surface.

Thus, alluvium of the Kuban River or relict alluvium from Pliocene deposits of abrasion shores was the source of quartz sands. This material accumulated to the North-West of the Blagoveshchensky butte during the Phanagorian regression. During the Nympheum transgression, the long-shore migration of sand to the South-East began. The main factor for sand lifting on the butte was a reduction in terrace width. As the coastline was leveled, the sand inflow decreased. To date, most of the sand has been carried to the South-East or accumulated in beach dune deposits. The growth of dunes on Blagoveshchensky butte has stopped, and their transformation is observed.

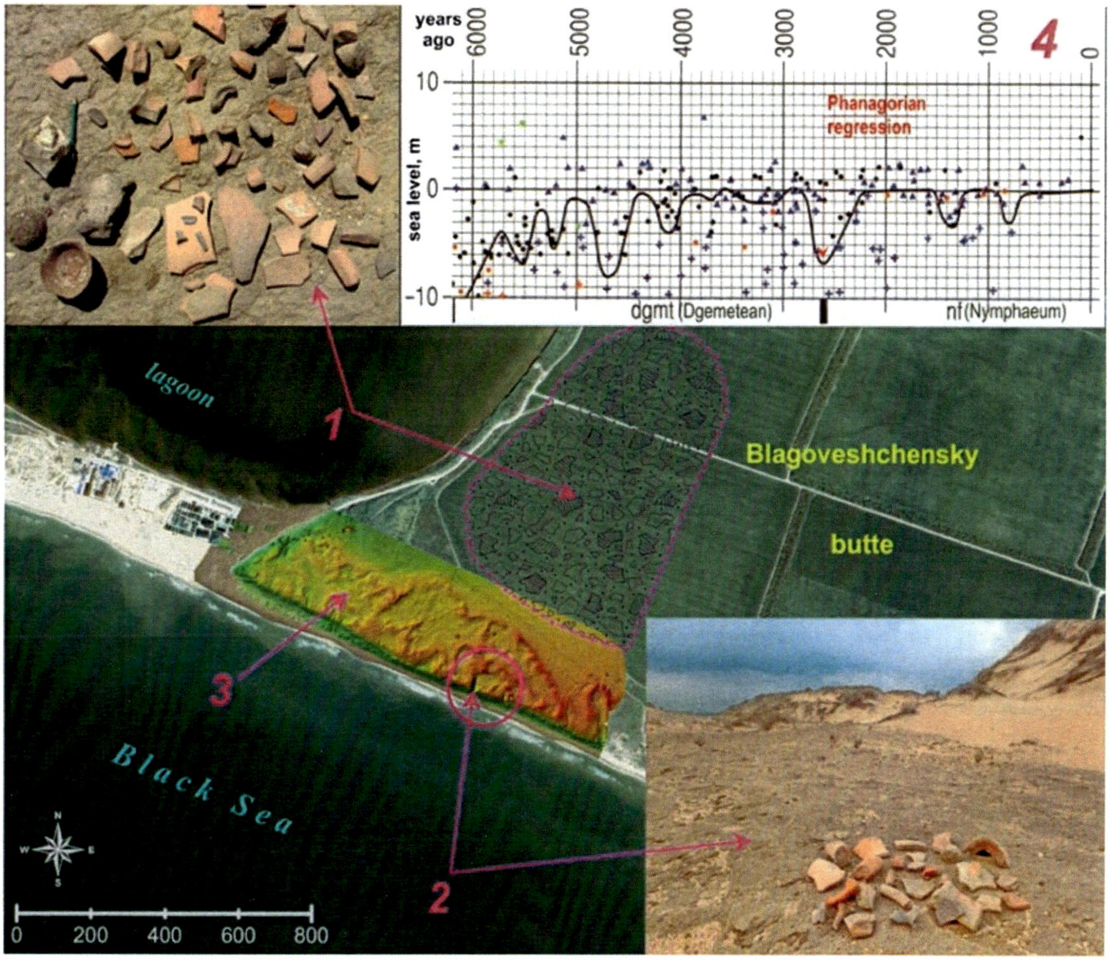

Fig. 2 Studies of the dune field age. *1*—ancient settlement Blagoveshchenskoye; *2*—ceramic fragments found under blowing dunes; *3*—digital elevation model (DEM) of the dune field; *4*—sea-level fluctuations (Balabanov, 2009)

5 Conclusions

As a result of the research, possible sources of sand for the formation of aeolian forms of Blagoveshchensky butte have been identified. In addition, the time period when these forms could arise has been clarified. The data show that there is currently no new influx and accumulation of sand. Thus, there are prerequisites for the irreversible destruction of the dune field with unique vegetation in the event of increased natural or anthropogenic impacts. In addition, the mechanism and conditions of the formation of the Blagoveshchensky butte dune field allows clarifying the chronology of Anapa bay-bar evolution.

The sediment flow studies with support of Russian Science Foundation (20-17-00,060). The sea-level fluctuations and morphological structure was supported by RFBR (18-05-00,333, 19-45-230,004).

References

Balabanov, I. P. (2009). *Paleogeographical background to formation of modern natural conditions of the Caucasus littoral Holocene terraces* (352 p). Dalnauka.

Izmailov, Y. (2005). *Evolutionary geography of the coasts of the Azov and Black seas* (175 p). Lazarevskoe Printing.

Izmailov, Y., & Arslanov, K. (2006). *Formation of the Anapa sand-spit and Late Holocene fluctuation of the Black Sea level.* Proc. of the IGCP 521,Odessa, Astroprint, pp. 81–82.

Kaplin, P., Porotov, A., Yanina, T., Gorlov, Y., & Fuash, E. (2001). Age and formation conditions of the Bugazskaya bay-bar. *Vestnik Mosk. University, 4*, 87–95.

Krylenko, M., & Krylenko, V. (2019). Dynamics of eolian forms on Blagoveschensky butte. *Proc. of the 14th MEDCOAST Congress, 2*, 543–551.

Krylenko, V. (2015). Seashore dynamics of the Anapa bay-bar. *Oceanology, 55*, 742–749.

Krylenko, V., Kosyan, R., & Kochergin, A. (2011). Regularities of the formation of the granulometric composition of the bottom and beach deposits of the Anapa bay Bar. *Oceanology, 51*, 1061–1071.

Nevessky, E. (1957). The history of the development of the Anapa bay-bar. *Trudy of Institute of Oceanology, 8*, 35–45.

Porotov, A., Gorlov, Y., Yanina, T., & Fouache, E. (2004). Features of the development of the Black Sea coast of the Taman Peninsula in Holocene. *Geomorphologiya, 4*, 63–77.

Features of the Early Pleistocene Pedosediment and Soil Formation Processes of the Lori Basin, Armenia

Anna Revunova and Olga Khokhlova

Abstract

The presented work is a continuity of a long-term research in the Lory Basin, Armenia. It is important to expand the area of paleosol research in the Lory Basin, because some habitats of an ancient people were found, living here periodically, advancing and retreating, during migration from Africa to Eurasia. We studied two sites, Kurtan-IV and Yagdan. No human-processed artifacts were seen, but the sites were dated and reconstructed. Paleoclimatic and paleolandscape data were used for clarification of ecological conditions around the hominids. The used methods include magnetic susceptibility, particle size distribution, major oxide and trace element determination, CHN, phytoliths and micro-morphological analyzes. Based on previous studies, we determine the formation time of the red-colored and pedosediments studied to be 1.9–1.4 Ma. The investigated pedosediments were formed in a changeable humid climate under savanna-like vegetation. Such conditions prevailed in the Lory Basin during the Pleistocene.

Keywords

Paleosol · Pedo-sediment · Paleo-reconstruction · Pleistocene

1 Introduction

Red-colored soils are representative of bright lithogenic characteristics due to the presence of hematite in the parent rock (Lesovaya, 2006), which determines the color of these soils. They are formed in humid and semi-humid tropics and subtropics under conditions of deep leaching water regimes (Inozemtsev, 2010; Kovda & Rozanov, 1988). In the Lori Basin of Armenia, signs of red-colored soil formation as the products of weathering of basalts were found during soil archaeological expeditions, in which the authors took part during 2011–2019.

Recently, paleosol studies of hominid tool sites in archaeological contexts have been carried out (Khokhlova & Sedov, 2017; Khokhlova et al., 2016a, 2016b; Lyubin et al., 2015; Sedov et al., 2011; Stolpnikova, 2017), and include sample dating using various instrumental methods (direct-SIMS U-Pb, K-Ar, $^{40}Ar/39Ar$, $^{13}C/^{12}C$, indirect-paleontological, magnetic, micro-morphological). Based on the previous studies, we can determine the formation time of the red-colored paleosols and pedosediments studied in this work. The purpose of this work is the morphological and genetic diagnostics of buried early Pleistocene red-colored paleosols and pedosediments of the Lori Basin and reconstruction of the paleoclimatic and paleolandscape conditions of their formation.

2 Materials and Methods

The Lori Basin, located in the northern part of Armenia, is a direct reflection of the co-named synclinorium. To the North and South, it is bounded by the Somkheti and Bazum fold ridges, and to the West by the volcanic Javakheti ridge. The bottom and sides of the basin are filled with volcanic weakly alkaline basalts and basaltic andesites with an age of ~2 million years. Colluvial and lacustrine sediments lie on top.

A. Revunova (✉)
Institution of Earth Sciences, Saint-Petersburg State University, 199034 Saint-Petersburg, Russia
e-mail: reina_abc1@mail.ru

O. Khokhlova
Institute of Physicochemical and Biological Problems of Soil Science, Russian Academy of Sciences, 142290 Pushchino, Russia

F. Lucci et al. (eds.), *Selected Studies in Geomorphology, Sedimentology, and Geochemistry*, Advances in Science, Technology & Innovation, https://doi.org/10.1007/978-3-031-43744-1_18

This paper describes two sections: Kurtan-IV (N 40°58'02.8″, E 44°31'32.9″, h = 1309 m) and Yagdan (N 41°00'27.8″, E 44°30'50.8″, h = 1277 m) (Fig. 1). The dating of the ash in the central part of one of the walls of the quarry, in which the Kurtan-IV section is located, using the uranium-lead method (SIMS U-Pb) showed an age of 1.432 ± 0.028 Ma (Presnyakov et al., 2012). For the upper horizon of pumice sands, there are SIMS U–Pb dates of 1.495 ± 0.026 and 1.496 ± 0.021 Ma (Trifonov et al., 2017), and also $^{39}Ar/^{40}Ar$ dating gives an age of 1.49 ± 0.01 Ma (S. Hynek, p. c., cited from Belyaeva and Lyubin 2012). The sides of the canyons, where the Yagdan section is located, are covered with basalts and basaltic andesites dated to 2–2.5 Ma (Trifonov et al., 2017). The paleosols and pedosediments studied here were formed prior to burial, implicating that, their age is not younger.

3 Results

In the micro-morphological sample from the Kurtan-IV section, the soil mass is structured; the clots of finely dispersed material contain rounded aggregates of ferruginous-clayey composition (Fig. 2a). Some microzones contain carbonate concretions composed of micrite, masked by ferruginous-clayey material. Carbonates from the Kurtan-IV section, according to Khokhlova et al. (2016b), are of hydrogenic origin, formed during the existence of a lake here in the Holocene. In the Yagdan section, the following signs of soil formation are observed: ferruginization, optical orientation (Fig. 2b), and a high degree of weathering of primary minerals (in the lower horizons, the degree of weathering of minerals decreases). The lower horizons also show traces of biogenic grinding (Fig. 2c), and rare traces of meso-fauna.

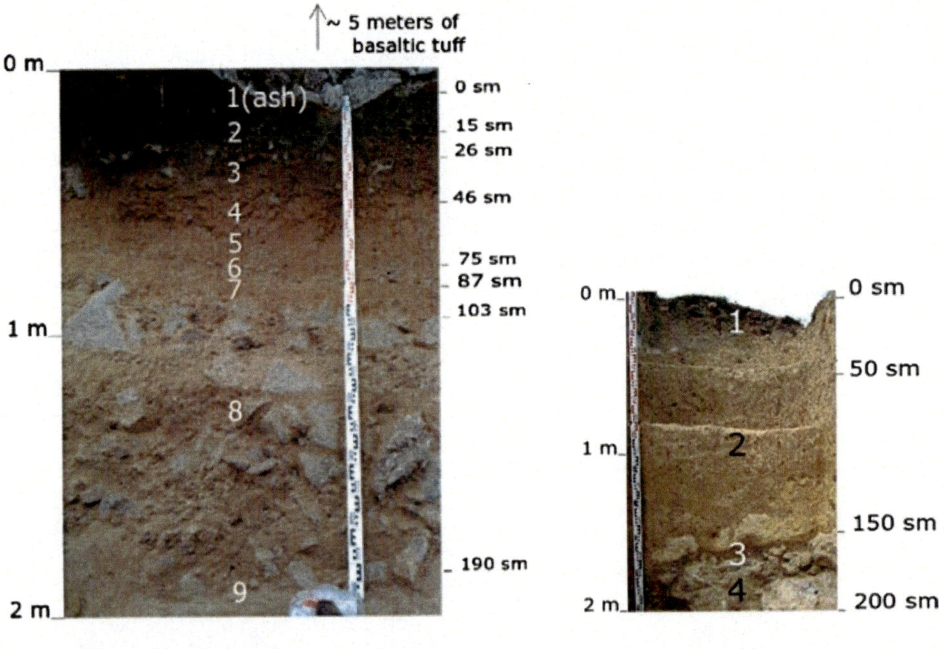

Fig. 1 Appearance of sites Yagdan and Kurtan-IV

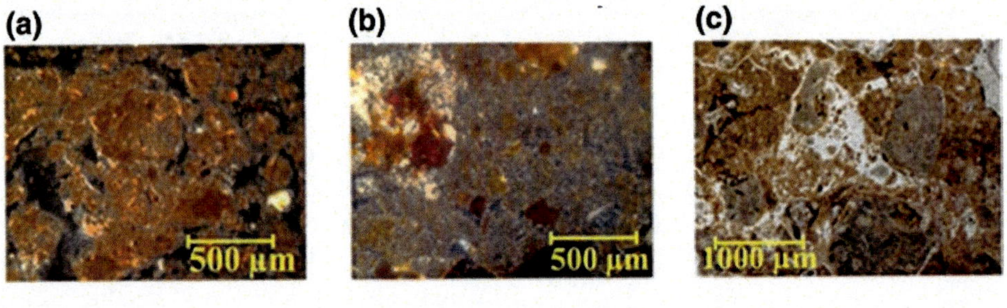

Fig. 2 Photos of micro-morphological sections: **a** Kurtan-IV, iron spots, traces of meso-fauna activity, around aggregate orientation of ferruginous-clay material; **b** Yagdan 1st layer: scaly, fibrous type optical orientation of finely dispersed material; **c** Yagdan 6th layer: biogenic grinding of soil mass

In the Yagdan section, from top to bottom, the amount of physical sand decreases and the amount of physical clay increases. Violation of this regularity due to fractions of coarse and medium sand is shown in layer 3. The increase in the percentage of physical clay occurs due to an increase in fine dust; in layers 2 and 7 the content of the silty fraction also increases. We measured nitrogen and total carbon in order to isolate layers with the lowest C:N ratios, i.e., the layers most developed by soil processes, in which the introduction of inorganic carbon due to volcanic processes was minimal. The nitrogen content in the Kurtan-IV section is 0.05%, and carbon is 0.18%. The nitrogen content in the layers of the Yagdan section varies from 0.3% in the upper layer to 0.6–0.7% on average, and drops to 0.5% in the lower layer. The carbon content is maximal in the second layer at 0.45%, and sharply decreases to 0.1% toward the upper and lower layers. Geochemical coefficients based on gross composition data show that in the Yagdan section the highest layer differs from the others. It contains the highest level of total weathering (4.2, while underlying layers have values of 2.2–2.5), oxidation (0.7, underlying layers 0.4–0.5), Ti and Al weathering (0.067, underlying layers 0.04–0.05) and the lowest degree of chemical weathering (0.26 and smoothly increase to 0.85 in the lowest layer), carbonization (0.14, underlying layers 0.18–0.20) and salinity (0.025, underlying layers 0.046–0.069). In the Kurtan-IV section, the weathering coefficient is 2.20, oxidation is 0.44, Ti and Al weathering is 0.06, chemical weathering is 0.56, carbonization is 0.31 and salinity is 0.040. Distribution of micro-elements shows a dramatically increased number of volcanic elements, such as Cu, Ni and Co in the upper horizon. Biophilic elements (Zn and Sr) are concentrated in the underlying layers. Magnetic susceptibility is the highest in the upper layer of the Yagdan section at $1006 \times 10^{-8} m^3 kg^{-1}$. By the second horizon, the indicator drops to $527 \times 10{-}^8 m^3 kg{-}^1$ and continues to decrease below. The second maximum is in the 7th layer at $171 \times 10{-}^8 m^3 kg{-}^1$. Susceptibility in the Kurtan-IV section is $63 \times 10^{-8} m^3 kg^{-1}$. Phytoliths were found in both sections. Samples from the Yagdan section contain phytoliths characteristic of the modern vegetation of grass meadows and reflect the modern invasive vegetation in this layer, rather than the time of accumulation and functioning of this layer. The sample from the Kurtan-IV section also contains phytoliths from wormwood, conifers, and lacustrine plants, which are currently not found in the study area and cannot coexist in one sample.

4 Discussion

In summary, the pedosediments of the Lory Basin were formed in the pre-Holocene. This is directly evidenced by phytoliths and the features found during micro-morphological analysis, and indirect evidence from magnetic susceptibility, bulk composition, and CHN analysis results. The primary carbonates found in micro-morphological thin sections indicate an arid climate, while the overlying ferruginous films indicate a subsequent change in conditions to more humid ones (Gagarina, 2004). Iron neoplasms found both in the Kurtan-IV and Yagdan sections can be indicative of paleohydromorphism and/or humid climate (Gagarina, 2004). The nitrogen content in the studied layer of the Kurtan-IV section is 0.05%, and the carbon content is 0.18%. This ratio is overestimated (nitrogen makes up 36% of carbon at 10–20% typical for modern soils (Kovda & Rozanov, 1988)), probably due to the higher rate and completeness of ammonification than carbon mineralization. Weathering coefficients are high in the Kurtan-IV section. A high titanium modulus was recorded of 0.060. We assume that this is a consequence of addition by volcanic processes during soil formation. Due to the higher location of the Kurtan-IV section (Fig. 1), the soils here were not overlain by lavas, which ensured a longer soil formation time than in other studied areas. Sediments in the Kurtan-IV section have the lowest magnetic susceptibility index, indicating a long period of soil formation (Dergacheva, 1997). It was previously reconstructed that the area was covered by a lake in the Holocene, which could also lead to a decrease in susceptibility (Babanin et al., 1995) due to a percentage increase in the amount of SiO_2. The sample from the Kurtan section does not contain volcanic minerals. On the contrary, magnetic susceptibility index values in the Yagdan section are very high in the two highest layers because of interactions with lava. Then, it gradually decreases, which is typical for the soil profile. Zinc and strontium are biogenic elements that accumulated in the second and third horizons of the Yagdan section, along with an increased content of nitrogen and carbon, indicating soil-forming processes. The maximum indicator of the coefficient of weathering is in the upper layer of the Yagdan section, which was affected by hot flowing lava. Down section, the indicator is evenly distributed and equal to ~2.5, which indicates a fairly high degree of weathering. The chemical weathering index in the upper horizon of the Yagdan section is minimal and increases downward, which confirms the strong temperature effect on the soil profile from above. The oxidation coefficient in the upper layer of the Yagdan section is at a maximum, and it differs from those of the underlying layers, which emphasizes the specificity of the development of the upper layer. In general, the index down the Yagdan section varies slightly, which indicates the absence of an oxidative barrier. The extremely low salinity index emphasizes formation in a humid climate. In the upper layers of the Yagdan section, a relatively high titanium modulus is recorded. We assume that this is a consequence of the introduction of volcanic materials after formation of the soil. Another effect of lava influence is high content of Cu, Co, Ni, Va, and Cr.

A complex phytolithic complex, characteristic of the samples, with a synlithogenic input of material, was found. These are phytoliths of conifers and wormwood, meadow and arid grasses, and reeds. Of particular interest is the presence of conifers, which currently do not grow near the section. According to the palynological analysis carried out by A. Simakova for the Early and Middle Pleistocene sections of the Lori Basin (Trifonov et al., 2017), red-colored soils were formed at an early stage (early Pleistocene), when savanna-like vegetation dominated on the plains with a large proportion of meadow grasses and coniferous-deciduous forests grew on the mountain slopes.

5 Conclusions

By color, slightly acidic and close to a neutral exchange reaction, the absence of primary carbonates, and also by the types of plants that we diagnosed by phytoliths, we can tentatively attribute paleosols from the Yagdan section as Vertic cambisols, and those from the Kurtan section IV as Luvisols. The studied pedosediments are an archive that is necessary both for archaeologists and natural scientists to reconstruct environmental conditions at different time intervals in the past, to understand the history of landscape development, as well as the paleoecological environment of the penetration and movement of ancient people deep into Eurasia.

References

Babanin, V. F., Trukhin, V. I., Karpachevsky, L. O., Ivanov, A. V., & Morozov, V. V. (1995). *Soil magnetism* (223 p). Yaroslav Publishing House.

Belyaeva, E. V., & Lyubin, V. P. (2012). The first data on the early stages of Acheulean in the Caucasus. *CSIA, 227C*, 28–36.

Dergacheva, M. I. (1997). *Archaeological soil science* (228 p). Novosibirsk.

Gagarina, E. I. (2004). *Micromorphological method of soil research* (156 p). SPb, Publishing House of SPSU.

Inozemtsev, S. A., & Targulyan, V. O. (2010) *Upper Permian paleosols: Properties, processes, conditions of formation* (188 p). Geos.

Khokhlova, O. S., & Sedov, S. N. (2017). Signs of soil formation in multilayer archaeological sites of the lower pleistocene in northern Armenia and paleoclimatic reconstructions. In: N. O. Kovalevoy (Ed.), *Genesis, diagnostics and prevention of ecological crises* (pp. 90–112). MAKS Press.

Khokhlova, O. S., Khokhlov, A. A., Kuznetsova, A. M., Stolpnikova, E. M., Kovaleva, N. O., Lyubin, V. P., & Belyaeva, E. V. (2016a). Carbonate features in the uppermost layers of quaternary deposits, Northern Armenia, and their significance for paleoenvironmental reconstruction. *Quaternary International, 418*, 94–104.

Khokhlova, O. S., Sedov, S. N., Khokhlov, A. A., Belyaeva, E. V., & Lyubin, V. P. (2016b). Indications of pedogenesis in lower pleistocene tool-bearing sediments in Northern Armenia and regional paleoclimatic reconstruction. *Quaternary International, 469*, 68–84.

Kovda, R. A., & Rozanov, B. G. (1988). *Soil, in two parts* (488 p).

Lesovaya, S. N. (2006). *Genesis and geography of soils on red-colored rocks of the European territory of Russia*. Dissertation, D. Geogr. Sci. SPb.

Lyubin V. P., Belyaeva, E. V., Trifonov, V. G., Simakova, A. N., Ozherelyev, D. V., Khokhlova, O. S., Nosova, A. A., Sazonova, L. V., Kolesnichenko, A. A., Golyeva, A. A., Trikhunkov, Y. I., Tesakov, A. S., Bachmanov, D. M., Shalaeva, E. A., & Frolov, P. D. (2015). *The dynamics of the natural environment and the formation of the earliest Early Paleolithic cultures of Southwest Asia*. All-Russian Scientific Conference, Moscow, Institute of Archeology RAS, pp. 45–49.

Presnyakov, S. L., Belyaeva, E. V., Lyubin, V. P., Radionov, N. V., Antonov, A. V., Saltykova, A. K., Berezhnaya, N. G., & Sergeev, S. A. (2012). Age of the earliest Paleolithic sites in the northern part of the Armenian Highland by SHRIMP-II U-Pb geochronology of zircons from volcanic ashes. *Gondwana Research, 21*, 929–938.

Sedov, S. N., Khokhlova, O. S., & Kuznetsova, A. M. (2011). Polygenesis of volcanic paleosols in Armenia and Mexico: Micromorphological annals of Quaternary climate changes. *Pochvovedenie, 7*, 832–847.

Stolpnikova, E. M. (2017). *Pleistocene soils of primitive man's sites as an archive of paleoecological information (on the example of the Lesser Caucasus, Inner Dagestan and the Central Russian Upland)*. Dissertation, Moscow State University.

Trifonov, V. G., Shalaeva, E. A., Bachmanov, D. M., Trikhunkov, Y. I., Simakova, A. N., Tesakov, A. S., Frolov, P. D., Kolesnichenko, A. A., Saakyan, L. K., Avagyan, A. V., Lyubin, V. P., Belyaeva, E. V., Lebedev, V. A., Latyshev, A. V., & Ozherelyev, D. V. (2017). Quaternary tectonics of recent basins in northwestern Armenia. *Geotectonics, 51*, 499–519.

Soil Erosion Susceptibility Mapping by Analyzing Morphometric Parameters Using ALOS PALSAR DEM and GIS Techniques for Rawanduz River Basins in Iraq

Waleed Mohammed Qader, Abbas Mohammed Noori, Fawzi M. Omer, Qahtan A. M. Al Nuaimy and Farman Galeb Saed

Abstract

Annually, soil erosion imposes significant hazards by raising soil fertility and filling reservoirs with sedimentation in the Rawanduz River basins, Iraq. The Rawanduz watershed consists of seven sub-basins. Therefore, it is important to consider areas in these basins that are vulnerable to soil erosion for preventive hazard action. ALOS PALSAR DEM 12.5 m-resolution data was used to analyze morphometric parameters in the study area. Linear, aerial, and relief morphometric parameters were determined using GIS techniques. A set of 15 parameters were studied, including slope (S), drainage density (Dd), texture ratio (T), relative relief (Bh), stream frequency (Fu), infiltration number (If), ruggedness number (Rn), length of overland flow (Lo), constant of channel maintenance (C), relief ratio (Rh), elongation ratio (Re), shape factor (Bs), form factor (Rf), circularity ratio (Rc), and compactness coefficient (Cc). Then, an Analytic Hierarchy Process (AHP) was used to classify sub-basins according to their susceptibility for soil erosion and important morphometric parameters that have an important influence on soil erosion. The results of the model show that erosion is highly affected by the slope and drainage density in the region of interest, mainly. The results of the AHP set model showed that sub-basin 2 was highly susceptible to soil erosion. We conclude that these results could be useful to various stakeholders such as farmers, water resource managers, conservation planners, and policymakers for better management and decision-making activities.

Keywords

Soil erosion · Rawanduz basins · Morphometric analysis · AHP · ALOS PALSAR · GIS

1 Introduction

Soil erosion is one of the most challenging issues in environmental and agricultural research (Arabameri et al., 2020a). Soil erosion has been a continual problem for agricultural productivity and the environment (Chalise et al., 2019; Mararakanye & Sumner, 2017; Vanmaercke et al., 2016). During one flood, millions of tons of new sediments are passing from the North of Iraq to Baghdad city in one day (Buringh, 1960). Watershed management includes proper use of a watershed's land, forest, water, and soil resources for best production with minimal risk of natural resources (Noori et al., 2019b). For this reason, predicting soil erosion could be a great factor in any decision-making and also support the policy formulation of environmental sustainability in conjunction with soil productivity. Therefore, it is key to predict the risk of soil erosion in the context of environmental and agricultural studies (Singh & Singh, 2018).

Several studies have recently shown that remote sensing and GIS tools have a high impact on the quality of watershed development, water conservation, soil management, and priority setting of sub-watersheds (Arabameri et al., 2020b; Kumar et al., 2016; Noori et al., 2019b). In addition, GIS techniques provide an efficient tool for evaluating, managing, and extracting information for a better understanding of spatial data (Ajaj et al., 2018; Hassan et al., 2019). Morphometric analysis can be performed through computing drainage network, linear, shape, and landscape parameters (Gajbhiye et al., 2014). Concerning this type of decision-making process, model optimization can be used rather than a single criterion. Several parameters have,

W. M. Qader
iMMAP Organization, Erbil, Iraq

A. M. Noori (✉) · F. M. Omer · Q. A. M. Al Nuaimy · F. G. Saed
Technical College of Kirkuk, Northern Technical University, Kirkuk 36001, Iraq
e-mail: abbasrsgis@gmail.com

F. Lucci et al. (eds.), *Selected Studies in Geomorphology, Sedimentology, and Geochemistry*,
Advances in Science, Technology & Innovation, https://doi.org/10.1007/978-3-031-43744-1_19

therefore, been applied to improve the precision of the decisions (Ajaj et al., 2019; Georgiou et al., 2015; Noori et al., 2019a). To identify areas affected by soil erosion, the use of multi-criteria decision-making (MCDM) models and morphometric parameters that are vulnerable to erosion would be crucial in developing new techniques in order to control soil erosion and make it more effective (Arabameri et al., 2020b).

Therefore, an Analytic Hierarchy Process (AHP) was used to classify sub-basins according to their susceptibility to soil erosion. It was also used to identify important morphometric parameters which have an important influence on soil erosion.

2 Materials and Methods

The study area is the Rawanduz River basins located in the North-East of Iraq (44°15'–44°30'E and 36°34'–36°'N). The area is bounded by a fold mountain on the northern side, while the elevation gradually decreases in the eastern part, where elevation varies between 424 and 2089 m asl. The major river is the Rwanduz River which originates from hills and flows toward the main river Greater Zab (Mirza & Rashid, 2018). The entire area is divided into seven sub-basins based on the distribution of tributaries, which produces various morphotectonic evidence (Fig. 1).

Topographical maps at 1:50,000 scale and ALOS PALSAR DEM with 12.5 m spatial resolution for 2017 were used to define seven sub-basins' boundaries. ArcGIS 10.5 software (ESRI, USA) coupled with ALOS PALSAR DEM data were used to analyze morphometric parameters such as linear, aerial, and relief morphometric parameters in the study area which were calculated based on Horton and Strahler (Arabameri et al., 2020b; Noori et al., 2019b). In addition, pairwise comparison was used to determine the relative significance of soil erosion morphometric factors, and identify and assign the relative weight of each parameter to the sub-basins. The AHP model was then used to produce a soil erosion susceptibility map (Fig. 2).

3 Results

The slope map is shown in Fig. 3a. The calculated number of streams with order shows that the number of streams decreases with increasing the order (Fig. 3b). Morphometric analysis of the Rawanduz River sub-watersheds was performed through the measurement of linear, areal, shape, and relief parameters using standard computational procedures.

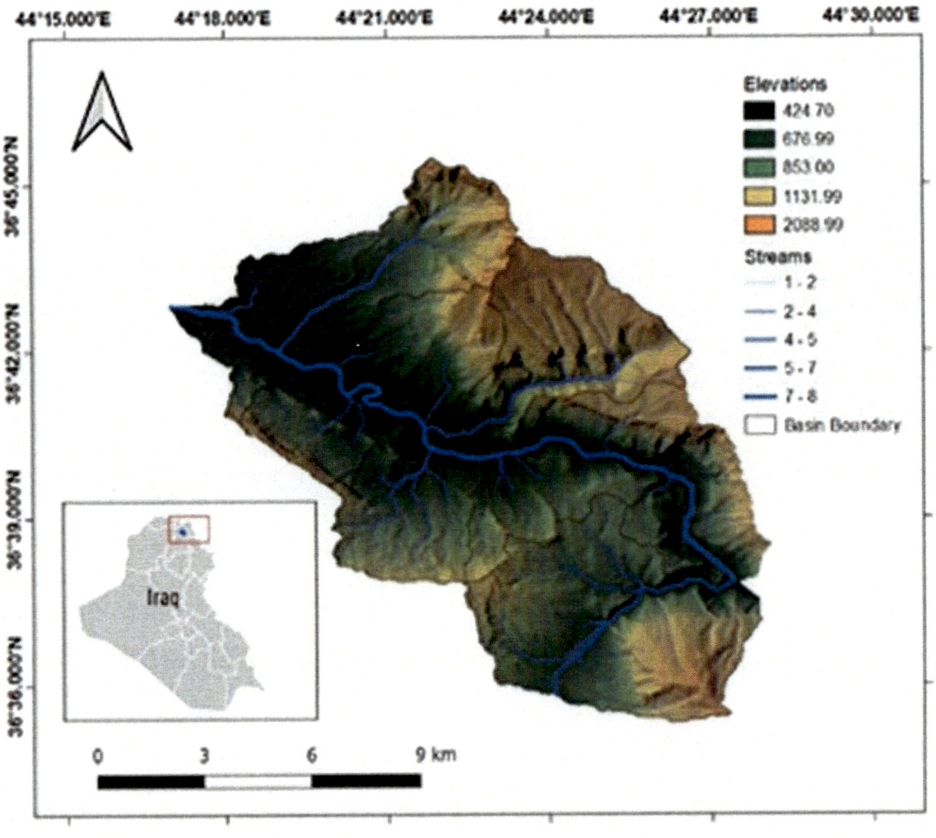

Fig. 1 Study area of Rawanduz River basin

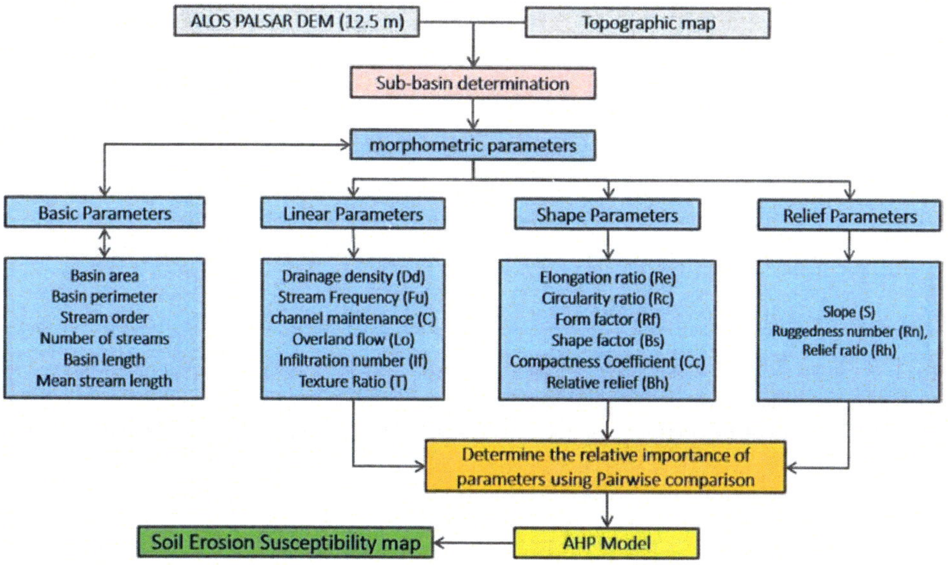

Fig. 2 Flowchart of the methodology

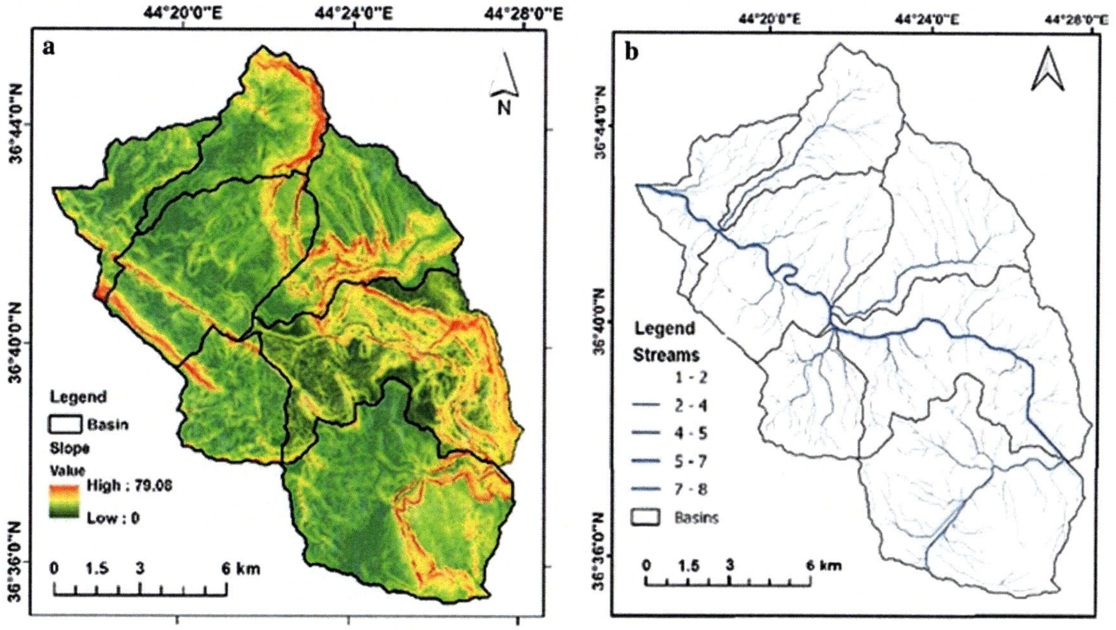

Fig. 3 **a** Slope of the study area, **b** stream order of the study area

Following the determination of the moral criteria, questionnaires were created for experts based on the determined factors, and classes from the literature review were determined. The experts offered input on the parameters and their relation to the vulnerability of soil erosion susceptibility. Then, the weights of every parameter were determined by geometrical mean in order to decide each parameter's final weight. The consistency ratio (CR) of this study was 0.04. If CR is less than 10%, the contrast between the parameters is appropriate; otherwise, the CR makes a reassessment. The results of the delineated watersheds based on the pairwise comparison matrix obtained are given in Table 1.

The morphometric basin defines the relationships between geomorphic, hydrological, geological, and landscape surface processes. Compound parameter values are determined, and a priority rating is given for the seven sub-watersheds of the Rawandus River catchment (Fig. 4). Sub-basin 5 obtained a very low rate of soil erosion and low-surface erosion classes are in sub-basins 1 and 7. The

Table 1 Pairwise comparison matrix for the calculation of weights of each morphometric attribute based on their degree of importance with soil erosion susceptibility

	(S)	(Dd)	(T)	(Bh)	(Fu)	(If)	(Rn)	(Lo)	(C)	(Rh)	(Re)	(Bs)	Rf)	(Rc)	(Cc)	Weights
(S)	1															0.202
(Dd)	0.45	1.00														0.137
(T)	0.32	0.50	1.00													0.115
(Bh)	0.29	0.43	0.45	1.00												0.098
(Fu)	0.27	0.38	0.42	0.50	1.00											0.084
(If)	0.25	0.33	0.38	0.43	0.50	1.00										0.071
(Rn)	0.23	0.31	0.34	0.38	0.43	0.50	1.00									0.060
(Lo)	0.22	0.29	0.32	0.34	0.38	0.43	0.50	1.00								0.050
(C)	0.21	0.26	0.29	0.29	0.34	0.38	0.43	0.50	1.00							0.042
(Rh)	0.19	0.24	0.27	0.27	0.29	0.34	0.38	0.43	0.50	1.00						0.035
(Re)	0.18	0.23	0.25	0.25	0.27	0.29	0.34	0.38	0.43	0.50	1.00					0.029
(Bs)	0.17	0.21	0.23	0.23	0.25	0.27	0.29	0.34	0.38	0.43	0.50	1.00				0.024
(Rf)	0.16	0.20	0.22	0.22	0.23	0.25	0.27	0.29	0.34	0.38	0.43	0.50	1.00			0.020
(Rc)	0.15	0.19	0.20	0.21	0.22	0.23	0.25	0.27	0.29	0.34	0.38	0.43	0.50	1.00		0.017
(Cc)	0.15	0.19	0.19	0.20	0.21	0.22	0.23	0.25	0.27	0.29	0.34	0.38	0.43	0.50	1.00	0.014

Slope (S), drainage density (Dd), texture ratio (T), relative relief (Bh), stream frequency (Fu), infiltration number (If), ruggedness number (Rn), length of overland flow (Lo), constant of channel maintenance (C), relief ratio (Rh), elongation ratio (Re), shape factor (Bs), form factor (Rf), circularity ratio (Rc), and compactness coefficient (Cc)

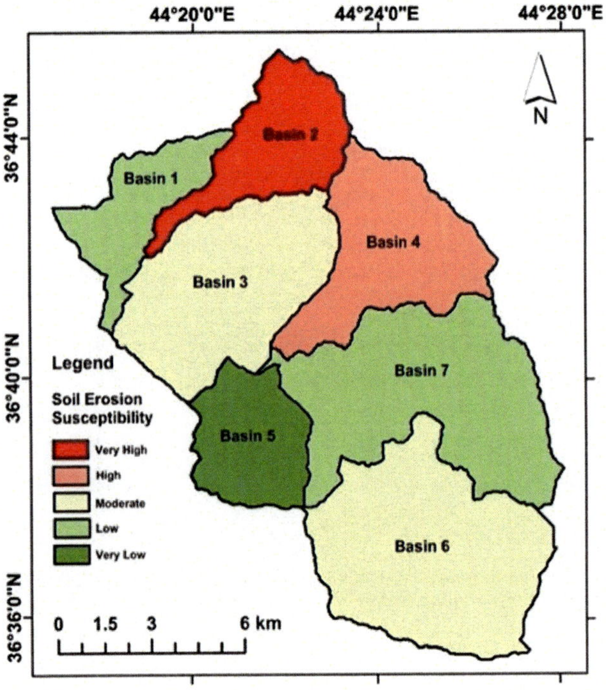

Fig. 4 Soil erosion susceptibility map of Rawanduz River basins

soil erosion capacity of sub-basins 3 and 6 is moderate, sub-basin 4 is highly likely to be vulnerable to soil erosion, and sub-basin is very highly likely.

4 Discussion

In any statistically dependent susceptibility method, the choice of determining factors for soil erosion is a significant starting point (Singh & Singh, 2018). This approach can be accomplished for producing soil erosion susceptibility maps with simple data like ALOS PALSAR DEM and morphometric parameters (Arabameri et al., 2020b). Based on results from the AHP model, it is found that the slope, drainage, and texture ratios in the study area have a significant impact on soil erosion, and these results agree with other scholars' (Arabameri et al., 2020b; Kumar et al., 2016). This approach constitutes a starting point to determine soil erosion susceptibility and can help evaluate environmental hazard risks in the future.

5 Conclusions

The current research prioritized the Rwanduz River in northern Iraq based on soil erosion susceptibility. The meaning of each of the 15 parameters was evaluated using an AHP. The results of this research showed that both slope and drainage densities were the main predictor factors of soil erosion. Sub-basins 2 and 4 are priorities for anti-erosive management, and a complementary study to take into account other determinant factors is necessary.

References

Ajaj, Q. M., Shareef, M. A., Hassan, N. D., Hasan, S. F., & Noori, A. M. (2018). GIS based spatial modeling to mapping and estimation relative risk of different diseases using inverse distance weighting (IDW) interpolation algorithm and evidential belief function (EBF) (Case study: Minor Part of Kirkuk City, Iraq). *International Journal of Engineering Technology, 7*, 185–191.

Ajaj, Q. M., Shareef, M. A., Jasim, A. T., Hasan, S. F., Noori, A. M., & Hassan, N. D. (2019). *An AHP-based GIS for a new hospital site selection in the Kirkuk Governorate*. 2nd International Conference Electrical, Communication, Computer, Power and Control Engineering, ICECCPCE, pp. 176–181.

Arabameri, A., Chen, W., Lombardo, L., Blaschke, T., & Bui, D. T. (2020). Hybrid computational intelligence models for improvement gully erosion assessment. *Remote Sensor, 12*, 140. https://doi.org/10.3390/rs12010140

Arabameri, A., Tiefenbacher, J. P., Blaschke, T., Pradhan, B., & Bui, D. T.: Morphometric analysis for soil erosion susceptibility mapping using novel GIS-based ensemble model. *Remote Sensing, 12*, 874. https://doi.org/10.3390/rs12050874

Buringh, P. D. (1960). *Soils and soil conditions in Iraq*. Ministry of Agriculture.

Chalise, D., Kumar, L., & Kristiansen, P. (2019). Land degradation by soil erosion in Nepal: A review. *Soil System, 3*, 12. https://doi.org/10.3390/soilsystems3010012

Gajbhiye, S., Mishra, S. K., & Pandey, A. (2014). Prioritizing erosion-prone area through morphometric analysis: An RS and GIS perspective. *Applied Water Science, 4*, 51–61.

Georgiou, D., Mohammed, E. S., & Rozakis, S. (2015). Multi-criteria decision making on the energy supply configuration of autonomous desalination units. *Renewable Energy, 75*, 459–467.

Hassan, N. D., Noori, A. M., Hasan, S. F., Shareef, M. A., & Ajaj, Q. M. (2019). *Cadastral mapping accuracy assessment using various surveying techniques and high-resolution satellites images*. 2nd International Conference of Electrical, Communication, Computer, Power and Control Engineering, ICECCPCE, pp. 182–187.

Kumar, R., Ashraf, I., Khan, T., Nyreen, H., Khan, J. N., & Bhat Owais Ahmad, R. D. (2016). Watershed based drainage morphometric analysis using Geographical Information System and Remote Sensing of Kashmir Valley. *Journal of Soil Water Conservation, 15*, 265–272.

Mararakanye, N., & Sumner, P. D. (2017). Gully erosion: A comparison of contributing factors in two catchments in South Africa. *Geomorphology, 288*, 99–110.

Mirza, T. A., & Rashid, S. G. (2018). Mineralogy, Fluid inclusions and stable isotopes study constraints on genesis of sulfide ore mineral, Qaladiza area Qandil Series, Iraqi Kurdistan Region. *Arabian Journal of Geoscience, 11*, 146. https://doi.org/10.1007/s12517-018-3490-x

Noori, A. M., Pradhan, B., & Ajaj, Q. M. (2019a). Dam site suitability assessment at the Greater Zab River in northern Iraq using remote sensing data and GIS. *Journal of Hydrology, 574*, 964–979.

Noori, A. M., Qader, W. M., Saed, F. G., & Hamdany, Z. A. L. (2019b). Quantification of morphometric parameters to analyze the watershed characteristics: A case study of Rosti Watershed, Iraq. *International Journal of Advance Science and Technology, 28*, 273–289.

Singh, O., & Singh, J. (2018). Soil erosion susceptibility assessment of the lower Himachal Himalayan Watershed. *Journal of the Geological Society of India, 92*, 157–165.

Vanmaercke, M., Poesen, J., Van Mele, B., Demuzere, M., Bruynseels, A., Golosov, V., Rodrigues Bezerra, J. F., Bolysov, S., Dvinskih, A., Frankl, A., Fuseina, Y., Teixeira Guerra, A. J., Haregeweyn, N., Ionita, I., Imwangana, F. M., Moeyersons, J., Moshe, I., Samani, A. N., Niacsu, L., … Yermolaev, O. (2016). How fast do gully headcuts retreat? *Earth Science Reviews, 154*, 336–355.

Comparison of Ten Regression Equations to Estimate Rainfall Erosivity Using a Long-Term Precipitation Dataset and Spatial Regionalization of *R* Factor in Tunisia

Insaf Cherif, Taoufik Hermassi and Mohamed Mechergui

Abstract

Water erosion is one of the major threats to soils in Tunisia, affecting 49% of the total area. Rainfall erosivity is one of the five factors considered in the Universal Soil Loss Equation to quantify the climate effect on the soil loss rate. Rainfall data are not available in adequate temporal and spatial resolution anywhere to compile rainfall erosivity using the empirical equations. This study aims to evaluate ten regression equations in Tunisia to calculate rainfall erosivity and to select the most appropriate one, for each bio-climatic region. This work is based on 115 years precipitation dataset for 31 pluviometry stations. The erosivity map resulting from the pre-selected equations can be used to develop the USLE erosion map. The result shows that it is crucial to select an appropriate equation that can be adapted to local conditions, based on climatic regions. In fact, for Saharan and arid region, the Ferro equation is the most appropriate to compile erosivity factor. For semi-arid region, the Rongo SI equation provides the closer erosivity, and for humid region, the Bollinne equation is most adapted one.

Keywords

Rainfall erosivity · Regression equation · Spatial distribution · Tunisia · Erosivity map

I. Cherif (✉) · M. Mechergui
National Agronomic Institute of Tunisia, 1082 Tunis, Tunisia
e-mail: cherifinsaf.inat@gmail.com

I. Cherif · T. Hermassi
Water and Forests, National Research Institute of Rural Engineering, 2080 Tunis, Tunisia

1 Introduction

Erosive processes depend on a multiplicity of interacting factors include the soil, land use, topography, and climate. The rainfall erosivity factor (*R*) represents the effect that rainfall has on erosion. Wischmeier and Smith (1978) have proposed a relation between the total kinetic energy E and the intensity i. For many regions worldwide, rainfall data are not available in an enough and adequate spatial and temporal resolution. To overcome this restriction, several studies have aimed at the assessment rainfall erosivity using regression function and expressed *R* factor as a function of t annual precipitation (P) or Modified Fournier Index (MFI).

The objective of this work is to present a comparison of multiple regressions to estimate the rainfall erosivity, then, a mapping of *R* factor.

2 Material and Methods

In this study, to compile rainfall erosivity, we have chosen ten different regression equations using the annual precipitation (P) or the Modified Fournier Index (MFI), and different equations inspired by work of Arnoldus in similar Tunisian climatic condition like Tunisia's conditions (Table 1).

Arnoldus didn't specify the precipitation units used, he just identified the *R* factor units as metric units and mentioned that the conversion factor of 1.735 is given for Morocco (Arnoldus, 1977) and a rounded 1.74 is given for Africa and the Middle East (Arnoldus, 1980) which are different from those presented by Foster et al. (1981) and Wischmeier and Smith (1978). According to Renard and Freimund (1994), the units presumably used are the units of t-m.cm/ha.h.year. Based on those conversion factors, a new conversion factor for the two different metric units used (t-m.cm/ha.h.year and Mj.mm/ha.h.year) can be identified and it is around 9.8 (17.02/1.735 = 9.8). Chehlafi et al. (2014, 2019) and Hniad et al. (2018), working in Morocco,

F. Lucci et al. (eds.), *Selected Studies in Geomorphology, Sedimentology, and Geochemistry*,
Advances in Science, Technology & Innovation, https://doi.org/10.1007/978-3-031-43744-1_20

Table 1 Ten regression equations used for rainfall erosivity estimation

No.	Regression equations	Authors (year)	Study area	Units
1	$R = P * (0.5 \pm 0.05)$	Roose (1976)	West Africa	*R*: 100 of foot tonf.inch/acre.h.year/*P*: Inch
2	$R = 1.735 * 10^{(1.5 * \log \text{MFI} - 0.8188)}$	Arnoldus (1980)	Morocco	R: t-m.cm/ha.h.year
3	$R = 3.27 * \text{MFI} - 168.42$	Bolinne et al. (1980)	Belgium	*R*: 10^{-2} tm.cm/ha.h/*P*: mm
4	$\log R = 1.74 * \log \text{MFI} + 1.29$	Rango and Arnoldus (1987)	Morocco	R: t-m.cm/ha.h.year
5	$R = 0.0483 * \text{P}^{1.610}$	Renard and Freimund (1994)	United States	*R*: MJ.mm/ha.h.year/*P*: mm
6	$R = 0.0483 * \text{P}^{1.61}$	Yu and Rosewell (1996)	Australia	
7	$R = 21.56 * \text{MFI}^{0.927}$	Ferro et al. (1999)	Italy	
8	$R = 21.56 * \text{MFI}^{0.927}$	Apaydin et al. (2006)	Spain	
9	$R = 0.1215 * \text{MFI}^{2.2421}$	Irvem et al. (2007)	Turkey	
10	$R_{m-30} = 16.66 * MFI_m^{00.82}$	Belaid (2015)	Tunisia	

defined the precipitation unit used is inch. We managed to convert and modify the Arnoldus and Rango and Arnoldus equations to the SI metric as following respectively:

$$\boldsymbol{R = 17 * 10^{(1.5 * \log \text{MFI} - 2.92)}} \quad (1)$$

$$\boldsymbol{R = 10 * 10^{(1.74 * \log \text{MFI} - 1.15)}} \quad (2)$$

Tunisia is divided in to four bio-climatic class: humid in the North of the country, arid and semi-arid in the center and Saharan in the south. In this study, we used a long-term precipitation dataset for 31 pluviometry stations spatially distributed which are part of a national rainfall gauge networks managed by National Institute of Meteorology (NIM). The amount of data per station is 115 years, between 1900 and 2015.

In terms of our data availability, we decided to use the 9 regression equations to compile and compare them to those compiled using Arnoldus SI equation. In fact, according to Belaid (2015) and Cherif et al. (2020) from Tunisia, Irvem et al. (2007) from Turkey and Apaydin et al. (2006) from Spain, the Arnoldus SI equation provides statistically equivalent erosivity values to those that were computed using empirical equation.

3 Results

The annual and the mean annual rainfall erosivity were calculate for each station using the ten regression equations. The Belaid relationship underestimates the erosivity with an average of 53%. On the other hand, the Irvem equation underestimates the erosivity with an average of 36% for stations where located in Saharan area. However, the difference increases in arid and semi-arid, and the equation overestimates the erosivity in humid area (242%). The Renard and Freimun and the Yu and Rossewell equations predict significantly smaller rainfall erosivity with an average difference around 34% and 47%, respectively, for all stations.

We decided to present the result of comparison between the *R* factors obtained using 5 regression equations and Arnoldus SI equation of 4 stations spatially distributed based on bio-climatic class (Ain Drahem station for humid, Tunis-Carthage station for semi-arid, Kairouan station for arid, and Rmeda station for Saharan).

Figure 1 showed that for Ain Drahem station, The Bollinne equation provides the closer erosivity values to those computed using Arnoldus SI equation with a 2% difference on average. And, for the Tunis-Carthage station, the Rongo SI equation provides the closer erosivity values to those computed using Arnoldus SI equation with a 4% difference on average. For the Kairouan and Remda stations, the Ferro equation provides the closer erosivity values to those computed using Arnoldus SI equation with, respectively, 4% and 8% difference on average.

4 Discussion

Based on those results, we managed to adopt and used each regression equation for each bio-climatic class. In fact, for humid climate, we used Bollinne equation to compile rainfall erosivity value for 4 stations. However, for the semi-arid climate where 13 stations are located, we managed to apply the Rongo SI equation. On the other hand, the Ferro equation was used to obtain the rainfall erosivity for 14 stations in arid and Saharan climates. Then, we managed to construct rainfall erosivity map in Tunisia using GIS and kriging interpolation as showed in Fig. 2.

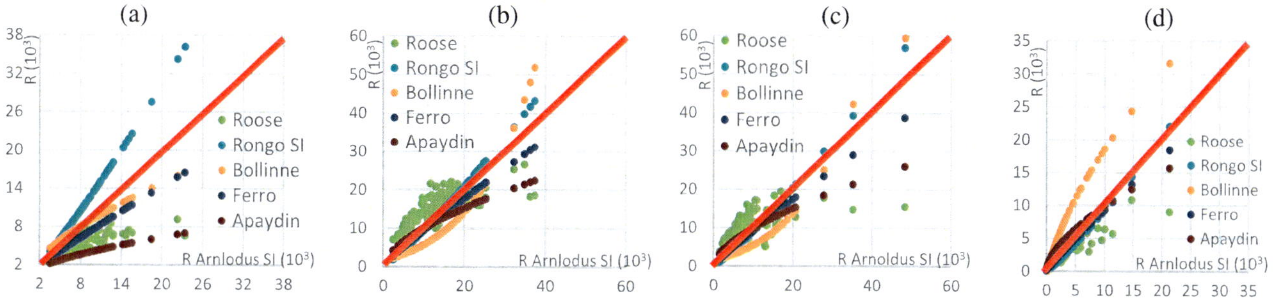

Fig. 1 Comparison between the *R* factors obtained using 5 regression equations and Arnoldus SI equation for **a** Ain Drahem Station, **b** Tunis-Carthage Station, **c** Kairouan station, **d** Rmeda Station

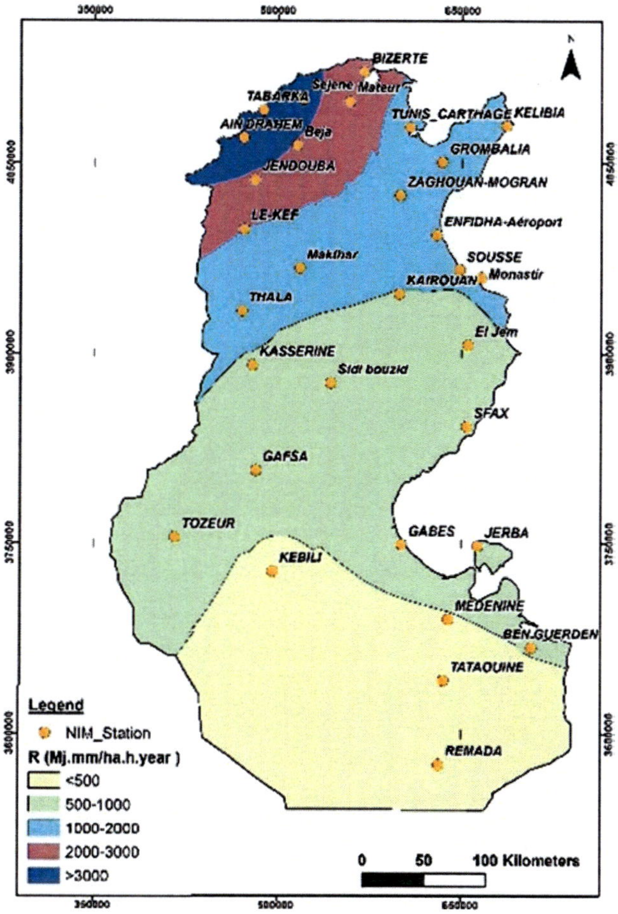

Fig. 2 Spatial distribution of the stations and rainfall erosivity map

5 Conclusions

This study showed that the rainfall erosivity estimates using regression equations can be highly dependent on the rainfall amount and the bio-climatic region. In fact, the erosivity estimates using the 6 equations were significantly different. This difference can be explained by the original climatic conditions of those equations. The results also showed that for Saharan and arid climates, the Ferro equation can be adopted to compile erosivity, for semi-arid, the Rongo SI equation provides the closer erosivity, and for humid climate, the Bollinne equation is more adapted. Finally, the spatial distribution is highly dependent on the number and the distribution of stations. Therefore, we manage to choose a largest number of stations with better distribution.

References

Apaydin, H., Ozturk, F., Merdun, H., & Aziz, N. M. (2006). Determination of the drainage basin characteristics using vector GIS. *Nordic Hydrology, 37*(2), 129–142.

Arnoldus, H. M. J. (1977). Methodology used to determine the maximum potential average annual soil loss due to sheet and rill erosion in Morocco. *FAO Soils Bulletin, 34*, 39–51.

Arnoldus, H. M. J. (1980). *An approximation of the rainfall factor in the universal soil loss equation, assessment of erosion* (pp. 127–131). Wiley

Belaid, H. (2015) *Contribution à l'etude des processus de l'érosion hydrique en milieu semi-aride Tunisien (Cas du bassin versant Sbaihia, Zaghouan)*. Ths, pp. 75–86.

Bolinne, A., Lauant, A., Rosseau, P., Pauwels, J. M., & Gabriels, D. (1980). *Aeltermann J provisional rain erosivity map of Belgium. D assessment of erosion* (pp. 111–120). Wiley.

Chehlafi, A., et al. (2014). Highway slopes protection through masonry or concreted arch. *Journées Nationales de Géotechnique et de Géologie de l'Ingénieur JNGG2014*, 4–6.

Chehlafi, A., et al. (2019). Highway cutting slopes with high rainfall erosion in Morocco: Evaluation of soil losses and erosion control using concrete arches. *Engineering Geology*, 3.

Cherif, I., Hermassi, T., & Mechergui, M. (2020). *Comparison of equations to estimate rainfall erosivity using a long-term precipitation dataset in Tunisia*. JISTEE.

Ferro, V., Porto, P., & Yu, B. (1999). A comparative study of rainfall erosivity estimation for southern Italy and southeastern Australia. *Hydrological Science Journal*, 8–11.

Foster, G. R., McCool, D. K., & Renard, K. G. (1981). Moldenhauer WC conversion of the universal soil loss equation to SI metric units. *Journal of Soil and Water Conservation, 36*, 355–359.

Hniad, O., Bahi, L., & Amgaad, S. (2018). Introduction of inclined open channels for the control of surface runoff of slopes in road structures. *MATEC Web of Conferences, 149*, 02098, p. 3.

İrvem, A., Topaloğlu, F., & Uygur, V. (2007). Estimating spatial distribution of soil loss over Seyhan River Basin in Turkey. *Journal of Hydrology, 336*, 30–37.

Rango, A., & Arnoldus, H. (1987). *Aménagement des bassins versants, Cahiers techniques de la FAO.* Food and Agriculture Organization of the United Nations.

Renard, K., & Freimund, J. R. (1994). Using monthly precipitation data to estimate R-factor in the revised USLE. *Journal of Hydrology, 157*, 287–306.

Roose. E. J. (1976). Use of the universal soil loss equation to predict erosion in West Africa. *Soil Conservative Society of America Ankeny, Iowa*, 60–74.

Wischmeier, W. H., & Smith, D. D. (1978). Predicting rainfall erosion losses. In *Agriculture handbook no. 537* (pp. 285–291). https://doi.org/10.1029/TR039i002p00285

Yu, B., & Rosewell, C. J. (1996). An assessment of daily erosivity model. *Australian Journal of Soil Research, 34*, 139–152.

Analysis of Hydro-sedimentary Balance Using ANSWER Model of Mellegue-2 Dam (Tunisia)

Olfa Hajji, Sahar Abidi, Nader Ben Rhouma and Lamia Laajili Ghezal

Abstract

In the present paper, the hydro-sediment balance of Mellegue-2 Dam was analyzed using Areal Non-point Source Watershed Environment Response Simulation (ANSWERS) distributed model. The adopted methodology is primarily based on hydrological data and physical state data. The integration of GIS data into the ANSWERS distributed parameter model requires some precision for some layers of information such as slope and flow directions. Three rainfall events are selected: January 2017 (81.3 mm and maximum rainfall intensity of 55 mm/day), February 2017 (161 mm and maximum rainfall intensity of 90 mm/day), and April 2017 (33.5 mm and maximum intensity of 16.5 mm/day). The simulated specific erosion rate was, respectively, for the three events, 6.5, 0.6, and 0.008 t/ha. Use of the model's results and GIS tools provided a soil erosion map with the classification of an erosion risk degree. A great spatial variability of erosion is observed for the three events; this variability is mainly related to the variation of the rainfall intensity as well as the state of the soil cover and the variation of the slope. The hydrological model ANSWERS is able to simulate the response of a watershed following a rain event while taking into account its spatial variability, to estimate runoff and average soil loss, and to provide the runoff hydrograph and the assessment of sediment concentration in runoff. In addition, the morphological, climatic, geological, and soil characteristics, as well as the vegetation, occupation, and tillage parameters of a watershed, have an important role in defining the runoff and soil loss processes. These factors must be added the data provided by GIS and by remote sensing, namely elevation maps, slopes, slope exposures, and lithological units which offer new possibilities of organization and prediction of natural phenomena of soil degradation.

Keywords

Soil erosion · ANSWERS · GIS · Rainfall · Runoff · Specific erosion

1 Introduction

Areal Non-point Source Watershed Environment Response Simulation was originally developed by Beasley, D.B. in 1977 (Beasley, 1977). The model has two major components: hydrology and upland erosion responses. The conceptual basis for the hydrologic model was taken from Huggins and Monke (1966) and for the erosion simulation from Foster and Meyer (1972a, 1972b).

According to Laajili (1998), the ANSWERS model (Beasley et al., 1980; Huggins & Monke, 1968; Silburn & Connolly, 1995) has been applied to simulate runoff for different rainfall events and at different scales of spatial units. According to Huggins and Monke (1968), the application of the model for two small watersheds indicates the need for additional research to better define the relationships for surface runoff and infiltration to improve the reliability of the simulated hydrographs. However, Beasley et al. (1980) confirm that ANSWERS may be applicable for watersheds without hydrologic measurements to obtain at least comparative results for different treatments or watershed management strategies.

O. Hajji · N. B. Rhouma
Engineers Higher School of Medjez El Beb, 9070 Beja, Tunisia

S. Abidi (✉)
Laboratory of Silvo-Pastoral Resources of Tabarka, 8110 Jendouba, Tunisia
e-mail: saharabidi01@gmail.com

L. L. Ghezal
Agriculture High School of Mograne, LR-SPADD, University of Carthage, Tunis, Tunisia

F. Lucci et al. (eds.), *Selected Studies in Geomorphology, Sedimentology, and Geochemistry*, Advances in Science, Technology & Innovation, https://doi.org/10.1007/978-3-031-43744-1_21

According to Connolly and Silburn (1995), the ANSWERS model, modified to include the infiltration equation of Green and Ampt, accurately predicts runoff from simulation plots and watersheds. Application of the modified ANSWERS model for ANSFOR forested areas (Thomas & Beasley, 1986a, 1986b) to watersheds in a foothills region of North Carolina indicates acceptable representations of maximum runoff and the time to reach it, but requires improvements in predicting total runoff.

Thus, the study of the upstream Mellegue watershed is based on the manipulation of a spatial database using a Geographic Information System, subsequently presenting a source of extraction of the input data necessary for the application of the hydrological model with physical parameters ANSWERS.

The objectives of this study are to simulate the water and sediment inputs at the outlet of the Mellegue watershed, which is part of the large watershed of the Medjerda.

2 Materials and Methods

In this present work, we are developing a spatialized modeling methodology for runoff and soil losses at the level of the Oued Mellegue watershed by applying the distributed hydrological model ANSWERS. This methodology is based on the use of physical state data (soil and plant cover), hydrological data (rainfall and hydrographic network), and especially data provided by GIS (DEM and all the maps derived from it: slope and aspect, soil texture map, and land cover map).

Three events were chosen for simulation: 07–09/01/2017, 07–07/02/2017, and 02–03/04/2017.

2.1 ANSWER Model

ANSWER is a conceptual model distributed by event, structured in different components: hydrological, sediment detachment/transport as well as different routines necessary to describe the movement of water on the surface of the plot, drainage, and the different phases of flow in the stream.

It operates on a square mesh grid; the geographic information system transforms the basic cartographic data into a set of distributed parameters. Simple attribute tables allow the user to provide the values of these input parameters in order to improve the management of the input parameters and to facilitate the manipulation of the different information planes.

2.2 Study Area

Wadi Mellegue is one of the most important tributaries of the Medjerda, located on its Upper Valley (Fig. 1).

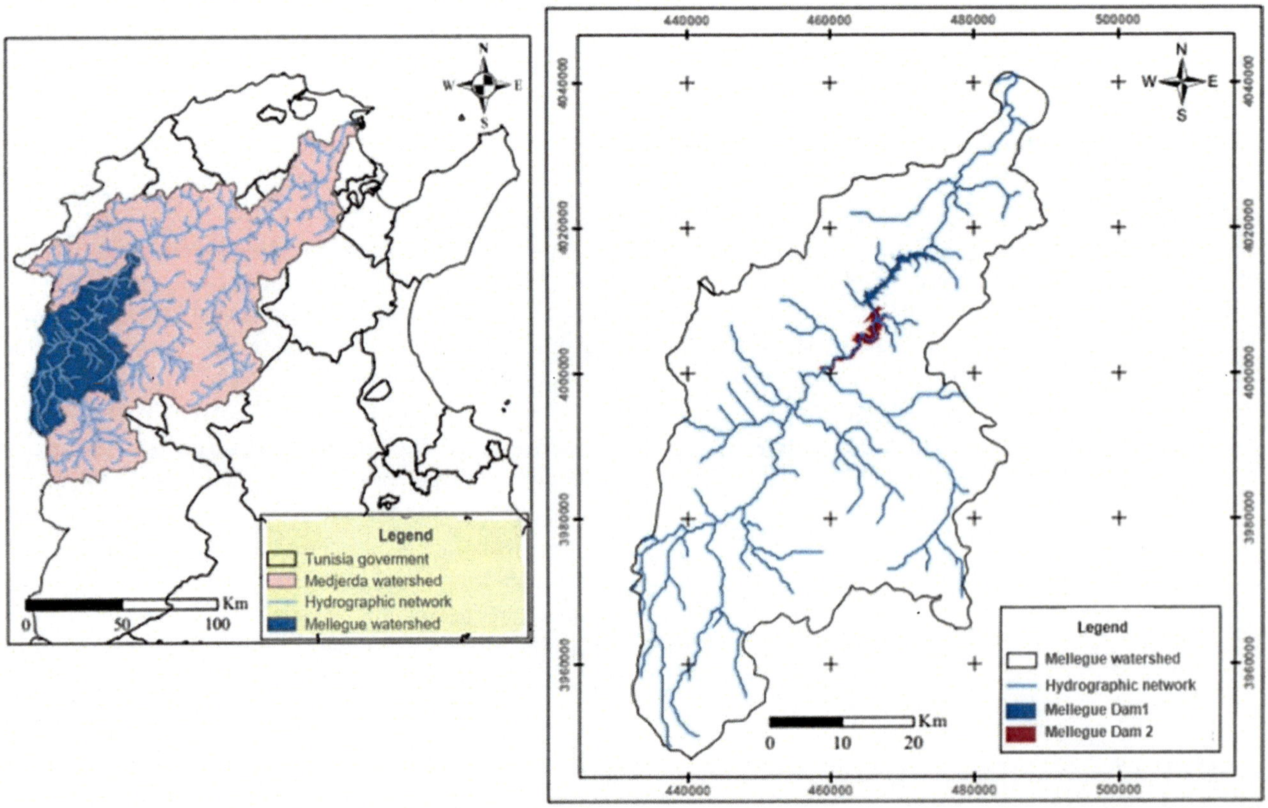

Fig. 1 Localization map of Mellegue watershed

The watershed is drained by a multiple vault dam with a retention capacity of around 300 million m^3. The main purpose of creating the dam is to prevent the flooding of the plain of Jendouba, as well as the irrigation of the lower valley of Medjerda, and the production of electricity. Mellegue dam had significant solid contributions which reduced its capacity to 20% of the initial capacity (DGBGTH, 2019).

3 Results

3.1 ANSWER Simulation Results

Table 1 presents the hydrologic and erosion results of ANSWER simulation for the three events.

Based on the ANSWERS model output file data, the sedimentogram which represents the sediment concentration in (g/l) as a function of time was constructed (see Figs. 2 and 3).

3.2 Erosion Mapping

Firstly, we prepared a maps database to well understand and identify the erosion causes, such as digital elevation model, slope, soil cover, and pedological maps (Fig. 4).

After the application of the ANSWERS model, the erosion map is generated based on the sign of values as the following principle:

- If the value is positive or zero, the mesh is characterized by a deposit.
- If the value is negative, the mesh is subject to erosion.

From this file, we can generate another Excel file comprising the column numbers, the row numbers, and the erosion values which allow establishing an erosion map of the watershed, which expresses the rate of detached or deposited sediments in t/ha (Fig. 5).

Table 1 Simulation results

Events	Hydrologic results					Erosion results			
	P (mm)	I_{max} (mm/h)	Q_{max} (m^3/s)	L_r (mm)	K_r (%)	V_r (10^6 m^3)	E_s (t/ha)	C_{max} (g/l)	T_s (10^6 t)
07–09/01/2017	81.3	2.29	117.64	5.489	6.75	15.094	6.535	159.121	1.478
07–09/02/2017	161	4.08	33.61	1.583	98.3	4.353	0.576	276.723	0.130
02–03/04/2017	33.5	1.12	25.21	0.055	16.41	0.151	0.008	502.172	0.002

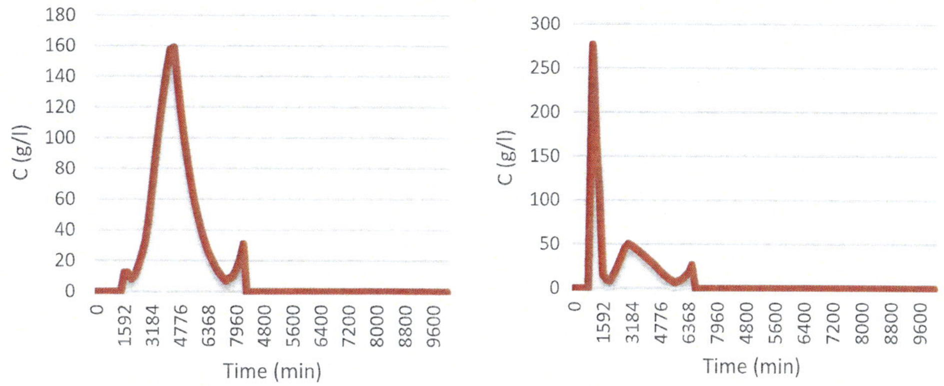

Fig. 2 Sedimentogram of the 07–09/01/2017 and 07–09/02/2017 events

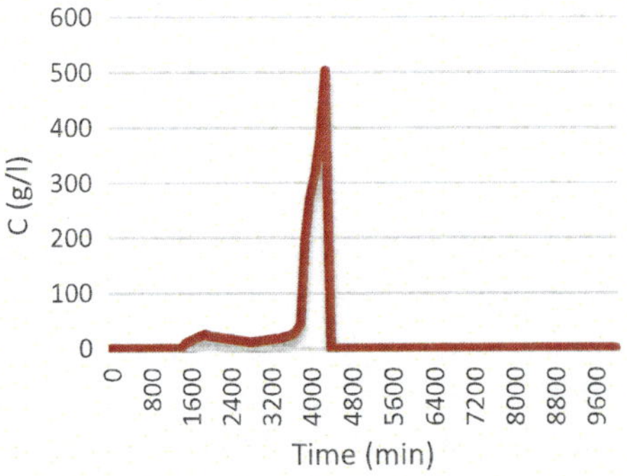

Fig. 3 Sedimentogram of the 02–03/05/2017 event

4 Discussion

From the results of ANSWERS simulation for the three selected events, it can be concluded that the average soil loss varies from 1849 t to 1.5 Million t which corresponds to a specific solid input varying from 0.008 to 6.535 t/ha.

In addition, generated runoff volumes vary from 0.15 to 15.09 Million m^3 which leads to a fairly significant potential for transport by runoff.

For erosion mapping, it was noted that the detachment and deposition values depend on elevation, slope, flow direction, soil characteristics, vegetation, and hydrographic network.

From the results obtained by the ANSWERS model, one notices a great spatial variability of the erosion for the rainfall studied on the watershed of Mellegue; this variability is mainly linked to the variation in the intensity of rain, thus the state of soil cover and the variation of the slope, as shown on Fig. 4.

5 Final Remarks

The morphological, climatic, geological, and pedological characteristics as well as the vegetation, occupation, and tillage parameters of a watershed have a significant influence on the definition of runoff processes.

The ANSWERS hydrological and erosion model is able to simulate the response of a watershed following a rainfall event, while taking into account its spatial variability.

Thus, it made it possible to estimate the runoff and the average soil loss and to give the runoff hydrograph and the evaluation of the sediment concentration in the runoff.

Coupling the ANSWERS model with the GIS gave us an idea of the distribution of water erosion in Mellègue watershed.

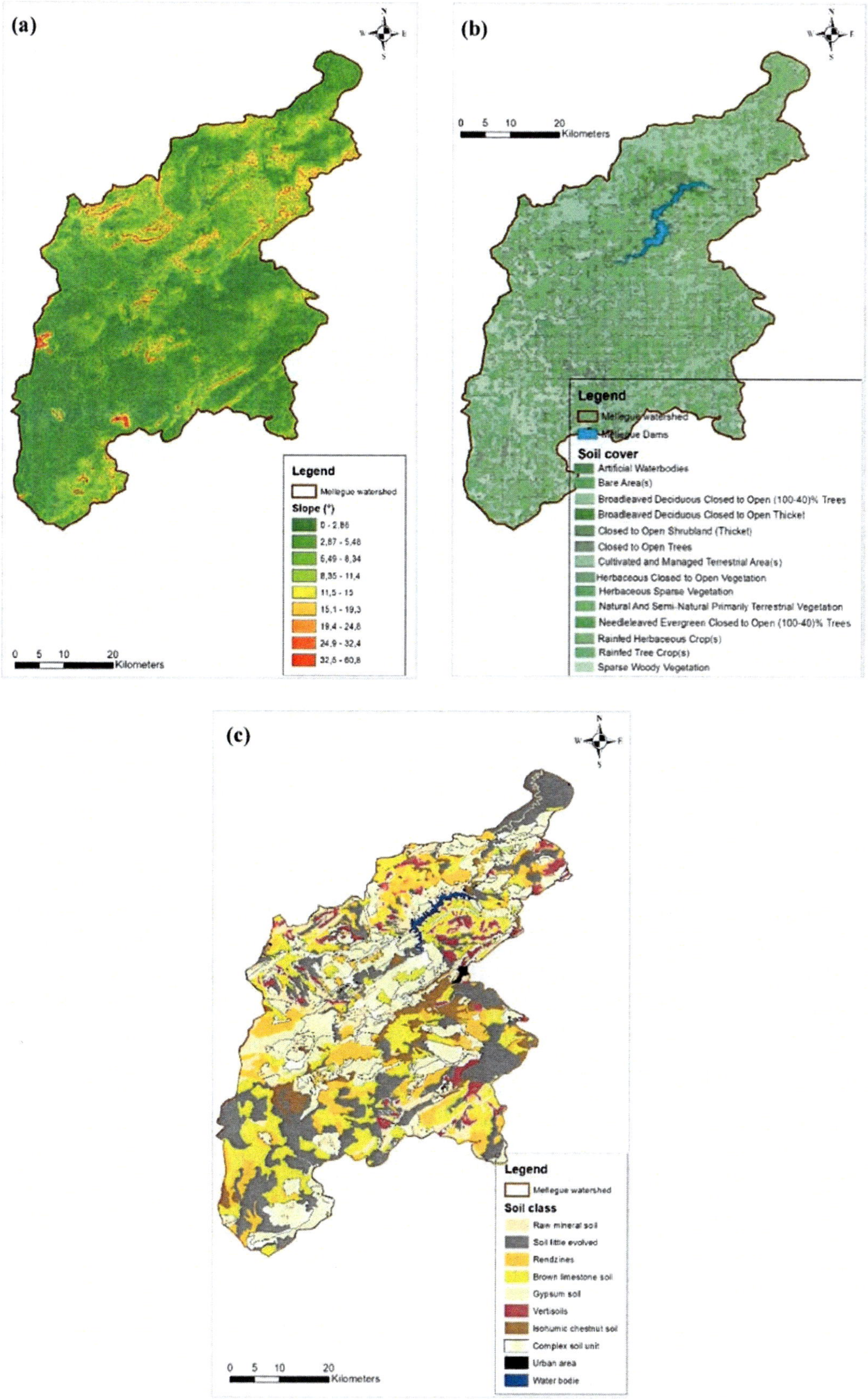

Fig. 4 **a** Slope map, **b** soil cover map, and **c** soil class map

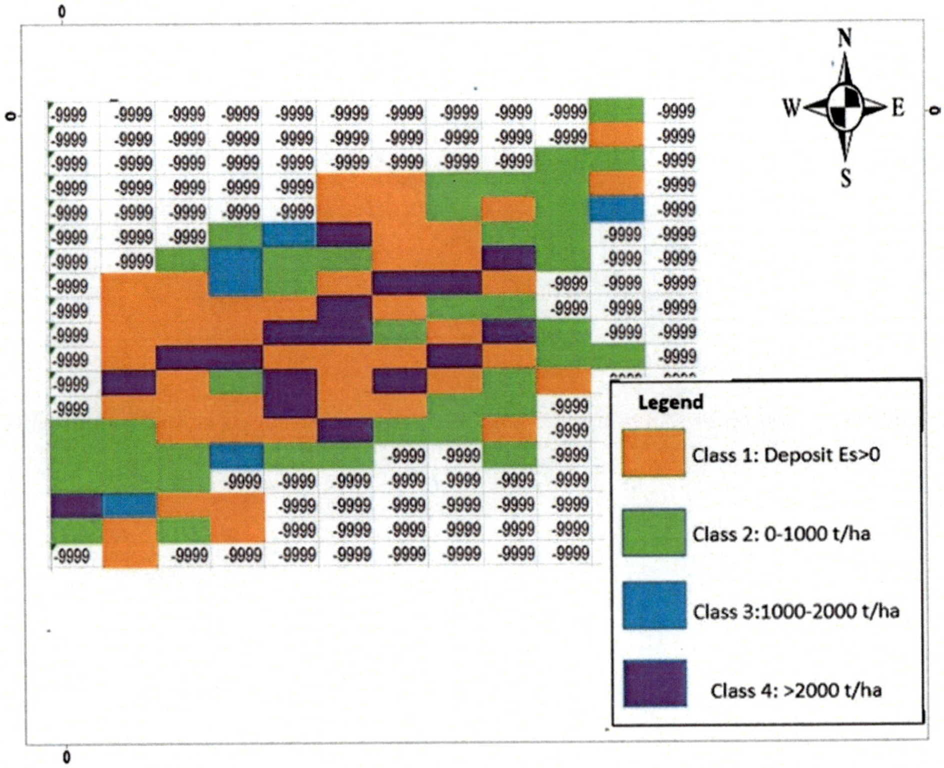

Fig. 5 Erosion map for the 07–09/01/2017 event

References

Beasley, D. B. (1977). ANSWERS: A mathematical model for simulating the effects of land use and management on water quality. Unpublished PhD. thesis, Purdue University, West Lafayette, 266 pp.

Beasley, D. B., Huggins, L. F., & Monke, E. J. (1980). ANSWERS: A model for watershed planning. American Society of Agricultural Engineers 0001-2351/80/2304-0938S02.0.0.

DGBGTH. (2019). *Bathymetric measurement of the Mellegue dam.*

Foster, G. R., & Meyer, L. D. (1972a). A closed-form soil erosion equation for upland areas. In H. W. Shen (editor and publisher), *Sedimentation (Einstein)* (Chap. 12). Colorado State University.

Foster, G. R., & Meyer, L. D. (1972b). Transport of soil particles by shallow flow. *Transactions of the ASAE, 15*(1), 99–102.

Huggins, L. F., & Monke, E. J. (1966). The mathematical simulation of the hydrology of small watersheds. Technical Report 1, Water Resources Research Center, Purdue University, West Lafayette, IN, 130 p.

Huggins, L. F., & Monke, E. J. (1968). A mathematical model for simulating the hydrologic response of a watershed. *Water Resources Research, 4*(3), 529–539. https://doi.org/10.1029/WR004i003p00529

Lajili, L. (1998). Water erosion in Tunisian semi-arid zone: Modeling, parameter estimation and application to anti-erosion management. Thesis presented for the degree of Doctor in Applied Biological Sciences (Land and Forest Management), Faculty of Agronomic and Applied Biological Sciences, Ghent University, Belgium, 215 pp.

Silburn, D. M., & Connolly, R. D. (1995). Distributed parameter hydrology model (ANSWERS) applied to a range of catchment scales using rainfall simulator data I: Infiltration modelling and parameter measurement. *Journal of Hydrology, 172*(1–4), 87–104.

Thomas, D. L., & Beasley, D. B. (1986a). A physically-based forest hydrology model I: Development and sensitivity of components. *Transactions of ASAE, 29*(4), 962–972.

Thomas, D. L., & Beasley, D. B. (1986b). A physically based forest hydrology model II: Evaluation under natural conditions. *Transactions of ASAE, 29*(4), 973–981.

Traces of Ancient Land Use in Cultural Landscapes of Desert Fringes: A Neglected Heritage and Archive

Bernhard Lucke, Jago J. Birk, Günther Schörner, Paula Kouki and Nizar Abu-Jaber

Abstract

Landscapes of desert fringes are often considered degraded: characterized by encroaching desertification and greatly diminished fertility due to poorly adapted land use. Deserted ancient cities have been interpreted as victims of past environmental mismanagement. However, recent research has made it increasingly clear that the man-made desertification narrative is largely fiction and partly a hegemonic discourse connected with colonial mandates and the establishment of modern statehood. Focusing on Jordan as an example, it is shown that numerous traces of past land use have been preserved in the desert fringes. These ancient land use features are an important part of the cultural heritage which is unfortunately rarely considered, although it provides important testimony to past land use practice and its environmental impact. This is illustrated with two examples: (1) In northern Jordan, some soils of the limestone plateaus are stable old land surfaces where remains of past land use accumulated, mainly in form of pottery. Surveys of off-site archaeological material, combined with systematic soil analysis, proved suited to reconstruct patterns of ancient land use. (2) Near Petra, the trained eye can observe a multitude of terrace remains, and there may have not been one slope that was not terraced at some time during antiquity. A case study of several terrace systems found that they are on the one hand suited to control runoff and thus minimize the risk of flash floods, and on the other hand collected sufficient water for runoff-irrigated agriculture. Feces biomarkers indicate manuring with human excrements. Jordan's cultural landscapes preserved numerous traces of ancient land use activities which are potential archives of environmental and cultural changes. The available evidence suggests that past human impact did in many cases not lead to land degradation, but created oases in the desert.

Keywords

Cultural landscape · Ancient land use · Terracing · Manuring · Environmental reconstruction

B. Lucke (✉)
Institute of Geography, FAU Erlangen-Nürnberg, Wetterkreuz 15, 91058 Erlangen, Germany
e-mail: bernhard.lucke@fau.de

J. J. Birk
Institute of Geography, University of Mainz, 55099 Mainz, Germany

G. Schörner
Institute of Classical Archaeology, University of Vienna, 1190 Vienna, Austria

P. Kouki
Ancient Near Eastern Empires Centre of Excellence, University of Helsinki, P.O. Box 3, 00014 Helsinki, Finland

N. Abu-Jaber
School of Natural Resources Engineering and Management, German Jordanian University, Madaba 11180, Amman, Jordan

1 Introduction

Well-preserved ruins of abandoned monumental cities from antiquity are common in dry lands characterized by marginal agricultural fertility, such as in present-day Jordan. One popular narrative explaining the desertion of these cities is that man-made soil erosion diminished the land's productivity (Lowdermilk, 1953). The current vegetation-poor state of these landscapes is seen as result of degradation (and largely 'empty' of heritage), and thus incapable to still supply the ancient population centers.

However, the spatial and temporal patterns of past agriculture are unknown, and there is mostly only circumstantial evidence of ancient farming strategies. In this

F. Lucci et al. (eds.), *Selected Studies in Geomorphology, Sedimentology, and Geochemistry*, Advances in Science, Technology & Innovation, https://doi.org/10.1007/978-3-031-43744-1_22

Fig. 1 Stars show the investigated hinterlands of the mentioned archaeological sites in Jordan. Circles mark nearby major cities. For detailed maps, see Lucke et al. (2019a, 2019b)

context, climate changes can have the same consequences as bad land management (Issar & Zohar, 2007). They could directly cause crop failures due to unfavorable weather, or impact yields indirectly by triggering soil erosion. In this context, soil erosion and sedimentation in northern Jordan was found to have been mainly connected with heavy rains inducing slope collapses, largely synchronous with the two Little Ice Ages of the sixth and fourteenth–fifteenth centuries CE (Lucke & Schmidt, 2017).

A means to reconstruct environmental history is the investigation of soils and sediments, which act as archives of environmental changes if they are deciphered correctly (Lucke, 2017). This contribution briefly presents two examples from seemingly 'empty' hinterlands in Jordan: fields at the site of Abila in Mediterranean climate in northern Jordan, and at Petra in rugged arid mountains in the South (Fig. 1).

2 Materials and Methods

We conducted intensive archaeological surveys of off-site pottery scatters on modern fields on the Terraecalcis soils of a flat limestone plateau near Abila, considered rather erosion-resistant, and on remains of ancient runoff-irrigated terraces built on slopes of the rugged sandstone mountains near Petra. The material culture was collected and soil samples were taken from profiles and the surface of current fields.

Abila is situated on a level of Eocene limestone plateau approximately 15 km North of Irbid, receiving ~450 mm of mean annual precipitation. Soil properties and depths on the limestone plateau are very heterogeneous, but a soil cover that suffices agriculture is present everywhere. The Wadi Queilbeh valley, which crosses the site, hosts an approximately 5 m deep fill of sediments (Lucke & Schmidt, 2017). The arid mountains surrounding Petra represent a harsh environment for agriculture, characterized by unpredictable, strongly varying rainfall and limited resources with regard to the availability of water and arable soils. The environment is prone to flash floods and extremely limited space for cultivation. However, virtually, all slopes surrounding the ancient city carry rests of runoff-irrigated terraces, which could represent the remains of a former agricultural hinterland (Lucke et al., submitted).

Laboratory analysis of soil samples included color, pH, electrical conductivity, organic matter, bulk density, contents of plant-available phosphorus, calcium carbonate, and chemical composition of major and trace elements according to standard methods. Grain sizes were determined by wet sieving and Sedigraph grain size analyzer without removal of calcium carbonate (Lucke et al., 2019b). Stanols, Δ5-sterols, and bile acids as feces biomarkers were extracted according to Birk et al. (2012).

3 Results

3.1 Pottery Distribution and Soil Properties Near Abila

A large but highly variable amount of material culture was found on fields surrounding the site, consisting mainly of pottery sherds. Its distribution did not follow spatial rules, such as steadily decreasing densities of pottery sherds with distance from the city center, but was more complex, suggesting other factors than distance to the city center to play a role for off-site pottery distribution. The strongly varying presence of material culture seems connected with diverse past land use practices (Lucke et al., 2019a), and supports earlier interpretations of soils around Abila as rather stable land surfaces, where remains of past land use accumulated (Lucke, 2017).

3.2 Soils of Ancient Runoff-Irrigated Terraces Near Petra

A hydrological model indicates that the terraces near Petra were suited for collecting sufficient water for agriculture (Lucke et al., submitted). Elevated contents of plant-extractable phosphorus and biomarkers indicative of human excrements suggest planned manuring, which is in agreement with references to systematic collection of sludge mentioned in the Petra papyri (Lucke et al., submitted). In difference to the level plateau at Abila, the material culture on the terraced slopes near Petra is strongly subject to erosion-sedimentation processes. However, redistribution seemed limited and surface pottery densities corresponded to other parameters considered indicative for manuring, such as plant-extractable phosphorus (Lucke et al., submitted).

4 Discussion

Results from our surveys show that remains of past land use practices are still omnipresent in Jordan. Material culture scatters indicate diverse activities, following the availability of water (Lucke et al., 2019a). At Abila, the limestone plateau was mostly stable, although short but intense occurrences of heavy rains (mostly during the two Little Ice Ages) led to drastic mass movements along valleys (Lucke, 2017). Similar sediment ages were found by other studies (Lichtenberger et al., 2019), but attributed to assume past land use changes as deposition processes were not considered. At Petra, terraces harvested not only runoff water, but also dust which provided the largest part of the substrate of terrace soils (Lucke et al., 2019b). Feces biomarkers suggest planned manuring (Lucke et al., submitted) and indicate intensive agriculture.

5 Conclusions

The barrenness of dry lands like in Jordan is misleading: a closer look reveals a vast, but understudied heritage of ancient land use remains. It testifies to sustainable land use, ingenious water management, and can be evaluated as environmental archive.

References

Birk, J. J., Dippold, M., Wiesenberg, G. L. B., & Glaser, B. (2012). Combined quantification of faecal sterols, stanols, stanones and bile acids in soils and terrestrial sediments by gas chromatography-mass spectrometry. *Journal of Chromatography, 1242*, 1–10.

Issar, A., & Zohar, M. (2007). *Climate change—Environment and civilisation in the Middle East* (2nd ed.). Springer.

Lichtenberger, A., Raja, R., Seland, E. H., Kinnaird, T., & Simpson, I. A. (2019). Urban-riverine hinterland synergies in semi-arid environments: Millennial-scale change, adaptations, and environmental responses at Gerasa/Jerash. *Journal of Field Archaeology, 44*, 333–351.

Lowdermilk, W. (1953). *Conquest of the land through 7000 years*. U.S. Department of Agriculture.

Lucke, B. (2017). *Landscape transformations in the context of soil development, land use, and climate: A comparison of marginal areas in Jordan, Mexico, and Germany*. Relief Boden Paläoklima 26. Schweitzerbart Science Publishers.

Lucke, B., & Schmidt, C. (2017). Debris flows, soil creep, and heavy rains in the Wadi Queilbeh, Northern Jordan. *Zeitschrift Für Geomorphologie, 61*(Suppl. 1), 159–201.

Lucke, B., al-Karaimeh, S., Schörner, G. (2019a). Channels, terraces, pottery, and sediments—a comparison of past irrigation systems along a climatic transect in Northern Jordan. *Journal of Arid Environments, 160*, 56–73.

Lucke, B., Sandler, A., Vanselow, K. A., Bruins, H. J., Abu-Jaber, N., Bäumler, R., Porat, N., & Kouki, P. (2019b). Composition of modern dust and Holocene Aeolian sediments in archaeological structures of the Southern Levant. *Atmosphere, 10*, 762.

Lucke, B., Birk, J. J., Hunt, C. O., Laparidou, S., Keilholz, P., Kouki, P., Abu-Jaber, N., Bäumler, R., Fiedler, S., Deckers, K., & Abudanah, F. (Submitted). The terraces of Petra, Jordan: Environmental archives of a lost agricultural hinterland. Geoarchaeology.

Marine Geosciences, Paleoceanography, Paleoclimatology

Element Geochemical Characteristics and Their Implications for Paleosalinity in the Paleogene Lishui Sag, East China Sea Shelf Basin

Long Sun, Jinliang Zhang, Zhu Zhu and Jinshui Liu

Abstract

The Lishui Sag in the East China Sea Shelf Basin has great oil and gas exploration potential but its research level is relatively low. In order to understand the geological background and sedimentary environment of the Paleocene in the Lishui Sag, this paper used X-ray diffraction clay mineral analysis, ICP-AES element analysis and other geochemical data, combined with the relationship between element geochemical characteristics and sedimentary environment. Various methods such as boron content, equivalent boron content, Adams formula, Couch formula, strontium-barium ratio, and boron-gallium ratio were used to distinguish paleosalinity. The results show that the Lishui Sag was a brackish-saltwater continental lake in the Early Paleoocene The basin was a salt water basin in the early Late Paleocene, and transitioned to a brackish water environment in the late period; the water temperature was relatively constant for a long time and was in a reducing environment for a long time.

Keywords

Paleosalinity · Element geochemical characteristics · Sedimentary environment · Lishui Sag · Paleogene

1 Introduction

As an important indicator of sedimentary environment analysis, water paleosalinity can provide a basis for studying sedimentary facies types. Currently, research on paleosalinity is mainly based on the geochemical methods of trace elements in muddy sediments. Generally, the boron content in marine muddy sediments is higher than that in land sediments, so the trace element boron is often used to indicate paleosalinity. Walker (1968) used the equivalent boron content to distinguish different paleosalinity environments. Couch (1971) used isotherm adsorption to fit a linear regression equation to quantitatively calculate paleosalinity. In addition, Sr/Ba and B/Ga ratios can also be used to analyze and distinguish the salinity of ancient water bodies.

2 Geology Settings and Methods

The Lishui Sag is located west of the Taipei Depression in the East China Sea Basin. It is a Mesozoic and Cenozoic rift-type basin with a northeast-southwest direction. Its western part is connected to the Fujian-Zhejiang uplift, the northeast is adjacent to the Jiaojiang Sag, and the southeast is the Yandang Uplift which is separated from the Fuzhou Depression (Fig. 1). The main strata encountered during the Paleocene in the study area are Mingyuefeng Formation, Lingfeng Formation, and Yueguifeng Formation.

In this study, some mudstone samples were collected in the Lishui-Jiaojiang Sag of the East China Sea Shelf Basin, mainly gray mudstone, gray-black mudstone, and dark brown mudstone. From the bottom of the Paleocene Yueguifeng Formation to the top of the Mingyuefeng

L. Sun (✉)
National Key Laboratory of Petroleum Resources and Engineering, China University of Petroleum (Beijing), Beijing 102249, China
e-mail: longsun163@163.com

College of Geosciences, China University of Petroleum (Beijing), Beijing 102249, China

J. Zhang
Faculty of Geographical Science, Beijing Normal University, Beijing 100875, China
e-mail: reser@cup.edu.cn

Z. Zhu
College of Global Change and Earth System Science, Beijing Normal University, Beijing 100875, China

J. Liu
CNOOC China Ltd, Shanghai 200335, China

F. Lucci et al. (eds.), *Selected Studies in Geomorphology, Sedimentology, and Geochemistry*,
Advances in Science, Technology & Innovation, https://doi.org/10.1007/978-3-031-43744-1_23

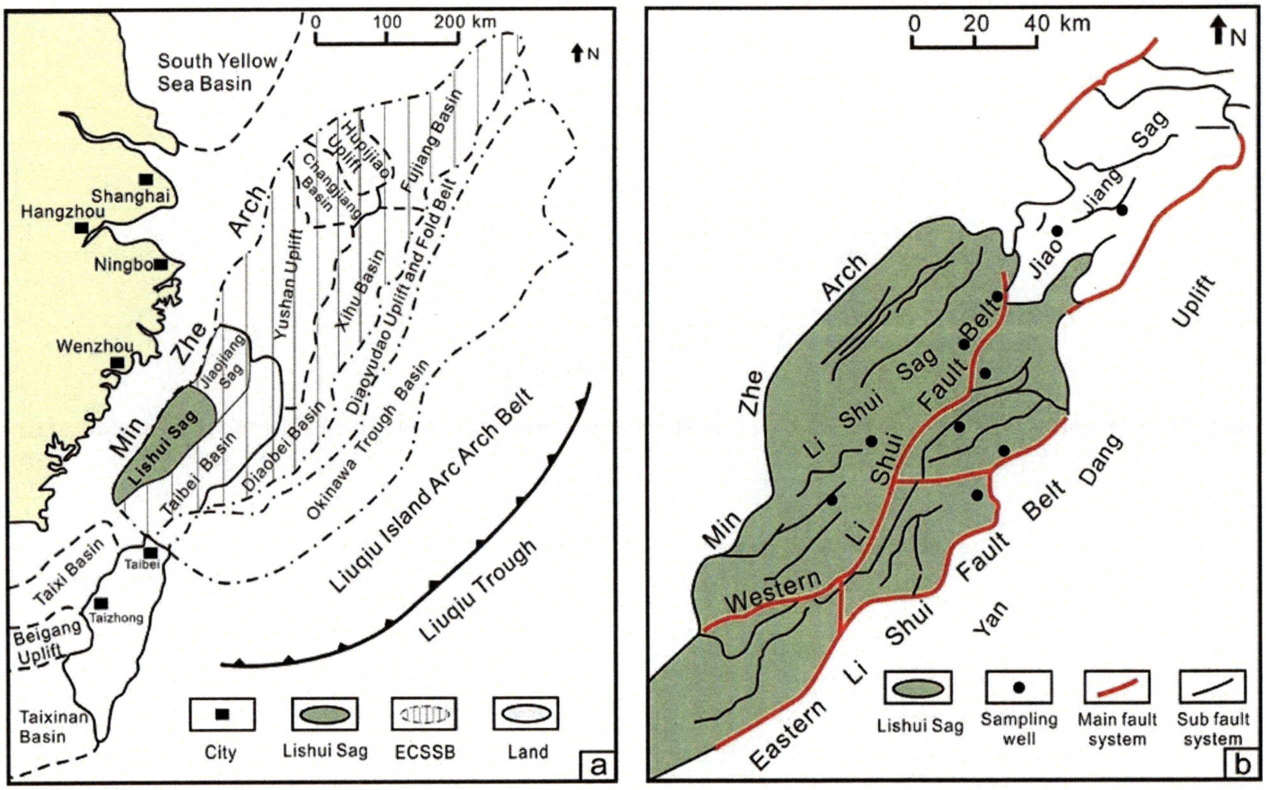

Fig. 1 Location of the study area in the Lishui Sag

Formation, six samples were collected systematically, including three from the Yueguifeng Formation, one from the Lingfeng Formation, and two from the Mingyuefeng Formation. Whole rock mineral X-ray diffraction analysis was carried out in the Geological Testing Center of the Exploration and Development Research Institute of Sinopec Shengli Oilfield, and the major and trace element contents of the samples were analyzed in the Stratigraphy Laboratory of the Shengli Oilfield Geological Research Institute. An ICP-MS, D/max-2500PC diffractometer, element analyzer (Vario EL III), and other related instruments were used for data collection.

3 Results

3.1 Sr/Ba Ratios

Generally, the Sr/Ba ratio of marine sediments (>1) is higher than that of terrestrial sediments (<1). A Sr/Ba ratio of >1 means the ancient water body is marine saline water. When the Sr/Ba ratio is 0.6–1.0, the ancient water body is brackish, and when the Sr/Ba ratio is <0.6, it is a freshwater environment. Using the element ratio method, the Paleocene Sr/Ba ratio in Lishui Sag was calculated. The achieved results show the average Sr/Ba ratio of the Yueguifeng Formation is 0.69, the highest is 0.78, and the lowest is 0.59, corresponding to a brackish water environment. The Lingfeng Formation has a ratio of 1.04, which belongs to the saltwater lake basin or brackish-salt water transitional environment. The Mingyuefeng Formation has a ratio of 0.68, with the highest of 0.94, and the lowest 0.42. The overall environment is brackish water. Because the highest value is close to 1, it may also be a brackish-salt water transitional environment. Overall using the Sr/Ba ratio method, the Lishui Sag was a brackish-salt water environment during the Paleocene.

3.2 B/Ga Ratios

The B/Ga ratio of marine facies is much higher than that of continental facies, while the B/Ga ratio of continental lakes is higher than that of river facies. When the B/Ga ratio is between 0.5 and 1.5, it is the freshwater phase, when the ratio is 1.5–3.0, it is a freshwater lake, when the ratio is 3.0–4.5, it is a brackish water phase, and when the ratio is greater than 4.5, it is a salt water phase.

B/Ga ratios of the Lishui Sag in the study area are generally high. The average ratio of the Yueguifeng Formation is 12.15 (highest 15.15, lowest 9.15), which is a salt water environment. The Lingfeng Formation has an average ratio of 5.54, which belongs to the salt water environment or transitional from brackish water to salt water. The average value of the Mingyue peak group is 8.33 (highest 11.74, lowest 4.91), suggesting a salt water environment, but the lowest value is close to 4.5 as the ratios span a large range, so it may also be a brackish-salt water transitional environment. From the B/Ga ratio as a whole, the Lishui Sag was in a brackish-salt water environment during the Paleocene, consistent with the Sr/Ba ratios.

3.3 Couch's Boron Content Analysis

The illite, kaolinite, and montmorillonite all contain boron, so Couch (1971) calculated the paleosalinity of Nigerian kaolin Tertiary shale based on the equivalent boron content and combined with the isotherm adsorption curve. The formula was re-calibrated, and the formula for calculating Couch paleosalinity was obtained as $\lg Sp = (\lg Bk - 0.11)/1.28$, where Sp is the ancient salinity (‰) to be determined, and Bk is the equivalent boron content (ppm). This formula was used to calculate paleosalinity values (‰) in the study area. This shows that the Yueguifeng Formation is 11.77‰ (highest 12.04‰, lowest 11.49‰), which is a brackish water environment. The Lingfeng Formation is 8.72‰, which is a brackish water environment. The Mingyuefeng Formation is 8.85‰ (highest 8.90‰, lowest 8.80‰), which is a brackish water environment. According to this method, the Lishui Sag was a brackish water environment during the Paleocene, consistent with the Sr/Ba and B/Ga ratios.

4 Discussion

Analysis of the above methods for judging paleosalinity and comparing their trends shows that the Sr/Ba ratio method has higher reliability and sensitivity than the Couch formula and B/Ga ratio. Using these methods, the Yueguifeng Formation was a brackish-saltwater basin environment; the Lingfeng Formation was a saltwater marine facies; and the Mingyuefeng Formation was a brackish-saltwater marine facies. The whole Paleocene of the Lishui Sag was a brackish water-salt water environment.

5 Conclusions

According to the discrimination results of boron content, equivalent boron content, Adams formula, Couch formula, and element ratio method, it is inferred that the Lishui Sag was a brackish-salty continental lake basin in the Early Paleocene, and a salt water basin in the early Late Paleocene.

References

Couch, E. L. (1971). Calculation of paleosalinities from boron and clay mineral data. *AAPG Bulletin, 55*(10), 1829–1837.

Walker, C. T. (1968). Evaluation of boron as a paleosalinity indicator and its application to offshore prospect. *AAPG Bulletin, 52*(1), 751–766.

Sedimentology, Stratigraphy, Paleontology, Geochronology

Origin, Paleoenvironment and Paleoclimate of the Paleocene-Eocene Clays of the BouCraa Deposit, Southern Morocco

Hanane El Boukhari, Mohammed Amine Nguidi, Mustapha Mouflih, Salem El Ouariti, Abdelmajid Benbouziane, Mohammed Aquite and Oussama Khadiri Yazami

Abstract

The Oued Eddahab basin, which is part of the large Tarfaya-Laayoune-Dakhla basin (TLDB) in southern Morocco, contains phosphate deposits at Izic, Laabadila, Imesleguen and BouCraa. The latter, currently exploited, is the most important due to the extension, thickness and richness of its phosphates. The BouCraa deposit consists of Upper Cretaceous-Paleogene intercalated clayey beds associated with phosphate mineralization. The objective of this work was to characterize the clays of the BouCraa deposit and determine the paleoenvironment, paleoclimate and origin of the clay beds. Analyses of the clayey beds by X-ray diffraction (XRD) and Fourier transform infrared spectrometry (FTIR) showed the presence of palygorskite-type clays, montmorillonites and illites in small quantities. Geochemical analyses confirm the XRD results by determining that the dominant constituent elements of the clays are silica, aluminum and magnesium. The Paleocene-Eocene clay minerals were interpreted as accumulated under a semi-arid climate with an alternation of dry (ion concentration) and wet phases (contribution of ions and detrital minerals). Montmorillonite is formed by neoformation from cations carried by solutions that migrate through the landscape during the wet season. These solutions come from the alteration of rocks exposed in the surrounding reliefs of the Anti-Atlas chain and the Reguibat shield. These solutions are loaded with silica and basic cations necessary to ensure such neoformation. The palygorskite is also neoformed but in a confined environment that was very rich in cations supplied from peri-marine water and the ions were concentrated by evaporation under drought conditions.

Keywords

Phosphate · Montmorillonite · Palygorskite · Neoformed clays · Paleocene · BouCraa

1 Introduction

The BouCraa deposit of southern Morocco consists of Upper Cretaceous-Paleogene intercalated clayey beds associated with phosphate mineralization. This deposit is the most important phosphate deposit in the Oued Eddahab basin, and is the middle of the onshore part of the Tarfaya-Laayoune-Boujdour basin. The Oued Eddehab basin, which is considered to be a subbasin within the greater Tarfaya-Laayoune-Boujdour basin, extends southeast from Laayoune to north of the town of Dakhla over a length of 650 km and a width of 200 km. Oued Eddehab basin is subdivided into four main sectors (Izic, Laabadila, Imesleguen and BouCraa) and three mining zones (Northern, Central and Southern) for mining of Paleogene phosphatic strata. Clay minerals are an important and widely distributed component in the phosphatic strata of the BouCraa deposit. This study characterized the clays of Mine A located in the central zone of the BouCraa deposit to identify the different types of clay minerals, their proportions and vertical variations. The characteristics of the clay minerals may be used for determining climatic conditions and the reconstruction

H. El Boukhari · M. Mouflih · S. E. Ouariti · A. Benbouziane
Dynamics of Sedimentary Basins and Geological Correlations Laboratory, Faculty of Sciences Ben M'Sik, Hassan II University of Casablanca, Casablanca, Morocco
e-mail: hanane.elboukhari-etu@etu.univh2c.ma

M. Aquite · O. K. Yazami
Geological Departments, Industrial Development Department, OCP s.a., 2, Rue Al Abtal Hay Erraha, 20200 Casablanca, Morocco

M. A. Nguidi (✉)
Geosciences Laboratory, Faculty of Sciences Aïn Chock, Hassan II University, Casablanca, Morocco
e-mail: nguidiaminemed@gmail.com

F. Lucci et al. (eds.), *Selected Studies in Geomorphology, Sedimentology, and Geochemistry*, Advances in Science, Technology & Innovation, https://doi.org/10.1007/978-3-031-43744-1_24

of the paleoenvironment during the Paleocene-Eocene (Chamley, 1989; Daoudi et al., 2008; Fürsich et al., 2005; Raucsik & Varga, 2008).

2 Materials and Methods

A lithostratigraphic section was surveyed in the central part of the BouCraa deposit and clays intercalated and/or associated with phosphate deposits were sampled. The clayey corteges contain associations of carbonates and organic matter. The impurities were removed by washing under a 50 µm sieve and with 5% hydrochloric acid and hydrogen peroxide. The mixture was then washed with distilled water until the liquid overlying the base remained opalescent. The clay, suspended in water, was then recovered in the first few centimeters by decanting after centrifugation. This operation made it possible to have granulometric fractions with a size of less than 2 µm. The preparations of the treated <2 µm fraction were analyzed by two methods: (1) from disoriented powders, and (2) from deposition on glass slides treated by saturation with ethylene glycol for 24 h and heated at 490 °C for 2 h. The diffractogram used was a Philips PANalytical DRX apparatus. This technique was added the Fourier transform infrared spectroscopy (FTIR), using a Perkin Elmer spectrometer equipped with an ATR-FTIR module. Percentages of chemical elements in clay minerals were evaluated by X-ray florescence (XRF) using an Epsilon 3XL type. Thermo-gravimetric analyses were carried out using a Setsys Ev 1750 type thermo-gravimetric analyzer to follow the mass loss as a function of heating temperature.

3 Results

3.1 Mineralogical Composition

The diffractograms of disoriented powder clays of the BouCraa deposit record the presence of quartz at $d = 3.34$ Å and $d = 4.27$ Å. Characteristic peaks of the clay minerals at 10.48 and 15.20 Å lattice spacing were observed in the majority of the samples analyzed (Fig. 1). The 10.48 Å distance is typical of palygorskite clay, whereas the 15.20 Å distance is typical of montmorillonite clay. Illites were present at about 4.45, 4.47, and 4.48 Å. When treated with ethylene glycol, all the samples showed that the line recorded at 15 Å increased to 17.1 Å, indicating that the clays were rich in montmorillonite clays. Reflections at 10.30 Å, and 6.44 Å confirmed the presence of fibrous phyllosilicates, corresponding to palygorskite. The persistence of a peak around 10 Å on all the difractograms confirmed the presence of illite (Brindley & Brown, 1980; Moore & Reynolds, 1989), proven by the peak at 3.33 Å which did not undergo any displacement following the treatment with ethylene glycol nor after heating. Quartz and fluoroapatite appear to be the minor mineral phases associated with the clays.

The examination of the infrared (IR) spectra obtained from the clays reveals several absorption bands (Fig. 2). The presence of bands between 3410 and 3480 cm^{-1}, relative to the OH groups vibrations, could indicate the penetration of water between the clay sheets. The presence of bands between 3410 and 3480 cm^{-1}, relative to the OH groupsvibrations, could indicate the penetration of water between the clay sheets. These bands are characteristic of hydrophilic 2/1 montmorillonite type materials (Hidalgo et al., 1995). The band at 1630 cm^{-1} is attributable to the bending of the H–OH bonds of the structural water molecules. The intense band around 1040 cm^{-1} that characterizes montmorillonite is attributed to the elongation vibrations of Si–O (Zaitan et al., 2008) and the band at 912 cm^{-1} is attributed to the bending vibrations of the Al–Al–OH and Al–Mg–OH groups, confirming the presence of montmorillonite (Caillere et al., 1982). According to Van Olphen and Fripiat (Olphen & Fripiat, 1979), the band around 750 cm^{-1} indicates the presence of illite.

The clays underwent heat treatment from 45 to 850 °C with a constant heating rate of 20 °C/min. The mass used for each sample ranged from 16 to 57 mg. Figure 3 depicts the analysis thermo-gravimetric/difference thermo-gravimetric ATG/DTG curves of clays. The thermogravimetry (TG) curves showed two main endothermic peaks. The first, between 60 and 290 °C, corresponds to the water loss and indicates the evaporation of surface water. This is confirmed by the loss of mass of about 12% in this temperature range. The second peak, between 500 and 800 °C, is due to dehydroxylation of the clay. The high intensity of this second peak indicates the presence of a swelling phase (montmorillonite) interbedded with a non-swelling phase (illite) (Srasra, 1987).

3.2 Geochemical Composition of the Clays

The results of the quantitative chemical analysis of the major elements of the clays indicate that the predominant constituents are SiO_2, Al_2O_3, CaO and MgO, with other oxides being traces of TiO_2 and K_2O. The content of SiO_2 ranges from 34.37 to 63.59%, and the content of Al_2O_3 from 3.75 to 8.17%. The predominant oxides (SiO_2, $A1_2O_3$) reveal the presence of aluminosilicates as montmorillonite, palygorskite and illite. Magnesium (MgO) confirms the presence of magnesian montmorillonite and palygorskite clays. The presence of alkaline K_2O in the chemical composition of the studied samples is very low, and reflects

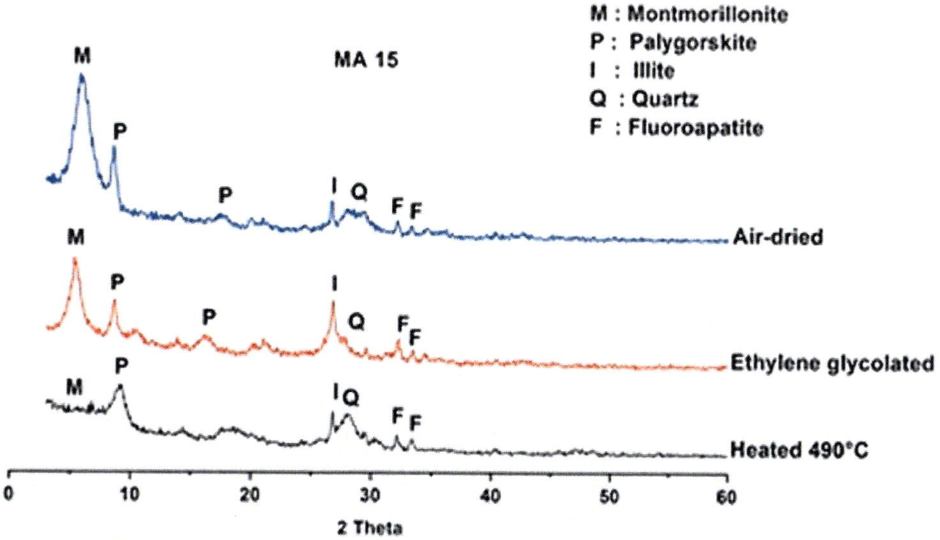

Fig. 1 X-ray diffractogram patterns obtained of the clay from the BouCraa deposit

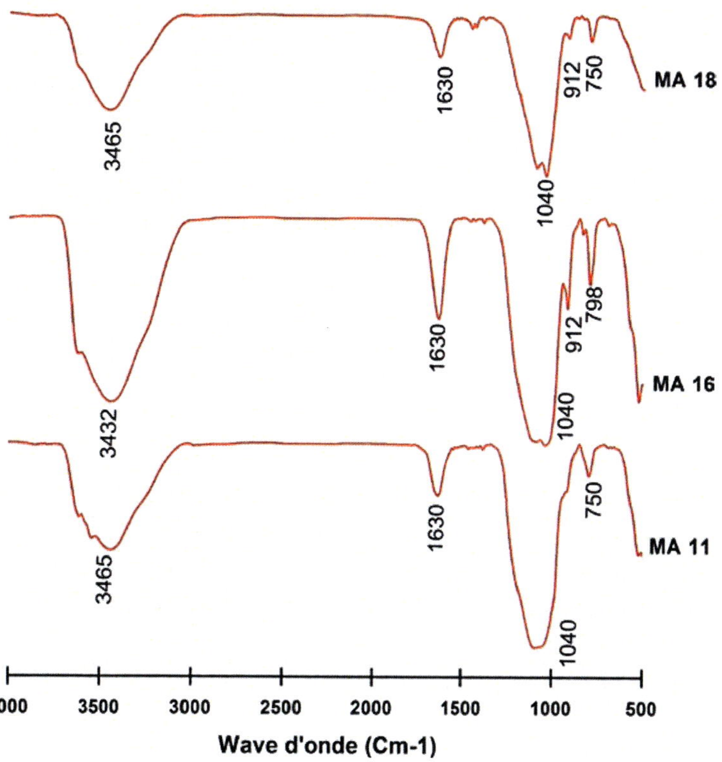

Fig. 2 Infrared spectrum of clay from the BouCraa deposit

the presence of a small proportion of illite, as revealed by XRD. Iron oxide (Fe_2O_3) is the main coloring agent in clay materials, being responsible for the reddish color after firing. According to Table 1, samples MA11 and MA13 have a higher amount of Fe_2O_3, which gives this type of clay a darker color after firing, whereas samples MA17 and MA18 have the lowest amount of Fe_2O_3, allowing a light color.

4 Discussion

The mineralogical and chemical analyses of clays from the BouCraa phosphate deposit show they consists of palygorskite (7–50%), montmorillonite (10–57%) and illite (7–65%). Clay minerals are phyllosilicates derived from weathering, which is controlled by multiple factors

Fig. 3 ATG/DTG curves of clay of the BouCraa deposit

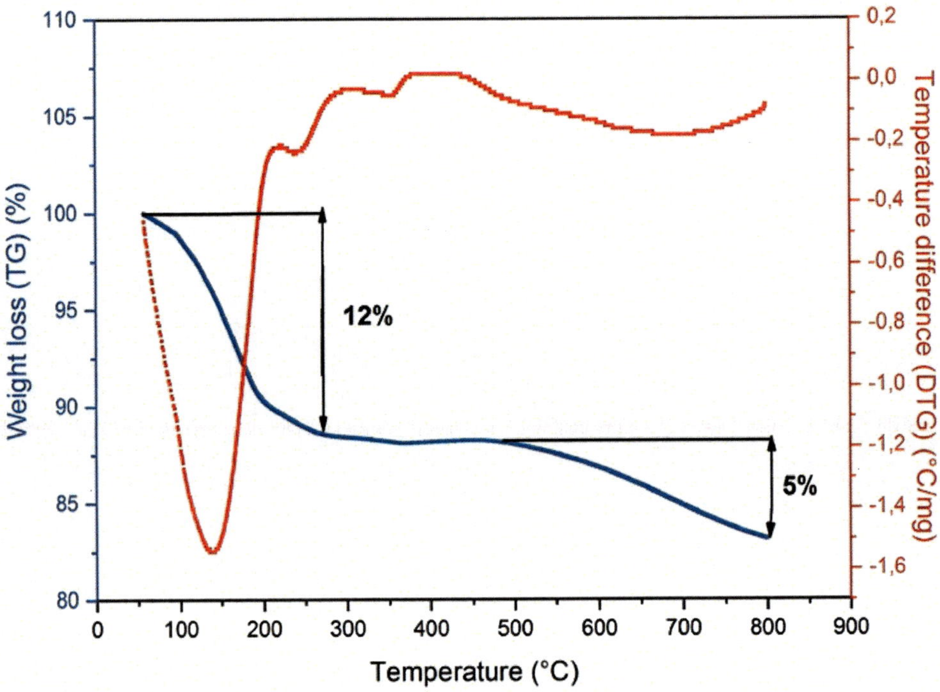

Table 1 Major chemical composition in oxides (%) of the clays of the BouCraa deposit

	SiO_2	Al_2O_3	MgO	CaO	Fe_2O_3	P_2O_5	TiO_2	K_2O	Loss to fire
MA10	34.67	3.75	1.350	28.81	4.11	12.75	0.47	0.76	15.64
MA11	61.25	6.07	1.92	0.88	8.15	1.21	0.92	2.85	16.27
MA13	36.71	8.17	1.84	8.68	7.40	3.84	0.97	1.07	19.27
MA17	63.59	4.97	1.14	1.36	2.65	0.78	0.55	0.83	11.11
MA18	42.16	6.38	1.42	16.89	3.78	8.66	0.6	0.75	19.04

including lithology of origin, paleoclimate and depositional environment (Chamley, 1989). Spatial and temporal distributions of clay minerals are therefore useful indicators of paleoenvironmental and paleoclimatic changes (Franke & Ehrmann, 2010; Liu et al., 2010).

Montmorillonite is one of the most common identified clay minerals. The phosphate series of the BouCraa deposit belongs to the Tarfaya-Laayoune-Dakhla basin. The latter is bordered to the east by the Reguibat Shield which consists of granitoids, migmatites, and metamorphic rocks of Precambrian basement and Paleozoic detrital deposits, and to the northeast by the Anti-Atlas Mountains formed by Precambrian granitoids and rhyolites and Paleozoic carbonates and detrital deposits (Michard et al., 2010). The montmorillonites come from the Western Anti-Atlas Mountains and the Reguibat Shield by the reworking and transport of soils and sediments loaded with silica and basic cations, exposed to a hot and semi-arid climate. Palygorskite represents the fibrous clay minerals associated with montmorillonites. In the BouCraa deposit, palygorskite and montmorillonite have the same temporal evolution. Thus, the two clay minerals were formed at the same time. Illite is the least abundant clay mineral. In this study, clay samples from progressively younger strata showed trends of increasing illite with decreasing montmorillonite. Therefore, illite is thought to have formed by the neoformation of montmorillonite into illite.

5 Conclusion

The results of the physical, mineralogical, chemical and thermal analyses of the clays of the Paleocene-Eocene BouCraa deposit (Oued Eddahab basin) show that the clays consist essentially of montmorillonite, palygorskite and illite as associated clay minerals. These clay compositions were controlled by variable climatic and tectonic conditions. The montmorillonite and palygorskite were formed by the reworking of detrital sediments of the Anti-Atlas Mountains and Reguibat Shield under arid to semi-arid climatic conditions, whereas the illite was formed by the neoformation of montmorillonite.

Acknowledgements The authors would like to acknowledge the support through the R&D Initiative–Appel à projets autour des phosphates APPHOS–sponsored by OCP (OCP Foundation, R&D OCP, Mohammed VI Polytechnic University, National Center of Scientific and technical Research CNRST, Ministry of Higher Education, Scientific Research and Professional Training of Morocco MESRSFC) under the project ID *GEO-MOU-01/2017*.

References

Brindley, G. W., & Brown, G. (1980). *Crystal structures of clay minerals and their x-ray identification*. Mineralogical Society.

Caillere, S., Henin, S., & Rautrureau, M. (1982). *Minéralogie des argiles* (Vol. I and II, 189 p.). Masson, Edition Masson et INRA.

Chamley, H. (1989). *Clay sedimentology* (623 p.). Springer.

Daoudi, L., Rocha, F., Ouajhain, B., Dinis, J. L., Chafiki, D., & Callapez, P. (2008). Palaeoenvironmental significance of clay minerals in Upper Cenomanian-Turonian sediments of the Western High Atlas Basin (Morocco). *Clay Minerals, 432*, 615–630.

Franke, D., & Ehrmann, W. (2010). Neogene clay mineral assemblages in the AND-2A drill core (McMurdo Sound, Antarctica) and their implications for environmental change. *Paleogeography, Paleoclimatology, Paleoecology, 286*, 55–65.

Fürsich, F. T., Singh, I. B., Joachimski, M., Krumm, S., Schlirf, M., & Schlirf, S. (2005). Palaeoclimate reconstructions of the Middle Jurassic of Kachchh (Western India): An integrated approach based on palaeoecological, oxygen isotopic, and clay mineralogical data. *Palaeogeography, Palaeoclimatology, Palaeoecology, 217*, 289–309.

Hidalgo, C., Bertaux, J., & Quantin, P. (1995). Forms of silica in indurated volcanic soils of the Mexico Valley. In G. C. Churchman, R. W. Fitzpatrick, & R. A. Eggleton (Eds.), *Proceedings of the 10th International Clay Conference* (pp. 487–493). CSIRO.

Liu, S., Zhang, W. G., He, Q., Li, D. J., Liu, H., & Yu, L. Z. (2010). Magnetic properties of East China Sea shelf sediments off the Yangtze Estuary: Influence of provenance and particle size. *Geomorphology, 119*, 212–220.

Michard, A., Soulaimani, A., Hoepffner, C., Ouanaimi, H., Baidder, L., Rjimati, E. C., & Saddiqi, O. (2010). The South-Western Branch of the Variscan Belt: Evidence from Morocco. *Tectonophysics, 492*, 1–2.

Moore, D. M., & Reynolds, R. C. (1989). *X-ray diffraction and the identification and analysis of clay minerals* (322 p.). Oxford University Press.

Raucsik, B., & Varga, A. (2008). Climato-environmental controls on clay mineralogy of the Hettangian-Bajocian successions of the Mecsek Mountains, Hungary: An evidence for extreme continental weathering during the early Toarcian oceanic anoxic event. *Palaeogeography, Palaeoclimatology, Palaeoecology, 265*, 1–13.

Srasra, E. (1987). Caractérisation minéralogique, propriété physicochimique et application des argiles du gisement Haidoudi. These de 3 eme cycle, Tunis. Univ. Tunis II, 166 p.

Van Olphen, H., & Fripiat, J. J. (1979). *Data handbook for clay materials and other non-metallic minerals*. Pergamon Press.

Zaitan, H., Bianchi, D., Achakb, O., & Chafikb, T. (2008). Comparative study of the adsorption and desorption of o-xylene onto bentonite clay and alumina. *Journal of Hazardous Materials, 153*, 852–859.

Sea-Level Variations, the Cause of Economic Phosphate During the Maastrichtian-Yprisan, BouCraa Deposit (Morocco)

Mohamed Amine Nguidi, Mustapha Mouflih, Salem El Ouariti, Hanane El Boukhari, Abdelmajid Benbouziane, Mohamed Aquit and Oussama Khadiri Yazami

Abstract

BouCraa deposit phosphorites accumulated in the central onshore part of the Tarfaya-Laayoune-Dakhla Basin in southern Morocco during the Upper Cretaceous-Paleogene phosphogenic event, producing the southernmost and westernmost deposits of the Atlantic-Tethyan Phosphorite Provinces. These deposits are interpreted as transgressive produced on an epeiric wave-dominated platform along the Atlantic coast. Lithofacies stacking patterns and stratigraphic surfaces record phosphogenesis through at least three sea-level cycles. The sequential division was established based on hardground and ravinement surfaces, sometimes containing Thassinoides bioturbations. The abrupt change of facies indicates lithological discontinuities. The granular phosphorites above a ravinement surface are interpreted as a transgressive surface and sequence boundary. The phosphorites were deposited during the transgressive systems tract. The maximum flooding surface/zone is shown by a condensed interval marked by a dense cm-scale coprolite and/or strong facies contrasts, pronounced deepening, and show burrowed levels associated with high authigenic mineral concentrations. The petrography, mineralogy and the presence of sulfur-oxidizing bacteria strongly suggest a reworked origin derived from a primary phosphate. The role that sea-level rise played in expanding the volume of accommodation on the platform has favored phosphogenesis by trapping terrigenous deposits in littoral environments. This relationship between relative sea-level rise, stratigraphic condensation, and the delivery of upwelled phosphorus is interpreted to have stimulated phosphogenesis. The winnowing and reworking of the phosphate mud were triggered during sea-level falls. Regressions exposed phosphorites to wave action; they winnowed and reworked them in situ. Marine transgressions allowed wave-induced currents to develop along the flooded margin, transporting winnowed sediments and concentrating them into economic phosphorites on ravinement surfaces. In the BouCraa deposit, storm currents have been a key agent in the enrichment and maturity of phosphorus sediments through cannibalism and dynamic sorting during transgressions.

Keywords

Cretaceous-Paleogene · BouCraa · Storm · Economic phosphorites · Reworking

1 Introduction

Sedimentary phosphate is a strategic resource in the Kingdom of Morocco with more than 70% of the worldwide phosphate reserves in sedimentary phosphorites, a marine biochemical sedimentary rock containing 18 wt% P_2O_5, the most important source of phosphorus (P). The main episodes of phosphorite accumulation are linked to variations in the biogeochemical cycle of P over time, which in turn are linked to changes in paleogeography, palaeoceanography and biological evolution (Pufahl & Groat, 2017). Phosphate sedimentary deposits are of primary or reworked origin, and sea-level fluctuations control this subdivision (Baioumy & Tada, 2005; Föllmi, 1996; Pufahl &

M. A. Nguidi (✉) · M. Mouflih · S. El Ouariti · H. El Boukhari · A. Benbouziane
Dynamics of Sedimentary Basins and Geological Correlations Laboratory, Faculty of Sciences Ben M'Sick, Hassan II University of Casablanca, Casablanca, Morocco
e-mail: Nguidiamine@gmail.com

M. Aquit · O. K. Yazami
Strategic Development Department, Industrial Development Department, OCP s.a., Casablanca, Morocco

F. Lucci et al. (eds.), *Selected Studies in Geomorphology, Sedimentology, and Geochemistry*,
Advances in Science, Technology & Innovation, https://doi.org/10.1007/978-3-031-43744-1_25

Groat, 2017). In Morocco, this sedimentation is very abundant and considerable in quantity and quality in its various basins. The Tarfaya-Laayoune-Dakhla Basin, also known as the Laayoune Basin, is a coastal basin in southern Morocco oriented NE–SW between the Atlantic Ocean and the Precambrian-Paleozoic Reguibat Shield. This basin contains a phosphatic sub-basin of Oued Eddahab, where the BouCraa deposit is located. Unlike the phosphate deposits in northern Morocco, the Oued Eddahab phosphate basin is poorly documented, especially when it comes to modern geoscientific views. Previous work has focused on the exploration of these phosphate deposits and mineral prospecting, which is now of historical interest only.

2 Materials and Methods

Stratigraphic sections of the BouCraa succession were described bed-by-bed in nine outcrops from mine sites. Systematic sampling included phosphate levels and associated facies (Fig. 1). Thin sections were made from over 70 representative samples. All the microfacies of the related phosphorites and rocks were observed under the petrographic microscope to examine their composition, texture, and grain size. Twenty representative samples were examined with a scanning electron microscope (SEM) HIROX SH-4000M for the texture, surface, architecture and internal fabric of the different types of phosphate grains. The X-ray powder diffraction (XRD) measurements of the different samples have been performed on an X'PERT Pro PANalytical diffractometer with CuKa radiation (1.5418 Å).

3 Results

3.1 Sedimentology and Sequence Stratigraphy

In the BouCraa deposit, sedimentological logs show a succession of Maastrichtian-Ypresian layers and comprise two phosphate-rich units, one of which is designated as L1, the economically most important interval. By contrast to other phosphate basins of Morocco, this sedimentary series record an exceptional rarity in carbonate facies that usually separates the loose phosphate beds (Mouflih, 2015a), while it has a remarkable silica richness with many silicified facies. The sequential division was established based on physical surfaces marked by hardened surfaces, and ravinements surfaces sometimes containing Thassinoides bioturbations. The abrupt change of facies indicates lithological discontinuities (Fig. 2).

The section of BouCraa starts with a falling stage systems tract of Maastrichtian age. It contains mainly quartz sand, clayey quartz sand and phosphatic sand, with cross-stratification structures.

SD1 is of Lower Paleocene (Danian) age, limited by two gully surfaces as a sequence boundary and transgressive surface. This sequence starts with phosphate layer L2A and L2B. The first is 2 m thick greenish-gray coprolitic fine phosphate and represents the transgressive system tract (TST), while the second is 70 cm thick silicified coprolitic fine phosphate that is bioturbated at the base. These phosphate levels are intercalated by a 1 m black silex level and a complex of chert and greenish clays and represent the

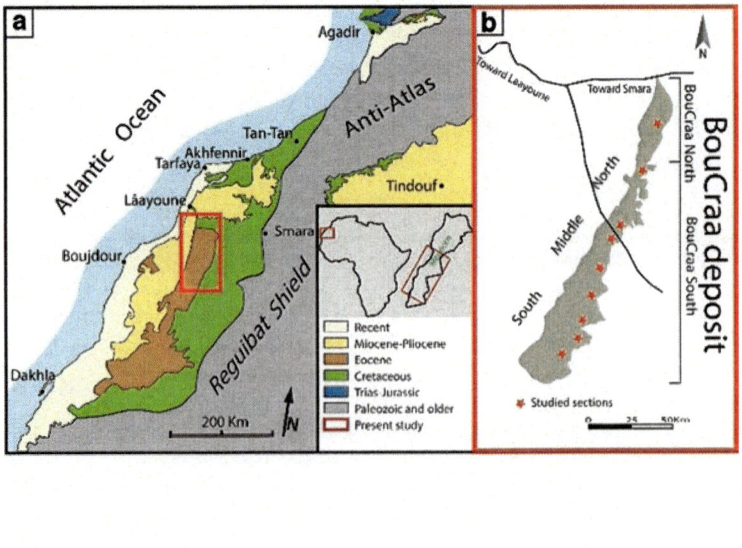

Fig. 1 Localities of the studied area: **a** simplified geological map of the TLDB with the locality of studied areas. **b** Map of the Oued Eddahab phosphate basin and the localities of measured sections

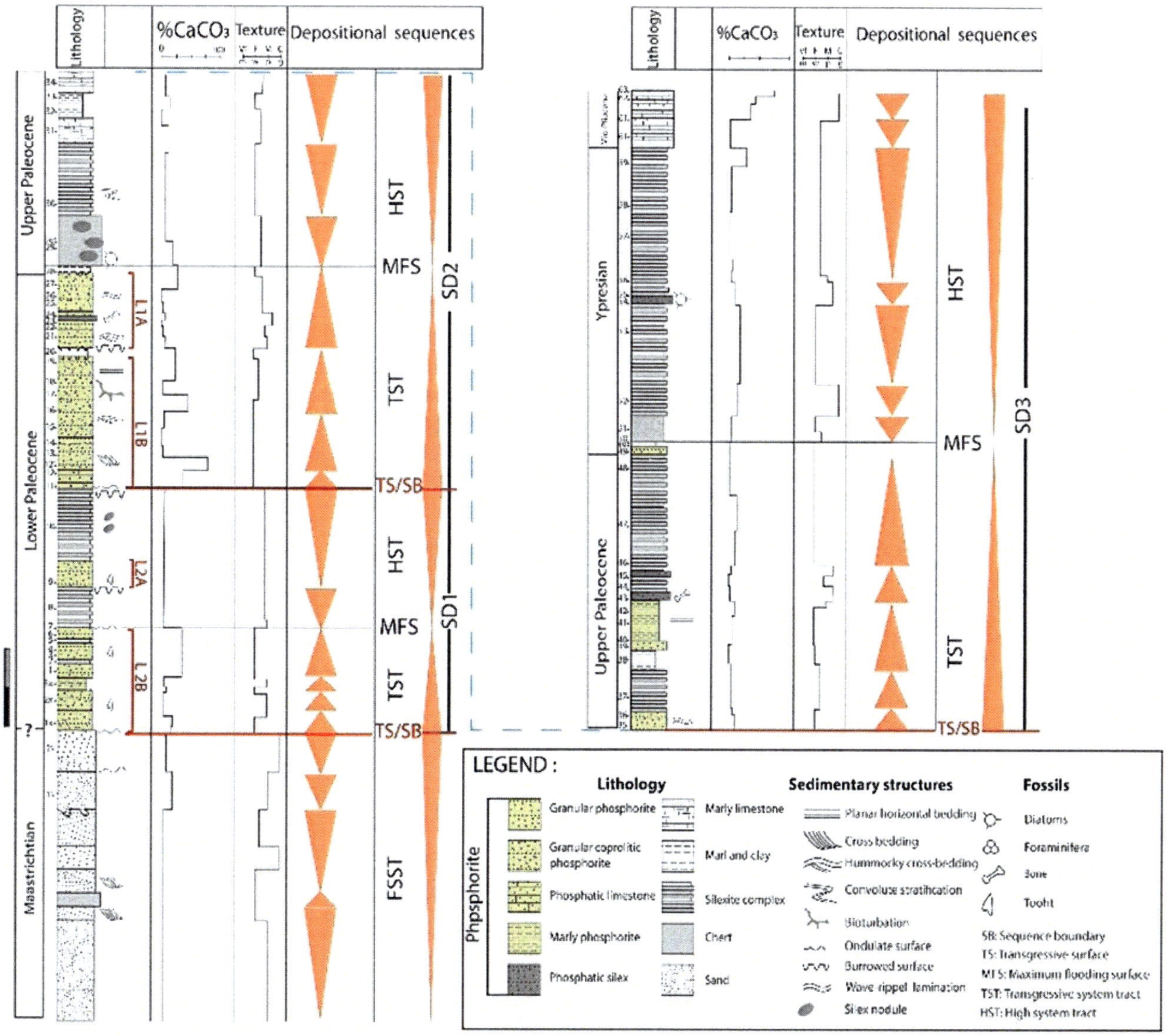

Fig. 2 Representative stratigraphic section of the BouCraa deposit showing the hierarchical arrangement of the different types of depositional cycles and sequences

high system tract (HST). Flooding surfaces in the transgressive system tract are unusually prominent and show strong facies contrasts, pronounced deepening and show burrowed levels associated with high authigenic mineral concentrations.

SD2, the TST, represents the wealthiest phosphate occurrence of the Danian age. It starts with L1 (L1B and L1A), of 4.5 m thick at the ravinement surface, and is interpreted as a transgressive surface (TS). L1B is composed of several levels of coprolitic granular phosphate, and presents tempestite structures, with cross-bedding and hummocky cross-bedding. This layer has intense bioturbation attributed to Ophiomorpha. The L1B and L1A levels are separated by 20 cm of bioturbated marls and clays at the base and roof. L1A, formed of granular, fining-up, wavy laminated, bioclast-rich, marly phosphate, displays sporadic Thalassinoides burrows filled with silt-sized phosphatic material. This level is characterized by the presence of scour marks and small channels, often filled with abraded coarse grains, shell debris, and coprolites. The MFZ is represented by a blackish silex rich in diatom. The HST comprises an alternation of 1.5 m of siliceous clays diatom rich, brown silex dm-scale nodules at the base followed by 2 m of whitish siliceous marls and silex. This intercalation is covered by two banks of marly limestone of 50 cm each, separated by a 50 cm thick level of yellowish and greenish marls and clays.

SD3, the HST is formed by an alternation of 1 m medium grayish phosphatic marls, greenish clayey marl, then a chert complex, and chalky facies evolving toward another clayey one. The HST is a siliceous complex

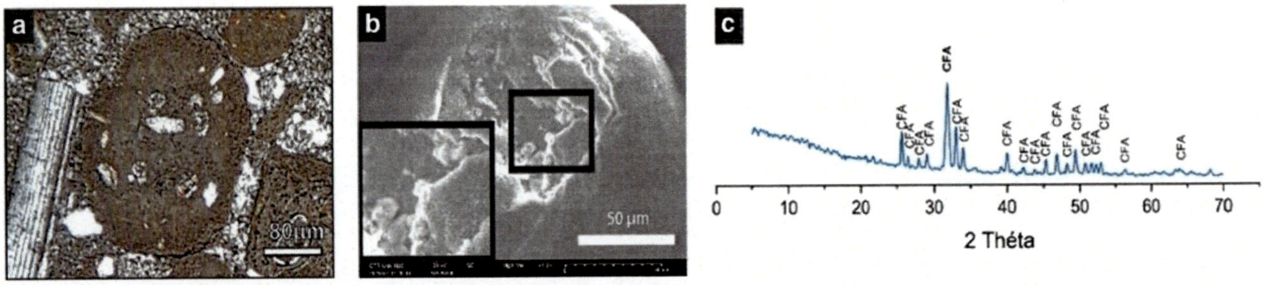

Fig. 3 **a** Well-rounded phosphatic mudclast containing detrital quartz grains. **b** Spherical phosphatic grain hosting spherical phosphatized bacteria. **c** XRD pattern of phosphorite samples showing the main mineralogical compositions of granular phosphorites

comprising alternating chert, clays and silex banks of the Ypresian age. The MFS is a condensed interval marked by a dense cm-scale coprolite.

3.2 Petrography and Mineralogy

Granular phosphorites are formed by phosphatic mudclasts (Fig. 3a). These are particles whose shapes and compositions suggest reworking, with detrital quartz, bone fragments and rounded to well-rounded shapes; their size is variable and often exceeds 200 µm. They also contain various microorganisms. The angular to well-rounded shape of the clasts indicate the degree of transport. Phosphatic mudclasts are generally associated with framboidal pyrite. Some grains have cavities showing bud structures (Fig. 3b); these cavities host phosphatized bacteria in the form of microspheres. The morphology of biomorphs of chemotrophic sulfuric bacteria observed in the pores and cavities of phosphate grains have sizes of less than 2 µm.

XRD analyses reveal that phosphorites consist mainly of carbonate fluorapatite (CFA). The highest peak intensity recognizes CFA at 2.784 Å spacing (Fig. 3c). Diffractograms revealed a very low crystallization state. Despite its diversity of appearance (bone, teeth, coprolites, phosphatic grains), apatite offers relatively constant diffractograms of common CFA with a carbonate content varying between 4.72 and 2%.

4 Discussion

Granular phosphorite formed mainly by phosphatic mudclasts with a cryptocrystalline apatite strongly suggests a reworked origin, derived from a primary phosphate. The presence of sulfur-oxidizing bacteria similar to those described in the Ganntour Basin (Mouflih, 1991a) and Oulad Abdoun Basin (Mouflih, 2015b). Framboidal pyrite which is usually related to the vital activity of sulfur-reducing bacteria like those of a Paleoproterozoic tidal flat in the 1.85 Ga Michigamme Formation (Mouflih, 1991b) suggests that this phosphate comes from an offshore reducing environment with high productivity. Glenn and Arthur (1990) (Cosmidis et al., 2013) proposed rivers as a source of P and Si because of the abundance of glauconite and pyrite in the phosphorites of southern Egypt, particularly in the Abu Tartur deposits. The absence of glauconite in the BouCraa phosphates shows that a river source for P and Si is excluded. Field observations revealed the deposition of reworked granular phosphorite in shallow marine settings. In addition to the role that sea-level rise played in expanding accommodation space on the shelves, it also promotes phosphogenesis by trapping terrigenous deposits in littoral environments, decreasing sediment accumulation rates on the shelves, promoting high productivity, intensification of the oxygen minimum zone and the consequent development of reducing environments. This relationship between relative sea-level rise and stratigraphic condensation also controlled the phosphogenesis (Hiatt et al., 2015).

Forced regression exposed the authigenic phosphorites to the influence of wave action which winnowed and reworked them. Storm-induced geostrophic fluxes are also crucial for the formation of economic phosphorite deposits and are probably responsible for the absence of pristine levels in the BouCraa series. The phosphorites in all sequences were deposited during the TST like, for example, the Eshidiyya phosphorite platform in Jordan (Glenn & Arthur, 1990). The formation and emplacement of the BouCraa phosphorites in their present arrangement can be described by three main events:

During HST, nutrient intake promotes high primary productivity. Upwelling currents provide P and Si from the depths. Higher bioproductivity, combined with increased mortality, resulted in high incorporation of P and Si into the sediments and led to the authigenic formation of phosphorites and cherts.
During LST, the authigenic phosphorites were above the storm wave base and the fair-weather wave base, so they were subject to erosion, reworking and remobilization by wave winnowing.
During TST, reworked phosphorites from the LST were transported, concentrated and scattered on the mainland by subsequent transgression. During this interval, the phosphorites matured in a shallow environment under storm actions over large areas and evolved to a high-grade ore of economic interest.

5 Conclusion

The Moroccan Saharan sedimentary phosphorites are located in the Oued Eddahab sub-basin, but the only phosphorite deposit that is currently being mined in this sub-basin is the BouCraa deposit. Like these northern Moroccan analogs, the Oued Eddahab phosphate series extends from the Upper Maastrichtian to the Lower Eocene and is divided into three sequences. The phosphorites are of reworked origin from a deeper, more anoxic, and high-productivity environment, and were deposited during the TST in a shallower, high-energy environment where the phosphorites matured under the effect of storms over large areas and evolved into a high-grade ore of economic interest.

Acknowledgements The authors would like to acknowledge the support of this study by the R&D Initiative–Appel à Projets autour des Phosphates APPHOS (sponsored by OCP, OCP Foundation, R&D OCP, Mohammed VI Polytechnic University, National Center of Scientific and Technical Research CNRST, Ministry of Higher Education, Scientific Research and Professional Training of Morocco MESRSFC) under the project ID: GEO-MOU-01/2017.

References

Baioumy, H., & Tada, R. (2005). Origin of Late Cretaceous phosphorites in Egypt. *Cretaceous Research, 26*, 261–275.

Cosmidis, J., Benzerara, K., Gheerbrant, E., Estève, I., Bouya, B., & Amaghzaz, M. (2013). Nanometer-scale characterization of exceptionally preserved bacterial fossils in Paleocene phosphorites from Ouled Abdoun (Morocco). *Geobiology, 11*, 139–153.

Föllmi, K. B. (1996). The phosphorus cycle, phosphogenesis and marine phosphate-rich deposits. *Earth-Science Reviews, 40*, 55–124.

Glenn, C. R., & Arthur, M. A. (1990). Anatomy and origin of a Cretaceous phosphorite-greensand giant, Egypt. *Sedimentology, 37*, 123–154.

Hiatt, E. E., Pufahl, P. K., & Edwards, C. T. (2015). Sedimentary phosphate and associated fossil bacteria in a Paleoproterozoic tidal flat in the 1.85 Ga Michigamme Formation, Michigan, USA. *Sedimentary Geology, 319*, 24–39.

Mouflih, M. (1991a). The phosphate series of the Benguerir deposit, facies sequences, sequence evolution, mineralogy, and geochemistry (Maastrichtian-Lutetian) (246 p.). Ph.D. Thesis, Cadi Ayyad University, Marrakesh, Morocco.

Mouflih, M. (1991b). La série phosphatée du gisement de Benguerir séquences de faciès, évolution séquentielle, minéralogie et géochimie (Maastrichtien-lutétien) (246 p.). Ph.D. Thesis, Université Cadi Ayyad, Marrakech, Maroc.

Mouflih, M. (2015a). Phosphates of central Morocco and Middle Atlas of Maastrichtian-Lutetian age: Sedimentology, sequence stratigraphy, genetic context, and beneficiation (340 p.). Ph.D. Thesis, Cadi Ayyad University, Marrakesh, Morocco.

Mouflih, M. (2015b). Les phosphates du Maroc Central et du Moyen-Atlas d'âge Maastrichtien-Lutétien: sédimentologie, stratigraphie séquentielle, contexte génétique et valorization (340 p.). Ph.D. Thesis, University Cadi Ayyad, Marrakech.

Pufahl, P. K., & Groat, L. A. (2017). Sedimentary and igneous phosphate deposits: Formation and exploration: An invited paper. *Economic Geology, 112*, 483–516.

Sedimentology and Palaeoenvironments of the Triassic Deposits of the Imini Basin (High Atlas of Marrakesh, Morocco)

Soukaina Obad, Rachid Essamoud and Abdelkrim Afenzar

Abstract

In this analytical work, we presented the facies and palaeoenvironments of the Triassic basin of Imini, in the southern sub-Atlasic area of the High Atlas of Marrakesh (80 km north of Ouarzazate). This basin was formed during the rifting phase that reactivated preexisting structures of the basement. The formations date to the Middle to Late Triassic. These are generally detrital and evaporitic sediments, capped by basaltic flows dating to the Late Triassic–Early Lias. The outcropping deposits are in discordance to Cambro–Ordovician schists of the basement. Our detailed sedimentological analysis identified facies associations with seven architectural elements (EA1 to EA7): CH (channels), GB (gravel bars and bedforms), SB (sandy bedforms), SG (sediment gravity flows), LS (laminated sand sheet), FF (overbank fines), and evaporites (Ev). Palaeoenvironments evolve from a proximal environment characterized by alluvial cones, to a braided fluvial environment, then to a distal fluvial environment with alluvial plains and playa lakes.

Keywords

Sedimentology · Triassic · Facies · Architectural elements · Palaeoenvironments

1 Introduction

The succession of Triassic deposits in the High Atlas of Marrakesh (Morocco) provides excellent opportunities to collect sedimentological and stratigraphic data. Within the scope of reconstructing depositional environments, several detailed sections have been recorded along with the identification of facies and architectural elements of the respective formations. The Imini basin is located about 80 km north of Ouarzazate city (road N9 Marrakesh–Ouarzazate). It belongs to the southern sub-Atlasic zone of the High Atlas of Marrakesh, bounded by the region of Tizi n'Tichka to the north, the Khela plateaus to the east, the first reliefs of Siroua to the south, and the ancient massif to the west (Fig. 1). It is a rift-type basin, the opening of which was initiated by the reactivation of preexisting structures in the basement. This basin is filled with red beds of Triassic age, unconformably resting on the Neoproterozoic Ouarzazate Group. Triassic deposits are exposed within a monoclinal structure.

2 Methodology

The study is based on a detailed sedimentological analysis of selected sections in different areas, allowing the identification of a wide range of sedimentary facies and characterization of their respective associations within architectural elements (Miall, 1985, 2006). Lithological characteristics, granulometry, texture, and sedimentary structures were observed in outcrops. In this study, the nomenclature of facies follows the scheme set by Miall (1977, 1978, 2006). The vertical and horizontal patterns of these facies help to characterize the architectural elements. These include the nature of lower and upper boundary surfaces, external geometry, scale, and internal geometry (Allen, 1983; Miall, 1985, 2006). The identification of palaeoenvironments considers the nature and types of lithofacies and their architectural elements.

S. Obad (✉) · R. Essamoud · A. Afenzar
Dynamics of Sedimentary Basins and Geological Correlations Laboratory Faculty of Sciences Ben M'Sick, Hassan II University of Casablanca, Casablanca, Morocco
e-mail: soukaina.obad@gmail.com

F. Lucci et al. (eds.), *Selected Studies in Geomorphology, Sedimentology, and Geochemistry*,
Advances in Science, Technology & Innovation, https://doi.org/10.1007/978-3-031-43744-1_26

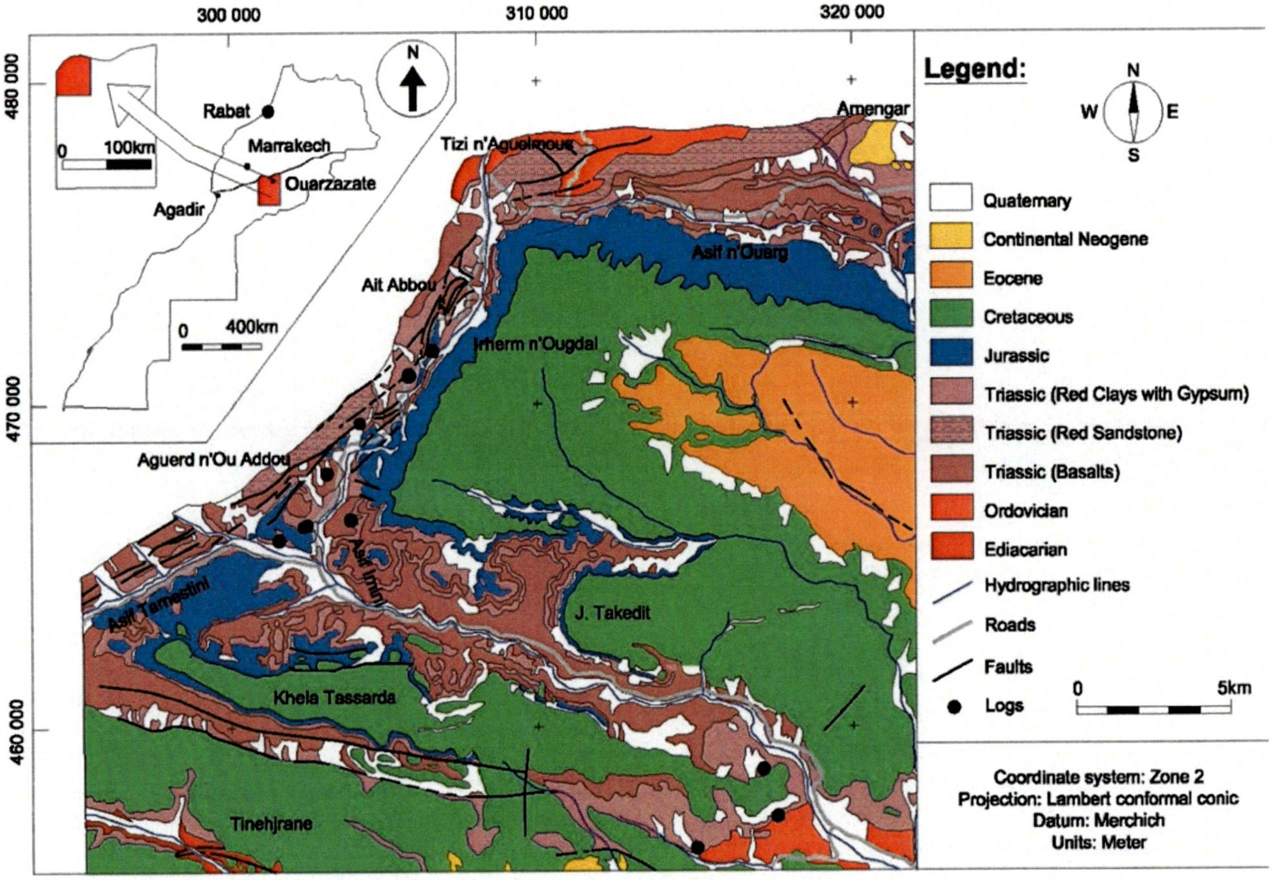

Fig. 1 Geographical and geological positions of the Imini basin with the respective studied outcrops (extract from the geological map of Ouarzazate 1/200.000)

3 Results

3.1 Facies Analysis

Conglomerates: In most studied outcrops, the conglomerates are in the basal part of the series (Timalizene Formation). Their color is reddish brown to purplish pink, the elements are sometimes angular, sub-angular, rounded or sub-rounded, and cm to dm scale, and the matrix consists of very fine sands and clays. These conglomerates are sometimes deposited without any stratification (facies Gmm: plastic debris flows; Gcm: pseudoplastic debris flows), and in other cases, the stratification is clearly visible with horizontal and imbricated bedding of pebbles (facies Gh), with planar cross beds (facies Gt), and/or oblique planes (facies Gp).

Sandstones: The sandstones are reddish brown to grayish green, outcropping in almost all the formations and sometimes interbedded with conglomerates and micro-conglomerates as well as siltstones and claystones. The sandstone bed forms range from cm scale to m scale (up to 18 m) and contain: horizontal (facies: Sh) and/or trough cross beds (facies: St), oblique planar beds (facies: Sp), massive (facies: Sm), different types of ripples (symmetrical, crescent, and climbing) (facies: Sr) as well as bioturbations. These sandstones are characterized by mottling at places (Table 1).

Siltstones: The siltstones are reddish in color, laminated to massive and around 40 m thick. These siltstones (facies: Fl) alternate with either sandstones or cm-scale evaporites.

Mudstones: The studied reddish mudstones in the basin are the predominant lithofacies within the Triassic series. These present marmorization spots and alternate with either sandstones or evaporites. This facies corresponds to facies Fm and can be interpreted as an overbank deposit in a floodplain or a deposit of abandoned channels (Miall, 1985; Einsele, 2000; EjiUba et al., 2005).

Gypsum: In the central and upper parts of these Triassic series, the gypsum facies is exposed and represented as horizontal, sub-vertical to sloping fibrous veins. This facies can originate and be preserved in relatively hot and humid environments.

Table 1 Identified facies in the study area (Miall, 1978, 2006, modified)

Facies code	Lithofacies	Sedimentary structures	Interpretation
Gmm	Matrix-supported, massive gravel	Weak grading	Plastic debris flow
Gcm	Clast-supported, massive gravel	–	Pseudoplastic debris flow
Gh	Clast-supported crudely bedded gravel	Horizontal bedding, imbrication	Longitudinal bars or sieve deposits
Gt	Gravel, stratified	Trough cross beds	Minor channel fills
Gp	Gravel, stratified	Planar cross beds	Linguoid bars
Sm	Coarse sandstone	Massive	Gravity deposits
St	Sand, medium to very coarse	Trough cross beds	Sinuous-crested and linguoid
Sp	Sand, medium to very coarse	Planar cross beds	Linguoid, transverse bars
Sr	Sand, very fine to coarse	Ripple cross-laminations	Ripples
Sh	Sand, very fine to very coarse	Horizontal lamination	Plane-bed flow
Fl	Siltstone	Fine lamination	Overbank or waning flood deposits
Fm	Mudstone	Laminated to massive	Overbank or drape deposits
Gypsum	Satin spar	Fibrous gypsum	Lake or lagoon deposits

4 Architectural Elements

In the study area, we characterized seven architectural elements (AE) (Fig. 2) referenced according to Miall (1985, 2006):

AE1: This element is formed by the assemblage of sandy facies St, Sp, and Sh. It is characterized by fourth-order base surfaces (erosion surface). These boundaries are slightly plane and sometimes upwardly concave with respect to erosion. AE1 may correspond to architectural element CH deposited as multiple channels.

AE2: Identified at the base and top of the studied series, this is formed by the association of interlocking and stratified coarse facies with oblique bedding planes Gh and Gp. Other minor facies such as Sh are also identified. This element corresponds to architectural element GB (gravel bar and bedform), interpreted as coarse deposits.

AE3: This is an association of the medium to fine sandstone facies St, Sp, and Sh. This element is characterized by first-, second- and third-order internal boundary surfaces and fourth-order external boundary surfaces. It corresponds to architectural element SB (sand bedform), deposited in crevice-channel and/or crevice-splay deposits.

AE4: This element lies at the base of the studied series and is formed by facies Gmm and Gcm. It corresponds to architectural element SG (sediment gravity flow) and is formed by gravity flow deposits, mainly pebbles and poorly sorted gravels, formed in the proximal regions of alluvial fans.

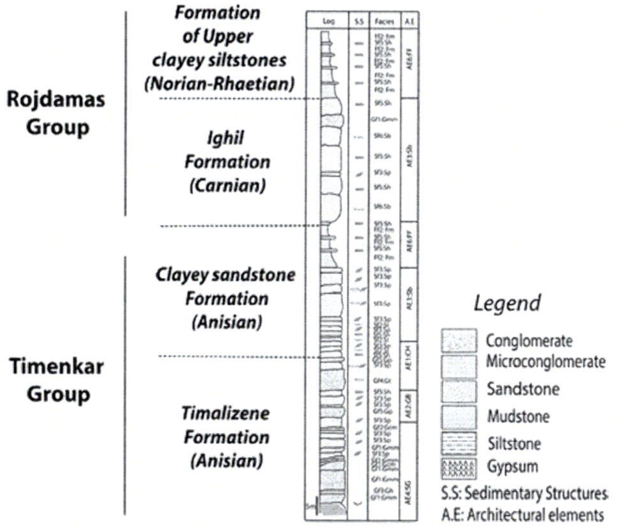

Fig. 2 Synthetic log of the respective formations observed in the Imini basin

AE5: It is formed by the assemblage of fine to very fine facies: Sh, Fl, and Fm with a thickness reaching 5 m. It corresponds to architectural element LS (laminated sand sheets). Deposits of this element are interpreted as the byproduct of flash floods (Miall, 1977, 1984; Rust, 1978; Tunbridge, 1981, 1984; Sneh, 1983).

AE6: This is an association of facies Fl and Fm, sometimes alternating with Sh. On the basis of their structure and architectural properties, it corresponds to architectural element FF which is interpreted as overbank fine deposits.

AE7: This architectural element is formed by the alternation of facies Fl, Fm, Sh, and sometimes gypsum. The alternation of sandstone and silt–clay facies (Ev) is interpreted as playa lake deposits.

5 Discussion

The detrital deposits of the Triassic Imini basin were deposited in a diverse river system. The first deposits in the succession are massive conglomerates formed by different alluvial cone structures, as architectural elements AE2 and AE4. These are similar to those of proximal alluvial cones with gravity flow lobes by Miall (1985). This environment then turned into a braided fluvial system characterized by coarse gravel deposit bars (AE2), and sometimes some fine deposits left in abandoned channels. This model corresponds to the braided fluvial model with gravel bars of Miall (1985, 2006). The fine and very fine deposits (siltstone and mudstone textures) were probably formed in a river system characterized by fine overbank deposits. The formation of these fine deposits is probably due to a retrograding clastic system. Finally, the river system was connected to a playa system characterized by evaporates and fine deposits.

6 Conclusion

The Triassic deposits of the Imini basin were deposited in a river system that evolved from a proximal system of alluvial cones marked by debris flow deposits to a braided river system rich in gravel bars and bedforms. These systems finally changed to an alluvial plain favorable for the accumulation of overbank fines associated with playa lakes. These evaporates were formed by the evaporation of saline water under a humid and arid climate in this basin.

References

Allen, J.R.L. (1983). Studies in fluviatile sedimentation: Implication of pedogenic carbonate units, lower old red sandstone Anglo-Welsh outcrop. *Geological Journal, 9*, 181–208.

Einsele, G. (2000). Sedimentary Basins: Evolution, facies, and sediment budget (p. 792). Springer.

EjiUba, C., Heubeck, C., & Hulka, C. (2005). Facies Analysis and basin architecture of the neogene sub and eanSynorogenic wedge, Southern Bolivia. *Sedimentary Geology, 180*, 91–123.

Miall, A.D. (1978). Lithofacies types and vertical profile models in braided river deposits: A summary. In: Miall, A.D. (Ed.) Fluvial sedimentology. Can Soc Petrol Geol Mem (vol. 5, pp. 597–604).

Miall, A.D. (1985). Architectural-element analysis: A new method of facies analysis applied to fluvial deposits. Earth-SciRev 22, 261–308.

Miall, A.D. (2006). The geology of fluvial deposits, sedimentary facies, Basin analysis and petroleum geology. Springer

Miall, A. D. (1977). A review of the braided river depositional environment. *Earth Science Reviews, 13*, 1–62.

Miall, A. D. (1984). Variations in fluvial style in the lower cenozoic synorogenic sediments of the Canadian Arctic Islands. *Sedimentary Geology, 138*, 499–523.

Rust, B.R. (1978). Depositional models for braided alluvium. In: Miall, A.D. (Ed.), Fluvial sedimentology. Can Soc Petrol Geol Mem (vol. 5, pp. 605–625).

Sneh, A. (1983). Desert stream sequences in the Sinai Peninsula. *Journal of Sedimentary Petrology, 53*, 1271–1280.

Tunbridge, I.P. (1981). Sandy high-energy flood sedimentation some criteria for recognition, with an example from the Devonian of S.W. England. Sediment Geology 28, 79–96.

Tunbridge, I.P. (1984). Facies model for a sandy ephemeral stream and clay playa complex; the middle Devonian Trentishoe formation of North Devon, U.K. Sedimentology 31, 697–715.

Sedimentological Characterization of Clastic Sediments of Anambra Basin: Implications for Provenance and Palaeoenvironments Sedimentological

Jimoh Ajadi, Yusuf Ayoola Jimoh and Adewole Adetoro Ajala

Abstract

Grain size and heavy mineral studies carried out on eleven samples from sandstone, siltstone and shale beds that are exposed along road cuts and river channels were used to decipher palaeodepositional conditions, provenance and tectonic setting within the Anambra Basin, south-eastern Nigeria. The study revealed three lithofacies belonging to the Lokoja-Basange and Mamu Formations. The sandstones mostly exhibit fining upward sequences with poorly sorted angular to subangular grains that have poor compositional maturity. These are characteristic of short transportation history and fluvial depositional environments. The siltstones and shale display a coarsening upward progradational pattern that is suggestive of deltaic sedimentation. Heavy mineral assemblages observed in the sandstone samples include zircon, rutile, magnetite, topaz and tourmaline which suggest igneous and metamorphic provenance derived from neighbouring rocks of the Precambrian Basement Complex of Nigeria.

Keywords

Anambra Basin · Lithofacies · Provenance · Palaeoenvironment

J. Ajadi (✉) · Y. A. Jimoh
Department of Geology and Mineral Science, Kwara State University, Malete, P.M.B. 1530, Ilorin, Nigeria
e-mail: jimoh.ajadi@kwasu.edu.ng

A. A. Ajala
Research Division, Stalwart Ventures LLC, Maryland, USA

1 Introduction

The Anambra Basin (Fig. 1) is one of the inland sedimentary basins in Nigeria known for its economic potential due to the presence of hydrocarbon and other mineral deposits. The westward displacement of depocenters in the Lower Benue Trough (southern part of the trough) led to the deposition of cyclic succession of coals, shales, siltstones and sandstones in the Cretaceous Anambra Basin. Recent studies in Anambra basin (Adeigbe & Salufu, 2009; Obi, 2000; Odunze & Obi, 2013) have led to the reconstruction of the stratigraphic framework within the basin. The present study covers Igarra, Auchi, Sabongida Ora, Uzebba and Ifon areas, showing good exposures of Cretaceous sediments along channels and road cuts. This study aimed at using sedimentological tools to provide geological information on provenance, transportation history and palaeoenvironments of the detrital sediments at different locations.

2 Materials and Methods

The methodology involves field study and laboratory analysis. Field studies were carried out on road cut exposures. Six lithostratigraphic sections which cover Lokoja-Basange Formation and Mamu Formation in the Anambra basin were logged and sampled (Fig. 2) with specific attention paid to differences in lithology, texture, colour, grain size, shape, sorting, bedding, thickness, weathering characteristics and structures. Samples were taken from a few cm below the surface to ensure sampling of in situ materials. A total of eleven rock samples were taken for grain size and heavy mineral analysis using different sieve sizes. In this method, 100 g of each sample was put in the upper sieve, the machine was allowed to shake for ten (10) minutes, and the sediments retained in each of the sieves and the bottom pan was weighed and recorded. Statistical parameters of graphic mean (Mz), graphic standard

F. Lucci et al. (eds.), *Selected Studies in Geomorphology, Sedimentology, and Geochemistry*,
Advances in Science, Technology & Innovation, https://doi.org/10.1007/978-3-031-43744-1_27

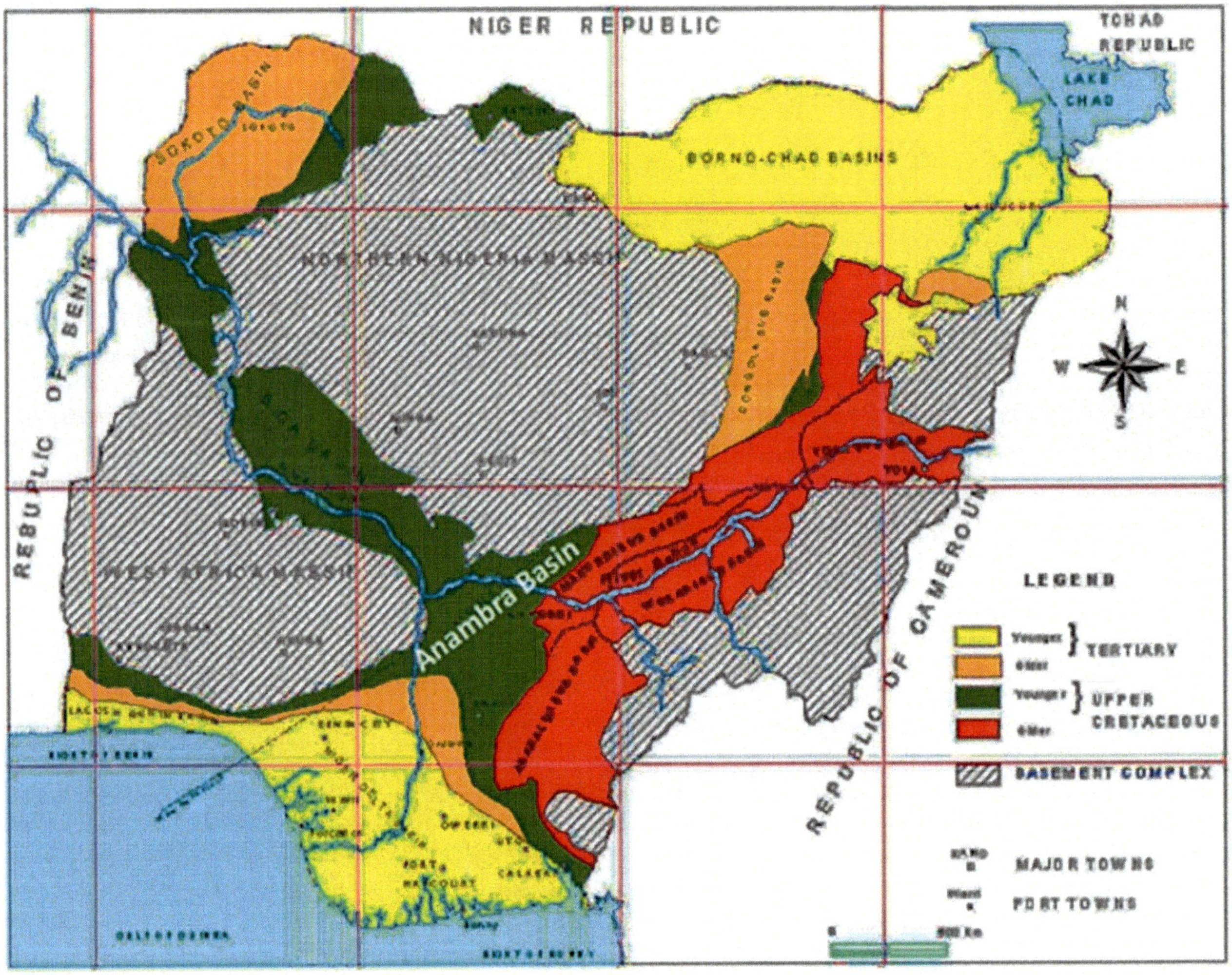

Fig. 1 Geological map of Nigeria showing the sedimentary basin and the study area (after Obaje et al., 2004)

deviation (sorting) (σ1), inclusive graphic skewness (Ski) and graphic kurtosis (KG) of each of the samples are presented in Table 1. Quantitative graphical values of percentiles from various samples obtained from the cumulative frequency curve were plotted for various samples as presented in Fig. 3a. The statistical parameters of the grain size frequency distribution were obtained and computed by the method of Folk and Ward (1957). Eleven samples were subjected to heavy mineral analyses using liquid bromoform (heavy liquid), acetone and filter paper. The disaggregated sample was poured into the bromoform in a separating funnel (with tap locked) and stirred vigorously with the glass rod. The heavy minerals were then drained into the receiving funnel with filter paper, washed with acetone and then dried. The conical flask at the bottom was used to collect the drained bromoform. The identification of minerals was done by using a binocular and polarizing microscope.

3 Results and Discussion

3.1 Lithostratigraphic Description of Outcrop Samples

Six locations were logged and described for facies characteristics. Sandstones, siltstones and shales were observed at various locations (Fig. 2) and each facies having distinct characteristics in terms of texture, structure, mineralogy and colour. The thickness of the sandstone beds varies between 0.3 and 2 m. The sandstones exhibit a fining upward (transgressive) sequence with thin laminations which may suggest fluvial deposition. There were no indications of preservation of fossil, and the grains were subangular to angular, indicating short transportation from the source area. At location 5, the exposure shows medium- to coarse-grained sandstone and form coarsening upward sequences with the underlying siltstone units.

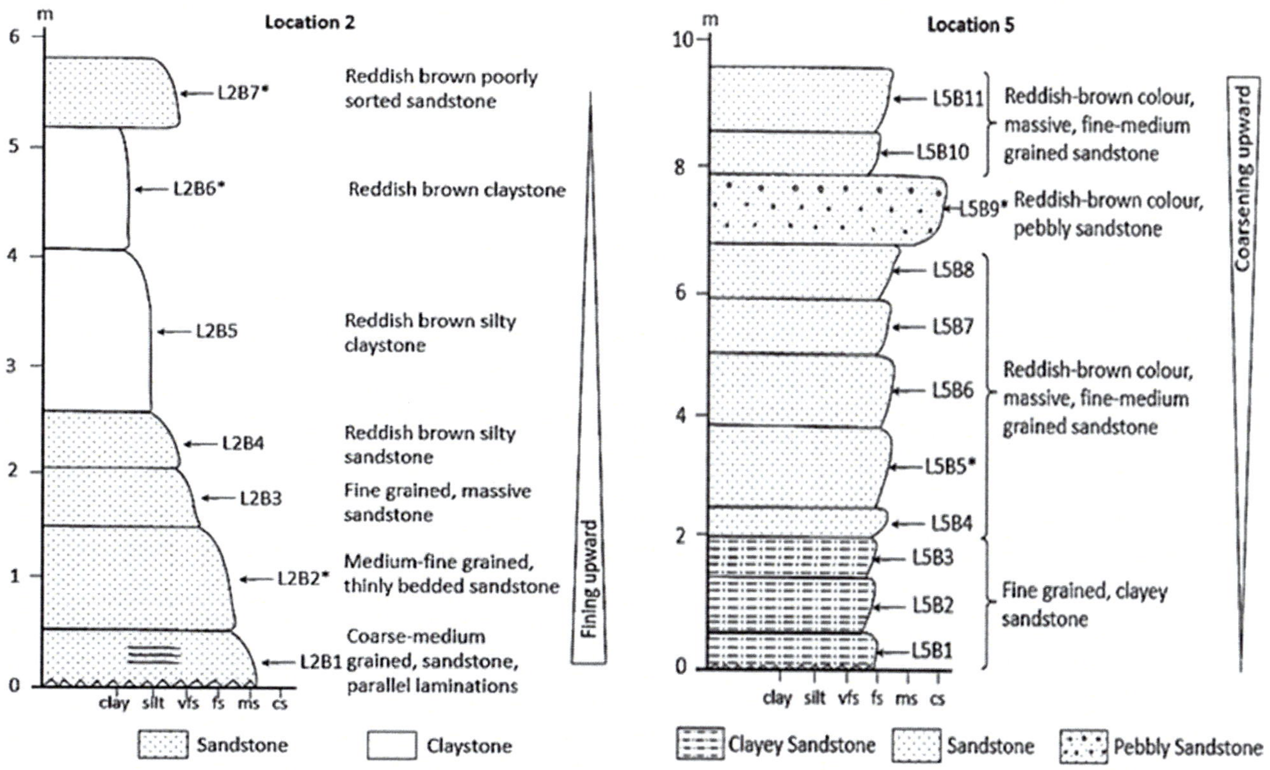

Fig. 2 Lithological succession of Lokoja-Basange Formation exposed at locations 2 and 5

Table 1 Statistical data obtained from grain size analysis (after Folk & Ward, 1957)

Sample	Sample code	Mean value	S.D	Skewness (Sk1)	Kurtosis (KG)
1	L2B2	0.07	1.22	0.38	1.61
2	L2B6	4.3	0.93	−1.18	0.41
3	L2B7	2.83	1.30	0.46	1.17
4	L5B5	2.47	0.78	0.23	4.10
5	L5B9	0.6	1.12	0.77	2.68
6	L7B1	3	1.23	0.40	1.64
7	L8B1	2.57	1.32	0.17	1.76
8	L9B8	3.9	1.06	−0.38	0.94
9	L14B1	1.47	2.20	0.42	0.53
10	L14B2	0.27	1.05	0.51	4.92
11	L15B1	1.53	2.15	0.47	0.48

The sandstone section here shows light to reddish brown coloured, ferruginous pebbly sandstones with a maximum thickness of 9.6 m. This unit exhibits coarsening upward sequences which get progressively pebblier towards the top. Elsewhere, the sandstone shows both cross stratification and flaser laminations. The underlying siltstones in locations 5 and 9 are light brown with beds varying in thicknesses of 0.4–2 m. The siltstone is composed mainly of quartz and clay as matrix and alternates with beds of shale and exhibits coarsening upward sequence (prograding) pattern which may suggest fluvio-deltaic deposition. These outcrops are characteristic of the Mamu Formation. The Mamu Formation is represented here, and the shale is exposed at location 9. The shale is fissile and has a thickness of 4 m. The carbonaceous shale has a dark grey colour due to concentration of organic matter. This suggests high level of biological productivity and low energy condition during deposition. It consists largely of clay minerals and micas. Fossils such as floral remains were present.

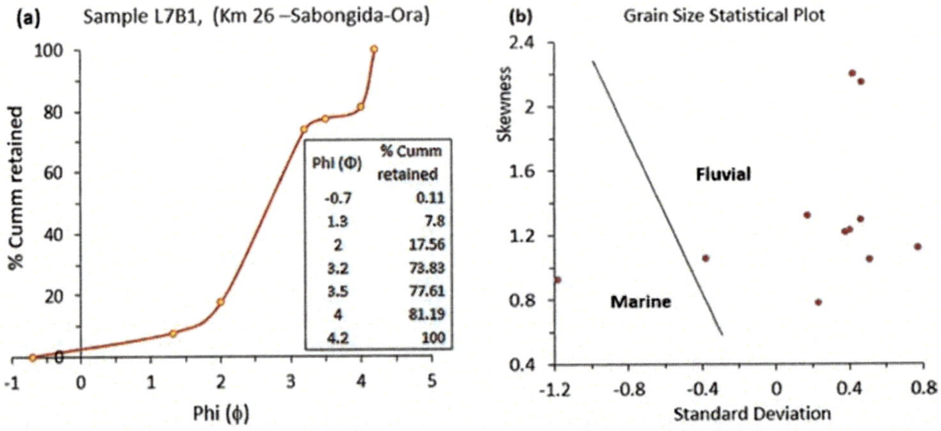

Fig. 3 **a** Cumulative frequency curve for sample L7B1 **b** Bivariate plots of skewness versus standard deviation

3.2 Grain Size Distribution and Palaeoenvironmental Reconstruction

Vischer (1969) and Friedman (1979) indicated the importance of using grain size distributions to deduce palaeoenvironments although it is suggested that particle size distribution for palaeoenvironmental reconstruction should be combined with other sedimentological parameters such as facies association and sedimentary structures (Amaral & Pyrol, 1977; Selley, 1985; Tucker, 1988). The statistical data obtained (Table 1) suggest that the sediments are characterized by fine-grained sand, medium-grained and coarse-grained sandstone whilst sample L2B6 contains silts.

3.3 Sorting and Transportation History

Standard deviation shows the effectiveness of the transporting medium in separating grains of different sizes. The sediments range from moderately to poorly sorted and very poorly sorted, and this is an indication that sediments have not been transported far from the source area before deposition. The poorly sorted sands indicate bedload transport, less winnowing and abrasion. The prevalence of poorly sorted sands and unimodal grain sizes suggest low energy unidirectional fluvial deposition. The skewness of a deposit is sensitive to the depositional environment. The sandstones are characterized by relatively poor sorting and positive skewness/strongly fine skewness which are indicative of river sand (Friedman, 1979). Okoro (1995) reported that negative skewness indicates marine reworking on continental shelf settings. However, in the samples analysed, there was no evidence of marine reworking. Kurtosis revealed that the sediments range from very and extremely leptokurtic to very platykurtic. The leptokurtic nature of the sediments reveals water-laid sediments and low energy conditions.

3.4 Provenance and Palaeoenvironments

Heavy minerals are used for provenance studies, correlation of sedimentary strata of one environment with another and to characterize facies of different localities. The provenance of the sediments is deduced from heavy mineral analyses. The opaque and non-opaque minerals tourmaline, magnetite, zircon, topaz, hematite, sphene, sillimanite and rutile were identified. The heavy minerals occur as detrital grains in the samples which suggests that the sediments are young and were derived from igneous and metamorphic rocks probably from the Basement Complex of south-western and north central Nigeria. The presence of staurolite, rutile, sillimanite and garnet suggests a metamorphic origin. The sediments are thought to have been derived from mixed sources. The presence of hematite suggests oxidation conditions during deposition in a fluvial setting. Palaeoenvironments were deduced from bivariate plots of the mean versus standard deviation and skewness versus standard deviation (sorting) (Fig. 3b) which indicate that the majority of sediments were deposited in a fluvial environment.

4 Conclusions

The grain size parameters (mean, standard deviation, skewness and kurtosis) indicate that the sandstones range from fine grained to coarse grained are poorly sorted, very fine skewed and highly leptokurtic. The angularities of the grains indicate a short transportation history. The range in texture suggests fluctuating energy conditions from low energy regime that deposited fine sand, silts and shale to high energy that deposited medium and coarse sandstone facies. The depositional patterns also suggest fluvio-deltaic sedimentation. The recovery of heavy minerals of zircon, rutile, tourmaline, sphene, sillimanite and staurolite from the sandstones indicates a mixed igneous and metamorphic sources that are probably derived from neighbouring rocks of the Basement Complex of Nigeria.

References

Adeigbe, O. C., & Salufu, A. E. (2009). Geology and depositional environment of Campano-Maastrichtian sediments in the Anambra Basin, Southeastern Nigeria: Evidence from field relationship and sedimentological study. *Earth Sciences Research Journal, 13*, 148–165.

Amaral, E. J., & Pryor, W. A. (1977). Depositional environment of St. Petters Sandstone deduce by textual analysis. *Journal of Sedimentary Petrology, 42*, 425–433.

Folk, R. L., & Ward, W. C. (1957). A study in the significance of grain size parameter. *Journal of Petrology, 37*, 327–354.

Friedman, G. M. (1979). Difference in size distribution of populations of particles among sands of various origins. *Sedimentology, 20*, 3–32.

Obaje, N. G., Wehner, H., & Scheeder, G. (2004). Hydrocarbon prospectivity of Nigeria's inland basins from the viewpoint of organic geochemistry and organic petrology. *American Association of Petroleum Geologists Bulletin, 88*, 325–353.

Obi, G.C. (2000). Depositional model for the Campanian-Maastrichtian Anambra Basin, Southern Nigeria. Ph.D. Thesis, University of Nigeria.

Odunze, S. O., & Obi, G. C. (2013). Sedimentology and sequence stratigraphy of the Nkporo Group (Campanian-Maastrichtian), Anambra Basin Nigeria. *Sequence Stratigraphy, 2*, 192–208.

Okoro, A. U. (1995). Petrology and depositional history of the sandstone facies of the Nkporo Formation (Campanian-Maastrichtian) in Leru area, Southeastern Nigeria. *Journal of Mining and Geology, 31*, 105–112.

Selley, R.C. (1985). Ancient sedimentary environment, 3rd ed., (p. 317). Cornell University Press.

Tucker, M.E. (1988). Sedimentary petrology: An introduction, (p. 252). Blackwell Scientific Publications.

Visher, G. S. (1969). Grain size distributions and depositional processes. *Journal of Sedimentary Petrology, 39*, 1074–1106.

Lithofacies and Palynofacies of the Miocene Deposits in the Paralic Basin of Tunisia (Basin of the Cap Bon and Basin of Saouaf): Reconstruction of Palaeoenvironments

Nadia Ben Moktar and Beya Mannaï-Tayech

Abstract

The Miocene micro-basins of north-east Tunisia are filled with molasse and fluvio-deltaic siliciclastic deposits, including organic matter-rich phytogenic deposits. A multidisciplinary study including sedimentological, palynofacies and palynological analyses was carried out in the Cap Bon and Saouaf basins. The study of palynofacies variation in correspondence with different facies associations allows us to evaluate the paleoenvironment of deposition. Our data set indicates an offshore-type environment during the Lower Burdigalian–Langhian. In the middle and upper Langhian, a shoreface environment evolved into a dominantly marine environment. The Langhian is known to be a phase of transgression in the Mediterranean. For palynofacies, sea level rise results in exceptional abundance of dinoflagellate cysts (*Impagidinium, Cordosphaeridium inodes minimus, Cleistosphareridium, Spiniferites* and others) and chitinous tests of foraminifera as well as dominance of amorphous organic matter. The base of the Serravallian is defined by a gradual change from a shallow submarine channel environment towards a coastal and, eventually, a deltaic plain. This change is followed by a regressive sequence that led to a decrease in amorphous organic matter in favour of non-oxidized and oxidized organic matter. Subsequently, a new transgression led to changes from lagoonal to a foreshore environment. These marine incursions are attested by an increase in amorphic organic matter, dinocysts and by the presence of foraminifera with chitinous tests. Finally, a deltaic deposit with a confined lagoon corresponds to the last regressive sequence. The environment is paralic in character (with twelve levels of lignite) and dominated by non-oxidized and oxidized organic matter. This is the last transgressive/regressive cycle of the Serravallian–Tortonian.

Keywords

Lithofacies · Palynofacies · Palaeoenvironment · Miocene · Tunisia

1 Introduction

Palynofacies are useful because they can help define different terrigenous/marine input in sedimentary rocks through the identification of spores, pollen, cysts, acritarchs, chitinozoans and scolecodonts etc. The Miocene series in Tunisia are characterized by a relative scarcity of fossils with chronostratigraphic value but they comprise phytogenic deposits that reflect a paralic nature (Mannaï-Tayech, 2004, 2006a). One of them is the Saouaf Formation which consists of clay levels with plant debris and lignites, intercalated in clays and sands, and is thus useful for palynofacies analysis. Numerous studies have discussed the geodynamics and palaeogeographic history of Tunisia. Here, Miocene micro-basins were characterized by the development of organic-rich deposits during the Serravallian–Tortonian. These deposits show alternations of sandy levels, sometimes marly clays and dm- to m-scale beds of phytogenic deposits rich in lignite. Previous studies have investigated the age, deposition mode, climatic conditions and tectonic regime of central and north-eastern Tunisia during the Miocene (Burollet, 1956; Salaj & Stranik, 1970; Abbes, 1983; Turki, 1985; Ben Salem, 1992; Bedir, 1995; Boujamaoui, 2000; Mannaï-Tayech, 2006a; Ben Moktar & Mannaï-Tayech, 2011, 2014). Stratigraphic studies show that the dating

N. B. Moktar (✉)
University of Gabes, 6072 Gabes, Tunisia
e-mail: nadiabenrhit34@gmail.com

B. Mannaï-Tayech
University Tunis El Manar, 2092 Tunis, Tunisia

F. Lucci et al. (eds.), *Selected Studies in Geomorphology, Sedimentology, and Geochemistry*,
Advances in Science, Technology & Innovation, https://doi.org/10.1007/978-3-031-43744-1_28

and stratigraphic subdivision of Miocene series are difficult except for those with marine characteristics (Solignac, 1927; Castany, 1951; Burollet, 1956; Jauzein, 1959b; Salaj & Stranik, 1970; Devolvet, 1974; Bismuth, 1984; Meddeb, 1984; Blondel, 1991; Bismuth & Hooyberghs, 1994; Saadi, 1997; Yaïch, 1997; Mannaï-Tayech, 2006a).

2 Methods

The Palynofacies analysis was carried out on 125 samples from different localities (Cap Bon, Grombalia, Jebel Jedidi, Jebel Maaza, Sidi Mahmoud, Saouaf) in central and north-eastern Tunisia. The chemical treatment for slide preparation follows Tayech (1984) for organic matter concentration. Mounting of slides was performed prior to HNO_3 application for the study of palynofacies. Pollens and spores, non-oxidized organic matter and non-oxidized woody debris, woody debris and oxidized organic matter, amorphous organic matter, dinoflagellate cysts, aquatic algae (Pediastrum) and late pyrite were counted (Fig. 1). The organic constituents ratios were also determined.

3 Results

The palynofacies analyses results indicate that samples from Takelsa, Grombalia and Jbibina show similar overall characteristics but with some differences. This attests for similar conditions of deposition and preservation of organic matter during the Burdigalian–Tortonian. The correlation of these sections allowed us to highlight eight palynofacies associations (Pf) (Fig. 2).

4 Discussion

To interpret palaeoenvironment evolution, correlation between sedimentary facies and palynofacies zones was established (Fig. 3). We considered three sections representative of the Takelsa, Grombalia and Jbibina basins which are placed along a NE–SW axis. The three represented sections are (1) the Douala section located near the village of Douala, 35 m thick, (2) the Grombalia section located near the village of Grombalia, 80 m thick and (3) the Saouaf section located near the village of Saouaf, over 1200 m thick.

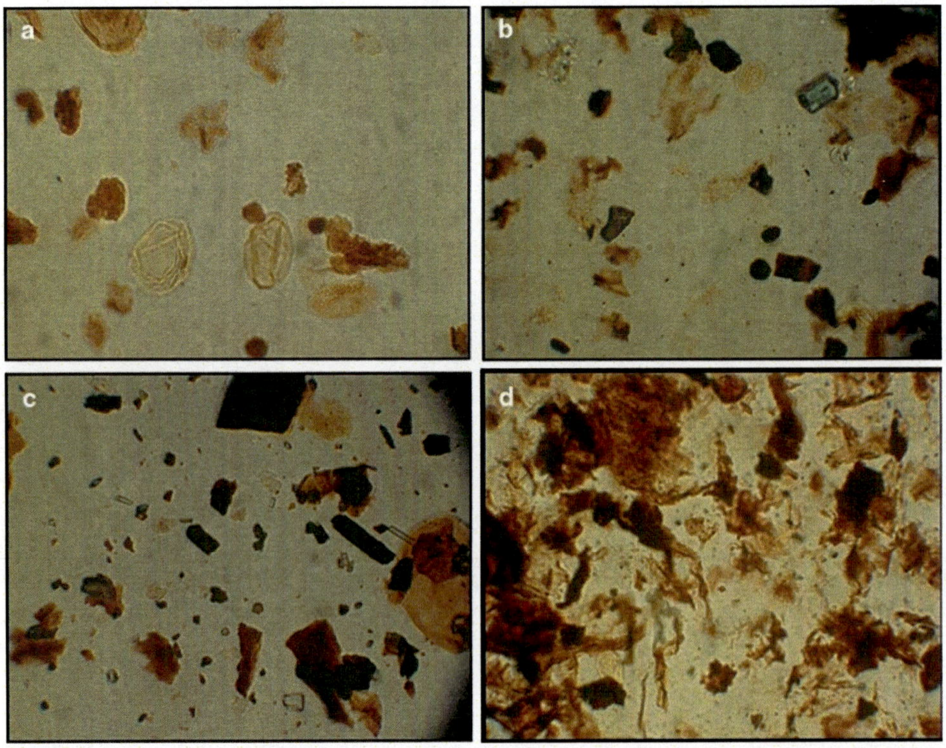

Fig. 1 The constituents of organic matter. **a** Palynofacies rich in pollen (16%), spores (30%) and non-oxidized and amorphous organic matter (42%), slide S53, × 40. **b** Heterogeneous palynofacies with pollen (3%), spores (14.6%), oxidized and non-oxidized organic matter (31.5%), amorphous organic matter (36.5%), pyrite and other minerals (13.7%) slide S38, × 40. **C** Palynofacies rich in oxidized woody debris (47.6%), slide SN25, × 40. **d** Palynofacies rich in amorphous organic matter (43%), slide S19, × 40

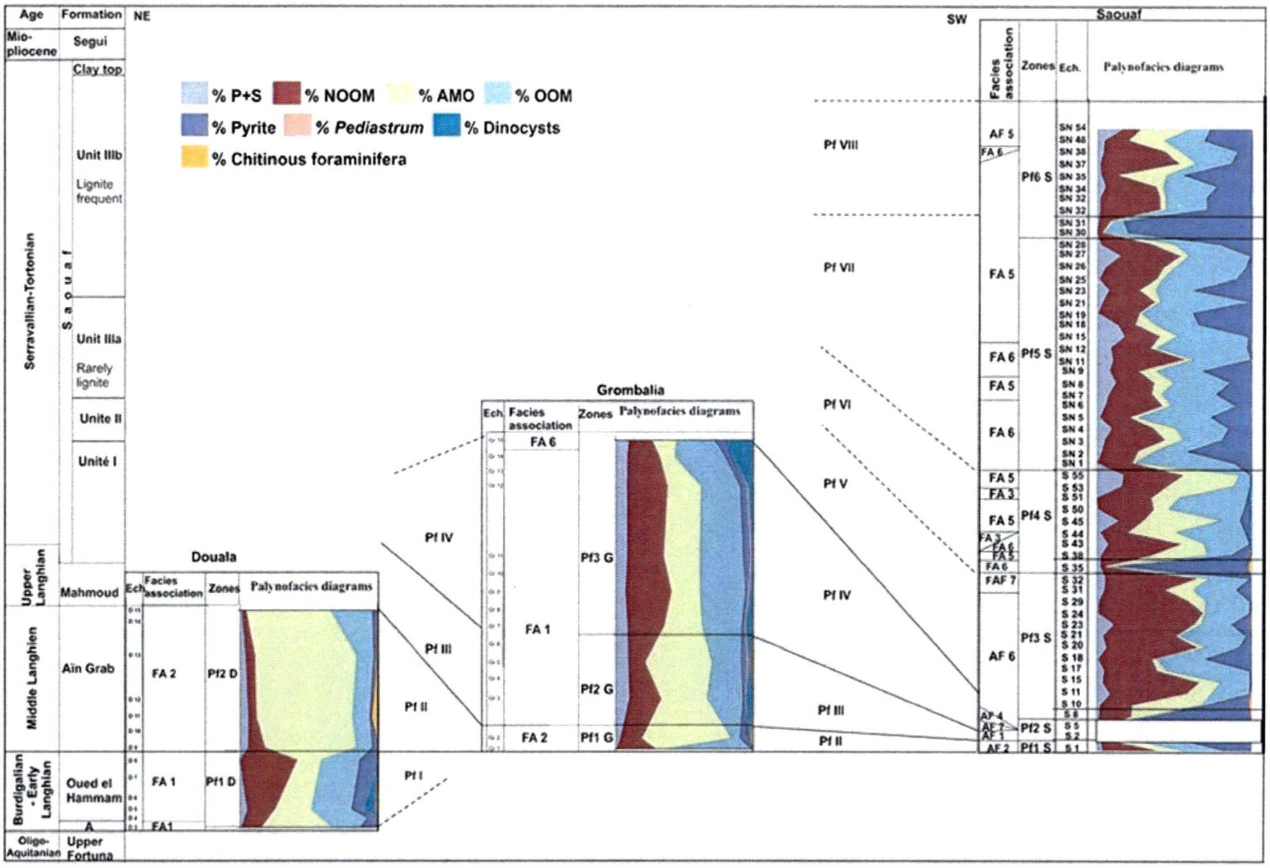

Fig. 2 Determination of Miocene palynofacies associations

5 Conclusion

Through the sedimentological analysis, facies associations were established, whilst the study of organic matter allowed us to define seven palynofacies. Correlation makes it possible to determine Miocene palaeoenvironmental evolution in north-eastern Tunisia. The facies associations are grouped into six sedimentary sequences which correspond to transgressive/regressive eustatic cycles (Ben Moktar & Mannaï-Tayech, 2014). The palynological analysis confirms the evolution of a Burdigalian–Langhian marine environment preceding the mainly marine environment of the Upper Langhian, then a deltaic channel milieu at the base of the Serravallian which evolves towards a more marine environment and finally in a deltaic environment in the Tortonian. It also allowed us to highlight two eustatic pulses during the Serravallian–Tortonian.

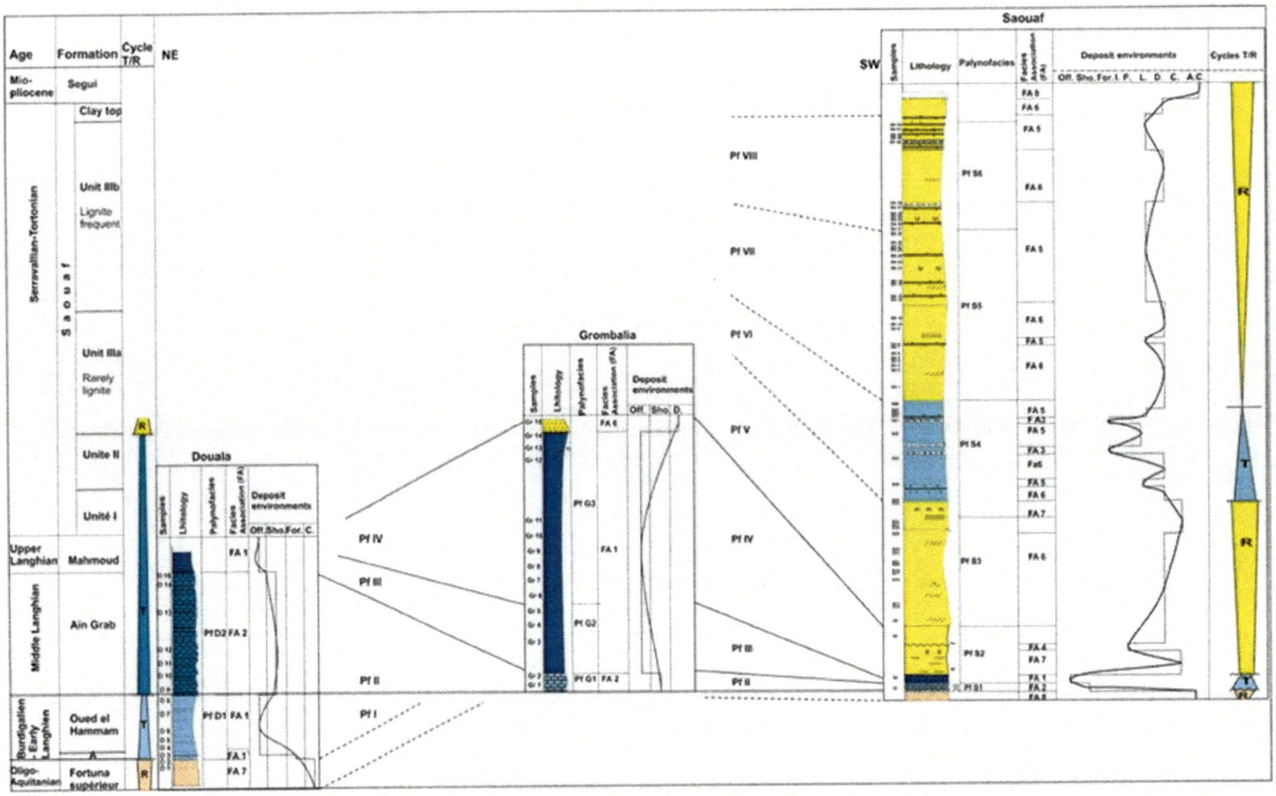

Fig. 3 Palaeoenvironmental evolution in the Miocene basins of north-east Tunisia

References

Abbes, C. (1983). Structural study of Jebel Touila. Northern extremity of the N-S Sidi Kralif-Nara-el Haouareb chain (p. 121). Thesis, Fac. Sc.

Bedir, M. (1995). Geodynamic mechanisms of basins associated with sliding corridors of the Atlas margin of Tunisia. Seismo-stratigraphy, seismo-tectonics and petroleum implications (p. 416). Ph.D. Thesis, Univ.

Ben Moktar, N., & Mannaï-Tayech, B. (2011). Reconstitution of the vegetation and climate during the Miocene in the Saouaf Basin (northeastern central Tunisia). *Geodiversitas, 34*, 445–456.

Ben Moktar, N., & Mannaï-Tayech, B. (2014). Palynology and sedimentology of the Miocene series in the north-east of Tunisia: The climatic and eustatic signature. *Arabian Journal of Geosciences, 7*, 385–396.

Ben Salem, H. (1992). *Contribution to the knowledge of the geology of Cap Bon: Stratigraphy, tectonics and sedimentology*. University Tunis.

Bismuth, H. (1984). Miocene lithostratigraphic units in eastern Tunisia. *Social Science*, p. 2.

Bismuth, H., Hooyberghs, H.J.F. (1994). Planktonic foraminifera and biostratigraphy of the Oligocene and Neogene in the Korba 1 borehole (Cap Bon, north-eastern Tunisia). *Bulletin des Centres de Recherches Exploration-Production Elf Aquitaine*, pp. 489–527.

Blondel, T. (1991). The series with a marine regressive tendency from the lower to middle Miocene in central Tunisia. Thesis, p. 487.

Boujamaoui, M. (2000). *Sequential and seismic stratigraphy of the Miocene facies of north-eastern Tunisia: Compilation of seismic, log and sedimentological data*. Univ. Tunis II.

Burollet, P.F. (1956). Contribution to the stratigraphic study from central Tunisia (p. 350). Ph.D. Thesis, Annal of Mines and Geologies.

Castany, G. (1951). Geological study of the Eastern Tunisian Atlas (pp. 151–160). Ph.D Thesis.

Devolvet, J.J. (1974) Structural study of the Enfidhaville region (eastern Tunisia) (p. 154). Thesis, Univ. Paul Sabatier.

Jauzein, A. (1959b). Revision of the Senonian from the surroundings of Tunis (vol. 18). Notes from the Geological Survey

Mannaï-Tayech, B. (2004). About the problem of the correlations of the Miocene series in Tunisia: a draft solution (p. 4). Earth and Water, Proceedings.

Mannaï-Tayech, B. (2006). The Miocene silico-clastic series from NE to SW Tunisia: An update. *Geobios, 39*, 71–84.

Meddeb, N. (1984). Petrographic and physicochemical examinations of Saouaf-Souar cores (p. 47). National Mines Office, unpublished report.

Saadi, M. (1997). Geodynamics of basins on Cretaceous-Cenozoic strike-slip relay and geometry of genetic sequences of the Saouaf basin (Centro-Eastern Tunisia) (p. 348). Thesis, Univ.

Salaj, J., Stranik, Z. (1970). Contribution to the Miocene stratigraphic study of the Saouaf syncline (J. Fkirine region, eastern Tunisia) vol. 3, pp. 79–82). Not. Geol. Surv.

Solignac, M. (1927). Dir. Gén. Trav. Publ., Tunis (p. 756). Ph.D. Thesis.

Tayech, B. (1984). Palynological study in the Neogene of Cap Bon (Tunisia) (p. 121). Thesis, University.

Turki, M.M. (1985). Polykinematics and associated sediment control on the Zaghouan-Nebhana scar (p. 252). Ph.D. Thesis.

Yaïch, C. (1997). Sedimentary dynamics, eustatism and tectonics during the Oligo–Miocene in Tunisia. Formation Fortuna, Messiouta and Grigma; Numedian and Greso–micaceous (p.479). Ph.D. Thesis, Univ.

Palaeoenvironmental Implications of Calcispheres in the Upper Hauterivian–Lower Barremian Bou Hedma Formation (Central Tunisia)

Mouna Amami, Besma Mardassi and Jamel Abdennaceur Ouali

Abstract

The Hauterivian–Barremian Bou Hedma deposits, cropping out in the Meloussi area (central Tunisia), consist of a mixed series. Facies stacking patterns allow the recognition of two distinct intervals. The lower interval is composed of siliciclastic deposits, encasing a few carbonates and evaporites. The upper interval shows the frequent occurrence of carbonates and evaporites at the expense of siliciclastics. Facies variations reflect palaeoenvironmental and climate changes. The abundant detrital input and scarce inner platform microfauna were previously interpreted as a shallow marine tidal flat environment for the lower interval. Sampling of the thickest series in this study can help describe well-preserved and abundant calcispheres recovered from well-bedded burrowed wackestones. These deposits overlie barren sandstones and clays. Locally, clays provide few miliolids and dwarf gastropods, indicating an inner platform context and pointing to stressful palaeoecological conditions. The discovery of calcisphere-rich facies provides new data to support reinterpretation of depositional environments and to better constrain eustasy. Their occurrence within the lower interval records sea level deepening and the transition from an inner platform to a deeper open marine shelf. The abundance of calcispheres associated with siliciclastic to carbonate sedimentary turnover highlights the role of climate change. Palaeoenvironment deepening could be linked to an abrupt warm and xeric episode, hampering continental inputs and favouring carbonate sedimentation.

Keywords

Hauterivian · Barremian · Calcispheres · Palaeoenvironment · Climate change · Open marine sedimentation

1 Introduction

Hauterivian–Barremian deposits, cropping out in central Tunisia, are marked by clear variations in thickness and facies. Each facies corresponds to an appropriate depositional area. Based on collected fauna and flora, previous works have assigned its lower part to inner shelf environments (M'Rabet, 1987). The discovery of calcisphere-rich facies within the lower part of the Bou Hedma Formation has brought new insight to refine environmental and climatic interpretations. Calcispheres have been interpreted as unicellular algae (Cayeux, 1935) but Cretaceous calcispheres have lately been considered as calcareous dinoflagellate algae (e.g., Flügel, 2004). Although some researchers consider them as dinoflagellates, calcispheres continue to be contested (e.g., Kietzmann et al., 2020). To resolve this problem, Versteegh et al. (2009) assigned organic calcareous cysts of uncertain origin in a new group, the Calcitarcha. Despite their uncertain origin, calcispheres are a reliable indicator of climate (Preto et al., 2013) and palaeoenvironmental proxies (Flügel, 2004).

The present work aims to examine the local occurrence and abundance of these open-shelf calcispheres within an inner marine context and to emphasize their palaeoenvironmental and climatic implications.

M. Amami (✉)
Faculty of Sciences of Sfax, Route de La Soukra Km 4, B.P, 802-3038 Sfax, Tunisia
e-mail: mouna.amami@gmail.com

B. Mardassi
Higher Institute of Biotechnology of Sfax, Route de La Soukra Km 4, B.P. 261 3000, Sfax, Tunisia

J. A. Ouali
National School of Engineers of Sfax, Route de La Soukra Km 3.5, B.P, 1173-3038 Sfax, Tunisia

F. Lucci et al. (eds.), *Selected Studies in Geomorphology, Sedimentology, and Geochemistry*,
Advances in Science, Technology & Innovation, https://doi.org/10.1007/978-3-031-43744-1_29

2 Materials and Methods

The study area is in the central-eastern Meloussi Anticline which is part of the Atlasic fold domain, located in central Tunisia (Fig. 1a). The upper Hauterivian–lower Barremian Bou Hedma Formation (Trabelsi et al., 2019), well-exposed in the northern flank of Jebel Meloussi (Fig. 1b), was measured and logged. This work focused on the first 80 m of the Bou Hedma Formation which spans the upper Hauterivian–lowermost Barremian (Trabelsi et al., 2019) (Fig. 2). Facies' investigation used petrographic and scanning electron microscopy, combined with clay mineralogy analysis.

3 Results

For clay mineralogy, the X-ray diffraction shows an upward increase in illite (38–92%) at the expense of kaolinite (62–8%). Illite shows an abrupt increase within clays bearing the calcisphere-rich bed, then slightly decreases within the overlying deposits. Petrographic analysis of the calcisphere bed from thin sections shows a calcisphere biomicrite with wackestone texture (Fig. 3a). Calcispheres appear as regular rounded, external calcitic-walled allochems 150–200 µm in diameter. Calcispheres are associated with some shell fragments. SEM analysis focused on small calcispheres < 10 µm in diameter. The ultrastructure shows spherical aggregates consisting of nested sub-micron crystals that are assigned to pristine calcispheres (Kazmierczak & Kremer, 2005) (Fig. 3b).

4 Discussion

The levels bounding the calcisphere-rich bed record changes in facies composition. The turnover from siliciclastic to carbonate sedimentation (Fig. 2) associated with an increase in illite (Table 1) outlines the transition from humid to xeric conditions (Adatte et al., 2002). The well-preserved calcispheres are indicative of an open marine environment (Flügel, 2004). Their position within inner shelf deposits points to a transgressive event leading to a slight deepening of the depositional setting. Calcisphere abundance records a transgressive event coeval with the Hauterivian–Barremian transition (Rehakova, 2000). This relationship is also supported by the latest dating of the studied series, now assigned to the upper Hauterivian–lowermost Barremian (Trabelsi et al., 2019). On the other hand, calcisphere abundance suggests major climate perturbations (Preto et al., 2013) as documented by a conspicuous change in sediment fabrics and clay paragenesis. Enrichment in illite suggests prevailing xeric conditions. The transgressive event was linked to a warmer and arid episode that favoured thin to well-bedded carbonate sedimentation.

5 Conclusion

Calcispheres could be considered as reliable tools to investigate sea level fluctuations and climate perturbations. Their occurrence coincides with a transgressive event of the Hauterivian–Barremian transition. This event was linked to a warm and xeric climate episode.

Fig. 1 Location map of the study area. **a** Jebel Meloussi location (yellow star), **b** logged section position (Google Earth)

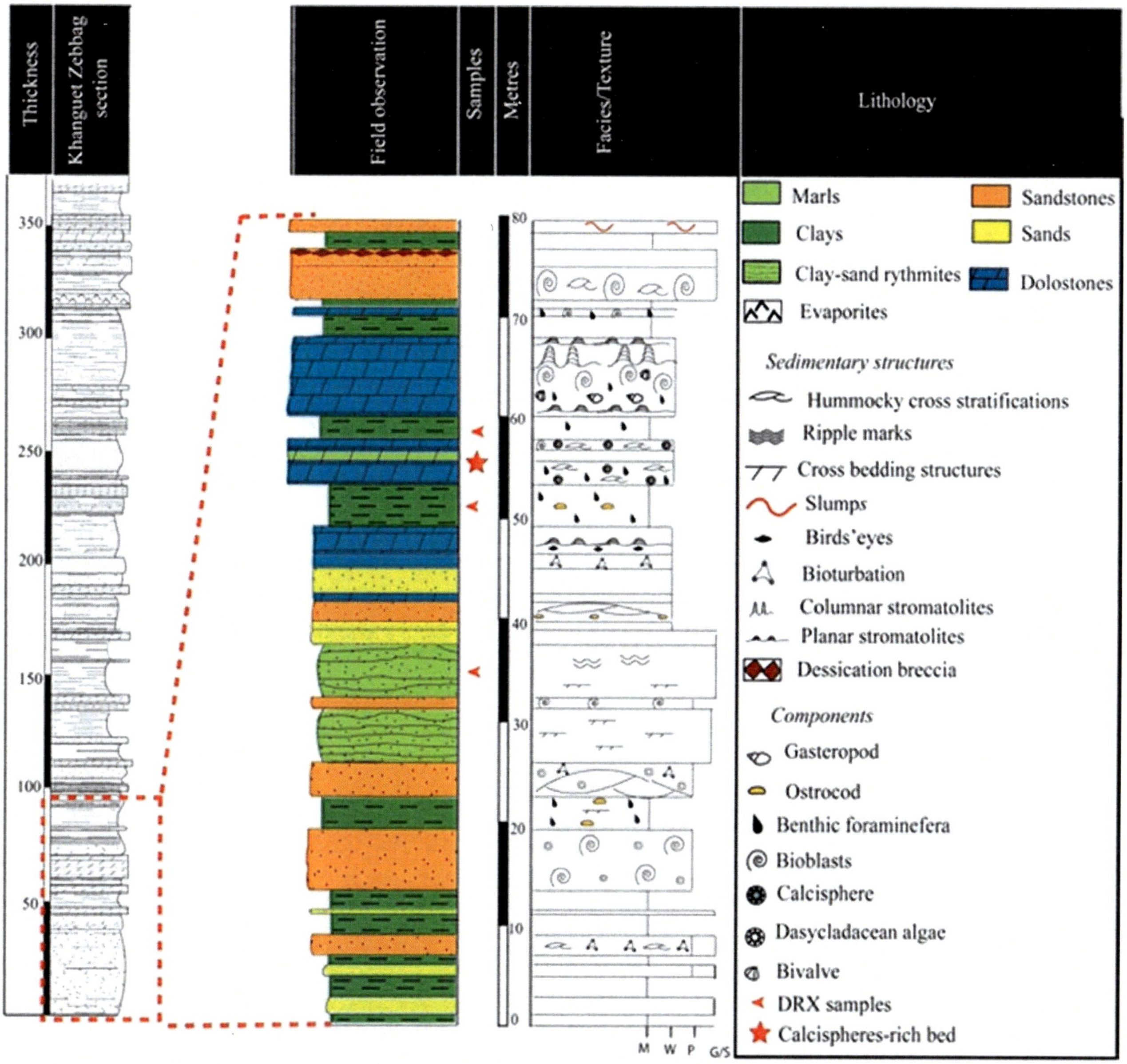

Fig. 2 Sedimentary characters of the Upper Hauterivian–lowermost Barremian in Meloussi area (central Tunisia)

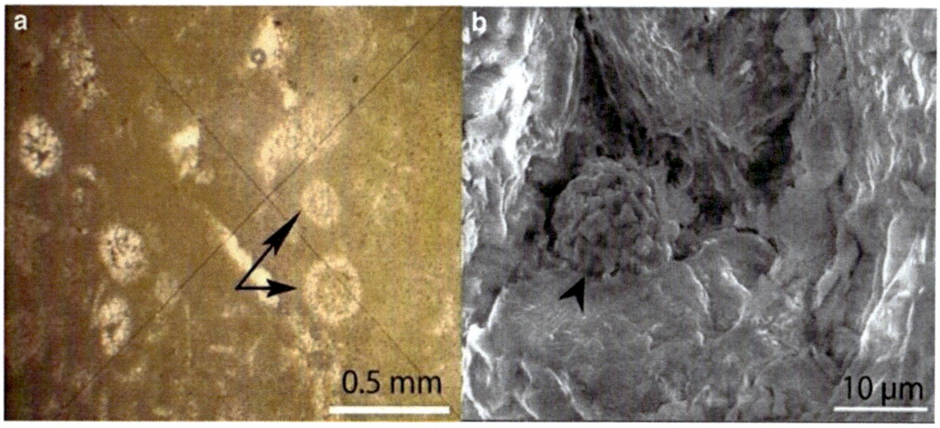

Fig. 3 Microphotographs of the calcisphere-rich bed **a** observed under a petrographic microscope. **b** SEM image showing the ultrastructure of calcispheres

Table 1 Relative abundance of identified minerals in the studied section

Samples	1 (%)	2 (%)	3 (%)
Illite	38	92	88
Kaolinite	62	8	12

References

Adatte, T., Keller, G., & Stinnesbeck, W. (2002). Late Cretaceous to Early Paleocene climate and sea-level fluctuations: The Tunisian record. *Palaeogeography, Palaeoclimatology, Palaeoecology, 178*, 165–196.

Cayeux, L. (1935). Les roches sédimentaires de France; roches carbonatées (calcaires et dolomies) (p. 463). Masson et Cie.

Flügel, E. (2004) Microfacies of carbonate rocks: Analysis, interpretation and application (p. 984). Springer.

Kazmierczak, J., & Kremer, B. (2005). Early post-mortem calcified Devonian acritarchs as a source of calcispheric structures. *Facies, 51*, 554–565.

Kietzmann, D., & Scasso, R. (2020). Jurassic to Cretaceous (upper Kimmeridgian–?lower Berriasian) calcispheres from high palaeolatitudes on the Antarctic Peninsula: Local stratigraphic significance and correlations across Southern Gondwana margin and the Tethyan realm. *Palaeogeography, Palaeoclimatology, Palaeoecology, 537*, 109419. https://doi.org/10.1016/j.palaeo.2019.109419

M'Rabet, A. (1987). Stratigraphie, sédimentation et diagenèse carbonatée des séries du Crétacé inférieur de Tunisie Centrale. Ph.D. Thesis, Paris-Sudcentre d'Orsay Univ.

Preto, N., Willems, H., Guaiumi, C., & Westphal, H. (2013). Onset of significant pelagic carbonate accumulation after the Carnian Pluvial Event (CPE) in the western Tethys. *Facies, 59*, 891–914.

Rehakova, D. (2000). Calcareous dinoflagellate and calpionellidbioevents versus sea-level fluctuations recorded in the West-Carpathian (Late Jurassic/Early Cretaceous) pelagic environments. *Geologica Carpathica, 51*, 229–243.

Versteegh, G. J. M., Servais, T., Streng, M., Munnecke Vachard, D. (2009). A discussion and proposal concerningthe use of the term calcispheres. Palaeontology, 52(2), 343–348. doi:10.1111/j.1475-4983.2009.00854.x

Trabelsi, K., Chnaina, M., Sames, B., Houla, Y., Soussi, M. (2019). New insights into the stratigraphy of the Bouhedma Formation (Lower Cretaceous) in Jebel Majoura (Central Tunisia). In 14th Tunisian Petroleum Exploration and Production Conference, abstracts (p. 99).

The Role of Eustasy and Local Tectonics in Jurassic Carbonate Sedimentation in the Blue Nile and Northwest Ogaden Basins of Ethiopia

Gilamichael Domenico and Hiroshi Kurita

Abstract

Microscopic, paleontologic and field studies were undertaken on the Middle to Upper Jurassic Antalo Limestone exposed in the Blue Nile and the northwest Ogaden basins of Ethiopia to determine the age, depositional environment and discuss their evolution. The study also explained the timing and formation of Mesozoic carbonate platforms in Ethiopia in detail for the first time. For the Blue Nile Basin, this work used a new age scheme by recent studies on Callovian–Oxfordian dinoflagellate cyst records, benthic foraminifers and green algae from the Antalo Limestone Formation. These allowed us to date the entire carbonate succession in the basin as Callovian to Early Kimmeridgian. A detailed facies analysis of the Antalo Limestone (ca. 550 m thick) in both basins reveals nearshore tidal flat (inner ramp), shelf (subtidal), open marine and barrier (shoal) facies associations. The second-order depositional sequence identified from the Blue Nile is characterized by more open marine facies and marked by a maximum flooding zone (MFZ) of Oxfordian age. In northwest Ogaden, the MFZ is represented by cherty and ammonoid-bearing spicular mudstones of Early Kimmeridgian age. Both local tectonics and eustasy are responsible for the evolution of the sediments. Differential subsidence formed a sub-basin in the Blue Nile area, whereas the northwest Ogaden could represent a paleohigh. Marine flooding could be at the peak stage during Late Oxfordian to Early Kimmeridgian, and this can be attributed to eustatic conditions during the opening of the Indian Ocean.

Keywords

Jurassic · Carbonates · Ethiopia · Local tectonics · Eustasy

1 Introduction

As a result of the Jurassic transgression of the Tethys Ocean on the continental margins of Gondwanaland, thick carbonate deposits were formed in Ethiopia (Bosellini, 1989) and Arabia (Althour, 1997; Zigler, 2001). The Middle to Upper Jurassic carbonate succession (Antalo Supersequence) (Russo et al. 1994) in the east (Ogaden Basin) and northwest Ethiopia (Blue Nile Basin) is part of this thick succession (Fig. 1). This study focused on the stratigraphy and facies distribution of carbonate beds of the Antalo Limestone in these two basins (Fig. 1). The age, depositional history and evolution of the sediments in relation to sea level change were also discussed. The study is based on field investigations, palynostratigraphy and microscopic examination.

2 Methods

Thirteen stratigraphic sections were measured and sampled. Microfacies studies were based on approximately 320 thin sections. Carbonate rocks were classified following Dunham (1962). Depositional sequences and inferred relative sea level changes were interpreted based on Flugel (2004). Second- and third-order depositional sequences were identified (Van Wagoner et al., 1988). Correlation of the depositional sequences in the two basins were related to the global sea level curve of Hallam (2001).

G. Domenico (✉)
East Los Angeles College, 1301 Avenida Cesar Chavez, Monterey Park CA91754, USA
e-mail: domenigk@elac.edu

H. Kurita
Faculty of Science, Niigata University, Niigata 950-2181, Japan

F. Lucci et al. (eds.), *Selected Studies in Geomorphology, Sedimentology, and Geochemistry*,
Advances in Science, Technology & Innovation, https://doi.org/10.1007/978-3-031-43744-1_30

Fig. 1 Map showing the location of the study areas in the African continent

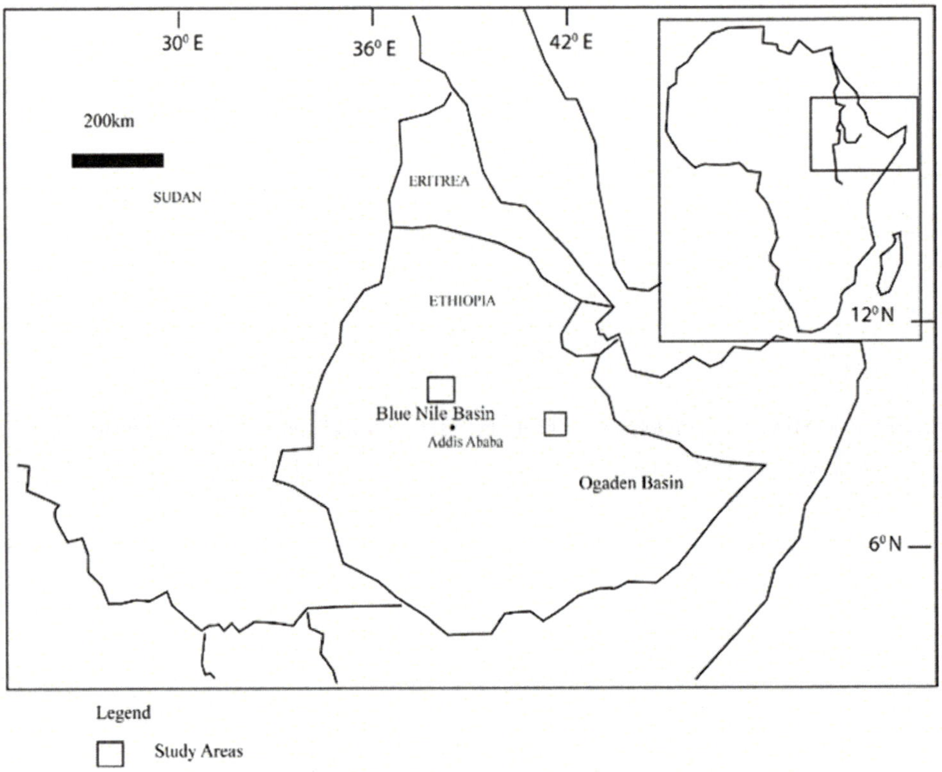

3 Results

Based on field studies (sedimentary structures) and microfacies studies, the depositional environment of the two basins includes tidal flat (nearshore), lagoonal (shelf) or subtidal, barrier and open marine (middle to lower ramp) facies associations.

3.1 Tidal Flat (Nearshore) Facies Association

Sandy mud to wackestone and dolomitic mudstone comprise this facies association. The Bathonian to Callovian basal beds of the Antalo Limestone in the sections of northwest Ogaden Basin (Fig. 2) are characterized by these interbedded facies (Domenico & Sano, 2006). Sandy mudstones and wackestones are generally non fossiliferous. However, some beds contain thin-shelled small bivalves, miliolid foraminifers and unidentified fauna are present.

3.2 Lagoonal (Inner Ramp) to Subtidal Facies Association

This facies association consists of bioclastic peloidal mud to wackestones. Bioclasts include benthic foraminifers, dasycladacean algae, corals, stromatoporoids and molluscs as well as brachiopods. In the northwest Ogaden Basin, beds of this facies association overlie basal dolomitic nonfossiliferous mudstones and sandy limestones. In the Blue Nile Basin, the lower part of the Antalo Limestone is composed of nodular bioclastic mud to wackestones and peloidal wackestones. These units also form medium to thick beds (0.2–1 m thick) at the top part of the succession overlying micritic limestones and oolitic granstones. Coral–stromatoporoid floatstones *(Cladocoropsis mirabilis)* (Domenico, 2010) are also common in northwest Ogaden sections and at the top of the Antalo Limestone of the Blue Nile Basin (Fig. 2).

3.3 Barrier (Shoal) Facies

Bioclastic peloidal, oolitic, bioclast ooid–peloid grainstones represent barrier (shoal) facies in both basins. Cliff forming beds of oolitic grainstones (1–2 m thick) are exposed at the top part of the Antalo Limestone in the Blue Nile Basin. The nuclei of the ooids in these facies contain dasycladacean algae and foraminifers.

3.4 Open Marine Facies

This facies association includes ammonitic lime mudstones, marls, spicular mudstones, coquinites, calciturbidites and

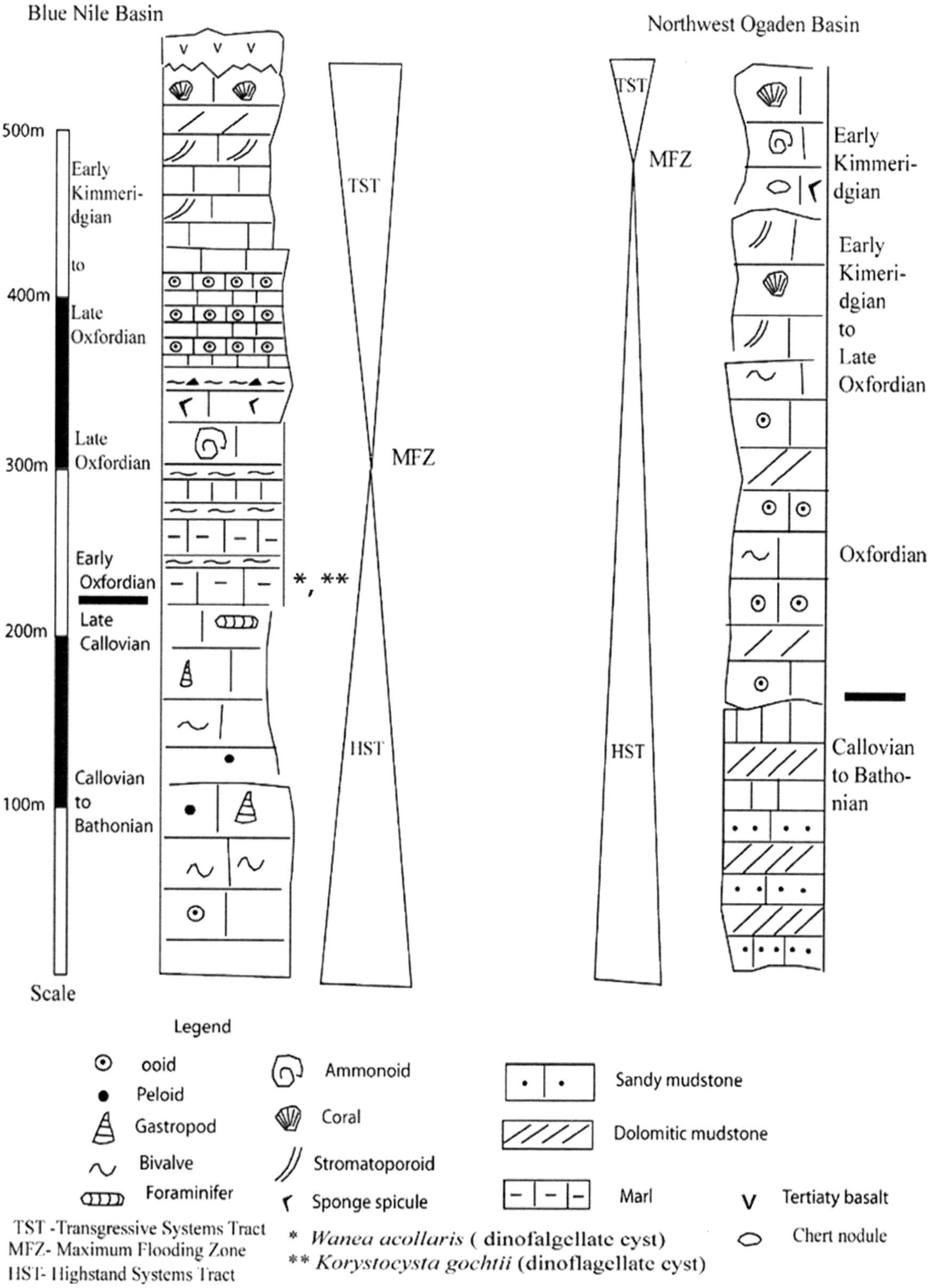

Fig. 2 Comparison of Jurassic sea level change patterns from Jurassic composite sections of the Antalo Limestone in the Blue Nile and northwest Ogaden basins

shelly packstones. Based on dinoflagellate cyst studies, these units mark Oxfordian beds in the Blue Nile (Kurita et al., 2009) and Lower Kimmeridgian beds in Northwest Ogaden Basin (Domenico, 2010) (Fig. 2).

4 Discussion

There is a transition from lagoonal and midshelf facies (Callovian) to more open marine conditions (Oxfordian) followed by a progradational sequence of Late Oxfordian to Early Kimmeridgian oolites, micrites, lime mudstones and coral–stromatoporoid mudstones in the Blue Nile Basin (Fig. 2). Likewise, in northwest Ogaden, the Antalo Limestone shows sea level rise as indicated by a nearshore facies sandy dolomitic mudstone at the bottom, which pass upwards to skeletal mud/wackestones, with coral–stromatoporoid floatstones and spicular mudstones with ammonites of Early Kimmeridgian age. This is overlain by a sequence of lime mudstones and reefal limestones marking a progradation stage (High Stand Systems Tract).

As shown in Fig. 2, more open marine conditions prevailed earlier (since Late Callovian to Early Oxfordian) in the Blue Nile Basin than in the northwest Ogaden Basin. This is evident from the timing of the maximum flooding zones, which is Oxfordian (from dinoflagellate cysts) (Kurita et al., 2009) in the former and Early Kimmeridgian in the latter. This indicates that the Blue Nile Basin represents a sub-basin (paleodepression), while northwest Ogaden was a paleohigh. However, the occurrence of Late Oxfordian and Early Kimmeridgian ammonites in the Blue Nile and northwest Ogaden basins, respectively, also indicates a global sea level rise of the Tethys Ocean during the Middle to Late Jurassic (Hallam, 2001), a time period which coincides with the opening of the Indian Ocean.

5 Conclusions

Nearshore (tidal flat), lagoonal to subtidal, open marine and barrier facies associations are recognized in the Bathonian to Kimmeridgian Antalo Limestone Formation of Ethiopia, in the Blue Nile and the northwestern part of the Ogaden basins. The prevalence of more open marine facies in the Blue Nile Basin indicates a paleodepression (sub-basin) which underwent differential subsidence. In contrast, northwest Ogaden was a paleohigh during the Middle to Late Jurassic. A maximum flooding zone (MFZ) of Oxfordian age in the Blue Nile sections and an Early Kimmeridgian MFZ in northwest Ogaden indicates global sea level rise during the Middle to Upper Jurassic culminating during the Late Oxfordian to Early Kimmeridgian. This can be attributed to the opening of the Indian Ocean. The difference of the timing of marine flooding between the basins is explained by the local tectonic overprints on the eustatic influence.

References

Althour, K. (1997). Facies sequences of the Middle to Upper Jurassic carbonate platform (Amran Group) in the Sana's region, Republic of Yemen. *Marine and Petroleum Geology, 14*, 643–660.

Bosellini, A. (1989). The continental margins of Somalia: Their structural evolution and sequence stratigraphy. *Memoire Di Scienze Geologiche, 16*, 373–458.

Domenico, G., Sano, H. (2006). Facies and biostratigraphy of Jurassic carbonate sediments of Bedessa area, eastern Ethiopia. In: 113th Annual meeting of the geological society of Japan, Geological Society of Japan (p. 80)

Domenico, G. (2010). Facies analysis of middle to upper Jurassic carbonates in eastern and north-central Ethiopia. Ph.D. thesis, Kyushu University.

Dunham, R. (1962). Classification of carbonate rocks according to their depositional texture. *AAPG Memoir, 1*, 108–121.

Flugel, E. (2004). *Microfacies of carbonate rocks*. Springer.

Hallam, A. (2001). A review of the broad pattern of Jurassic sea-level changes and their possible causes in the light of current knowledge. *Paaeogeography, Palaeoclimatology, Palaeoecology, 167*, 23–27.

Kurita, H., Domenico, G., Sano, H. (2009). Timing of Jurassic carbonate deposition in an East African aulacogen; dinoflagellate cysts from the Dejen section, Blue Nile Basin, north central Ethiopia. In: Abstracts 2009 Annual Meeting of Palaeontological Society of Japan.

Russo, A., Getaneh, A., & Balemwal, A. (1994). Sedimentary evolution of the Abay River (Blue Nile) Basin Ethiopia. *Neues Jahrbuch für Geologie und Paläontologie-Monatshefte, 5*, 291–308.

Van Wagoner, J., Posamentier, H., Mitchun, R., Vail, P., Sarg, J., Loutit, T., Hardenbol, J. (1988). An overview of sequence stratigraphy and key definitions. In C. Wilgus, B. Hastings, C. Kendall, H. Posamentier, C. Ross, J. Van Wagoner (Eds.), Sea level changes—an integrated approach (vol. 42, pp. 39–45. SEPM Special Publication.

Ziegler, M. (2001). Late Permian to Holocene paleofacies evolution of the Arabian Plate and its hydrocarbon occurrences. *GeoArabia, 6*, 445–504.

Facies Analysis of the Maastrichtian Taloka Formation in the Iullemmeden Basin, Nigeria: A Mixed Wave- and Tide-Influenced Depositional System

Mutari Lawal, Meor Hakif Amir Hassan, Wan Hasiah Abdullah, Lanre Asiwaju, Muhammad Abdullahi and Habib Adamu Ibrahim

Abstract

The Taloka Formation is the oldest member of the Rima Group, southeastern Iullemmeden Basin, Nigeria. The Iullemmeden Basin is a Late Jurassic–Early Cretaceous sag basin whose evolution is related to the West African Rift System (WARS) that is known to contain working petroleum systems. The possibility of a viable petroleum system in the Maastrichtian is being explored in the Iullemmeden Basin. However, the facies characteristics of the basin are still poorly documented. This study focused on the Taloka Formation (Maastrichtian) within the Basin. Outcrop studies including facies analysis and ichnology were here used to construct a depositional model for the study area. Six facies associations were recognized: (i) mangrove swamp/marsh; (ii) intertidal; (iii) tidal bar; (iv) tidal channel; (v) Bay/estuarine, and (vi) offshore transition-lower shoreface. Three facies successions were also identified: offshore-shoreface facies, bay–beach ridge, and tide-dominated delta front. Low/moderate degree of bioturbation and low ichnodiversity (*Thalassinoides*, *Cylindrichnus*, *Planolites*), representative of the *Skolithos* and *Cruziana* ichnofacies, and some occurrence of bivalve imprints were observed. The facies associations reflect deposition in a tide-dominated environment with some wave influence with brackish? water input. The Taloka Formation deposits share similarities with those of the Maastrichtian Bida Basin, Nigeria, and the modern Trusan Delta, Brunei Bay, based on fine-grained and tidal influence of the deposits. The facies arrangement is of dominantly regressive conditions in the basin during the Maastrichtian. The carbonaceous shale and lignitic mudstone of the offshore/delta-front facies and mangrove/intertidal facies, which range between 0.5 and 1.5 m in thickness and traceable for about 2 km, indicate a potential petroleum source and seal rocks, while the sandy bars and channels of delta-front facies may serve as reservoirs.

Keywords

Sokoto Sub-Basin · Tide-dominated · Coastal systems · Taloka formation · Facies analysis

1 Introduction

The Taloka Formation is the oldest member of the Rima Group in the Sokoto Sub-Basin, southeastern section of Iullemmeden Basin, Nigeria. The Iullemmeden Basin is part of the West African Rift System (WARS) that developed in the Late Jurassic–Early Cretaceous. The Sokoto Sub-Basin (Fig. 1) covers about 76,000 km^2. The Sub-Basin is NE-SW In trend and contains Upper Mesozoic-Eocene sediments. Its evolution is related to the Mesozoic–Early Cretaceous split-up of the Gondwana continent in a series of vertical (normal?) faulting and translational movements of the lithospheric block over a bulging asthenosphere (this process is regarded as epeirogenesis). The epeirogenesis resulted in the creation of shallow intracontinental depocenters within the Iullemmeden, one of which is the Sokoto Sub-Basin (Bellion, 1987; Guiraud et al., 2005; Lawal et al., 2023; Petters, 1979). Previous studies have discussed the characteristics of the sediments

M. Lawal (✉) · M. H. A. Hassan · W. H. Abdullah · L. Asiwaju
Department of Geology, University of Malaya, 50603 Kuala Lumpur, Malaysia
e-mail: mosalah1733@gmail.com

M. Lawal · H. A. Ibrahim
Department of Geology, Usmanu Danfodiyo University, Sokoto P.M.B. 2346, Nigeria

M. Abdullahi
Department of Geography, Shehu Shagari College of Education, Sokoto State P.M.B. 2346, Nigeria

F. Lucci et al. (eds.), *Selected Studies in Geomorphology, Sedimentology, and Geochemistry*, Advances in Science, Technology & Innovation, https://doi.org/10.1007/978-3-031-43744-1_31

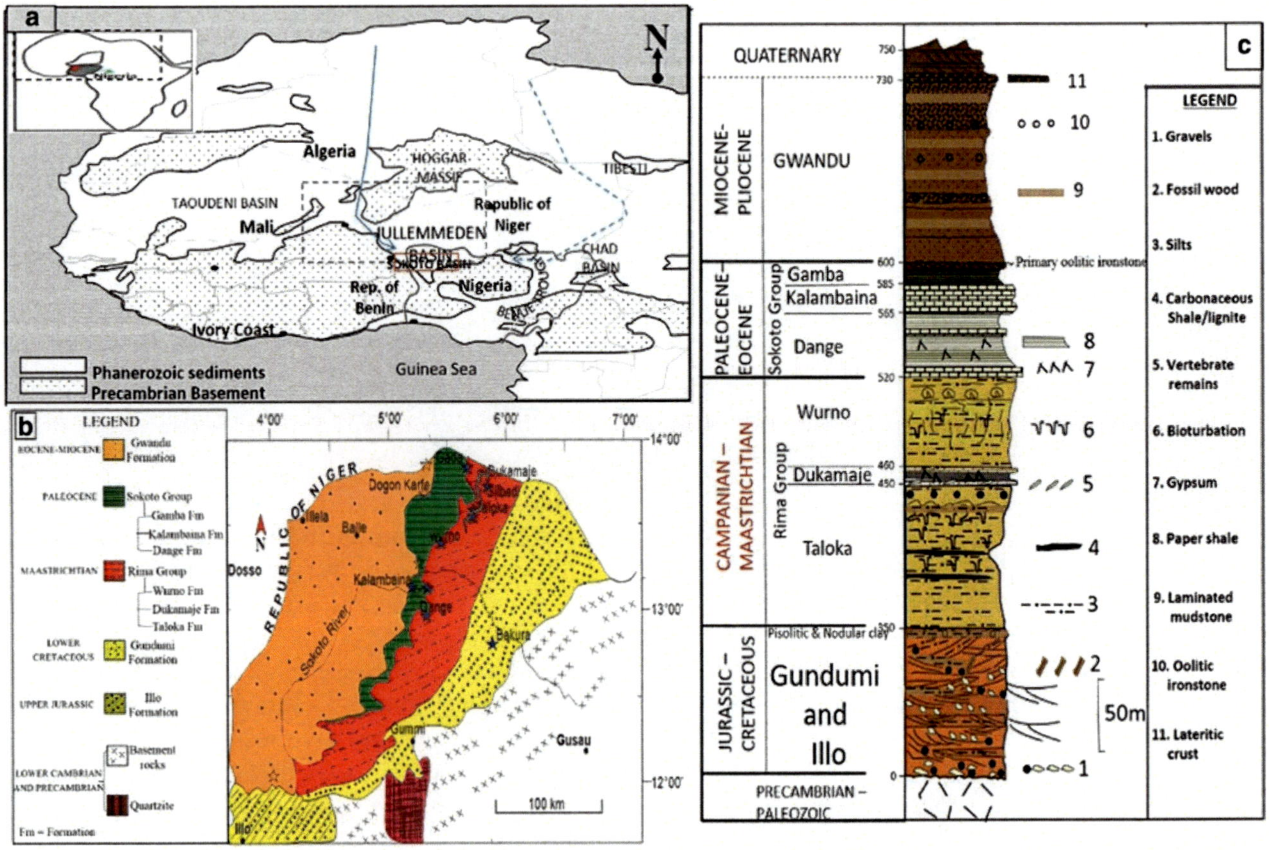

Fig. 1 Geological map **a** Africa and **b** the Sokoto Basin (modified after (Lawal et al., 2023)); **c** Stratigraphic section of the Sokoto Sub-Basin (Bellion, 1987). 1 Gravels; 2 Fossil wood; 3 Silts; 4 Lignites and carbonaceous shales; 5 Vertebrate remains; 6 Bioturbation; 7 Gypsum; 8 Paper shale; 9 Laminated mudstone; 10 Oolitic ironstone; 11 Lateritic crust

(as predominantly fine-grained tidal deposits) of the Late Cretaceous transgression into the Iullemmeden Basin, and their general hydrocarbon potential (Bellion, 1987; Kogbe, 1981; Whiteman, 1982). The Basin is filled with more than 2000 m of sedimentary successions that comprise continental fluvial deposits, coastal plain siliciclastics and shallow marine calcareous deposits (Bellion, 1987; Kogbe, 1981).

Coastal plain and shallow marine deposits of Campanian–Paleocene age are divided into the Rima Group and the Sokoto Group. The Taloka Formation (which is the focus of this study) forms the oldest member of the Rima Group. Previously, the Taloka Formation has been broadly interpreted as predominantly silty and muddy successions deposited in a coastal plain environment during the Maastrichtian (Kogbe, 1981). Here, depositional environments and processes and framework for the stratigraphy of the Taloka Formation are well constrained using outcrop investigations, sedimentary logging and ichnological attributes.

2 Methods

A total of over 100 m thick of outcrop sections were logged, measured and sampled. The sections were described in details in terms of lithology, grain size, texture, bed thicknesses and attitudes, sedimentary structures and ichnological properties. The logs were used to designate facies and facies associations. Facies successions were recognized and depositional environments were assigned. Strata patterns and relationships were applied to assign system tracts.

3 Results: Lithofacies and Facies Associations

Ten lithofacies were identified from the outcrop section: (L1) structureless siltstone and mudstone, (L2) laminated mudstone and siltstone, (L3) lignitic shale/mudstone, (L4) flaser-lenticular bedded siltstone, (L5) rippled siltstone and sandstone, (L6) hummocky cross-stratified heterolith, (L7) Crossbedded siltstones and sandstones, (L8) Thin-bedded Heterolith, (L9) Inclined siltstone and sandstone, and (L10) Phosphatized calcareous beds. Six facies associations were derived (Table 1).

4 Discussion

The facies characteristics of the Taloka Formation indicate a marginal marine depositional setting, with a mixture of wave, tide and fluvial influences. Hence, a mixed-energy depositional model, made up of two depositional systems, was proposed for the deposits of Taloka Formation: (i) a mixed wave- and tide-influenced deltaic system and (ii) a tide-dominated deltaic system. The latter constitutes the dominant environment.

Table 1 Facies associations and interpreted depositional environments of the Taloka formation

Facies association	Lithologic description	Structures	Fossil/ichnology	Interpretation
FA1 intertidal/marsh 	Dark, massive, bioturbated lignitic silty mudstone with organic debris; commonly sandwiched by relatively sandy units (L3; i.e., delta front); Or dark carbonaceous shale with gypsum at the top and dispersed organic material overlain hummocky beds. Sometimes occur as cm-thick muddy siltstones with mud drapes overlain by delta front/HCS beds. FA1 reaches 3 m thick; shale/lignite range between < 1 and 1.5 m thick)	Fissility; horizontal—wavy bedding and mud cracks	Low—moderate BI; bivalves imprints, *Skolithos* and rooted	Low energy, suspension deposits, mangrove/marsh, Tide-influenced subaqueous/intertidal marginal marine
FA2 delta front 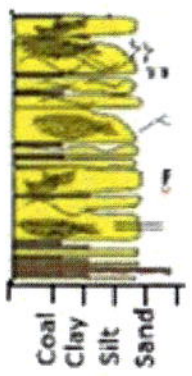	Very fine—fine-grained off white and light grey thin-bedded coarsening- and thickening-upwards rippled sandstone and siltstone/siltstone-mudstone heteroliths. Muddy at the base with brownish ferruginized pebbly mud layer (L5, L8). Associated with flaser—lenticular beds (L4) which cut into the heteroliths and rippled sandstone. FA2 is the sandiest part of the successions. It terminates in a thin pedogenic layer, overlain by tidal bar (FA3). FA2 is up to 12 m thick (10–20 cm thick beds)	Ripples, laminations, Flaser-lenticular beds	Weak BI, *Skolithos* ichnofacies	Moderate-high energy currents; stressed marine/marginal marine?; Delta front deposit probably on a distal position?
FA3 tidal/mouth Bar 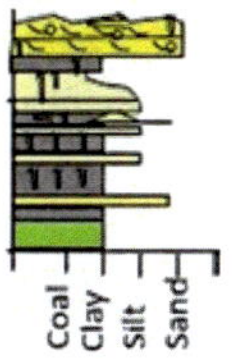	Coarsening- to fining-upward heterolithic sandy siltstone and mudstone. (i) 5 m thick basal bioturbated muddy siltstone and sandy mudstone crossbed with laterally extensive ferruginized, brown, off white and yellow-brown sandy mudstone (L7). (ii) Upper, 4 m thick near-horizontal / inclined thin bedded or laminated siltstone with mud drapes (L8 and L9); the heteriliths is about 2–5 m across, terminate in a churned (*Thalassinoides*) bed or ferruginzied calcareous "ledge" or by inclined plane beds/calcareous lag. FA3 is 9 m thick; <25–90 cm thick beds	Crossbeds, few slumps, inclined beds mudcracks	*Skolithos*, *Ophiomorpha*; *Cruziana* ichnofacies; bivalves?	Lateral/backward/forward accretion sediments of a tidal bar deposit shallow marine to brackish water

(continued)

Table 1 (continued)

Facies association	Lithologic description	Structures	Fossil/ichnology	Interpretation
FA4 tide—dominated channels 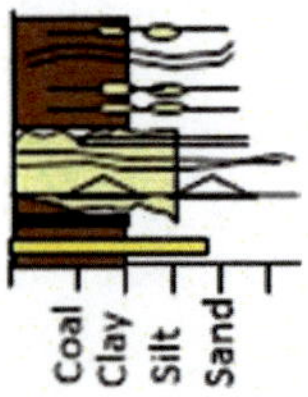	Fining upward dm-scale heterolithic successions of flaser—wavy-lenticular shaped siltstone bodies. Grey, off white/ yellowish brown. The lenticular beds have convex-up geometry with mud partings (L4) and bivalve imprints; homogenized. Mudcracked at the top. FA4 is interbedded with crossbedded siltstone (L7 and L8)/mudstone with carbonaceous debris. Overlain gradationally by FA5. 1–2 m thick. Laminated bed is 0.5–2 cm; lenticular beds<2–5 cm	Lenticular, wavy bedding; laminations and mudcracks	Weak to strong BI; increasing-up ichnodiversity (e.g., horizontal burrows)	Fluctuating low and high energy current deposition. Low energy tractions; Tide-dominated channel/ tidal flat deposits
FA5 bay/subtidal—intertidal flat 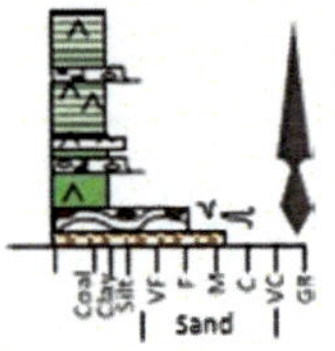	Fining upward offwhite strongly bioturbated/homogenized or bedded silt- and mud-dominated heteroliths (L1 and L4). Rarely thinly laminated and slightly inclined bedded siltstone (L8). Some minor lignitic debris (L3/FA1) also present. At the top, is parallel-wavy bedded mudcracked siltstone. Some lignitic mudstone infill the burrows as lenses. Networked calcified burrows (*Glossifungites?*) crosscut both the top of FA5 and the base of an overlying calcareous lag. Or the top is replaced by slightly inclined siltstone (L9), which is overlain by wavy calcareous lag. FA5 is up to 4 m thick	Mudcracks, horizontal—wavy bedding, inclined bedding	Moderate to intense BI; low ichnodiversity (*Thalassinoides*) (Stumps/roots?)	Tide—dominated (subtidal or intertidal environment) brackish water deposit/bay deposits?
FA6 lower shoreface 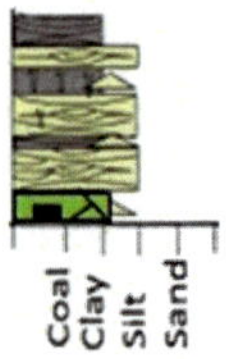	Very fine-grained grey, brown and reddish hummocky cross-stratification (HCS) siltstone/sandy siltstone (L6 and some L5) interstratified and cross laminated with strongly bioturbated mudstone (L2) and horizontal thin bedded siltstone. FA6 is associated/overlain gradationally or erosively by units FA2 or FA3. Bedding thickness ranges from 5 to 20 cm	HCS, asymmetric ripples, laminations	Low BI; *Skolithos*, *Planolites*? Or *Cruziana* species	Fluctuating energy, oscillatory flows (storm deposits) Lower shoreface

5 Conclusion

Based on predominance of tidally-influenced facies associations (FA 1–5) with a limited occurrence of wave influence (FA 6), and the stratal patterns of the recognized facies, a mixed tide-and wave-depositional model was proposed for the Taloka Formation (Ainsworth et al., 2011; Galloway & Broussard, 1975; Lawal & Hassan, 2023; Ta et al., 2002).

References

Ainsworth, R. B., Vakarelov, B. K., & Nanson, R. A. (2011). Dynamic spatial and temporal prediction of changes in depositional processes on clastic shorelines: Toward improved subsurface uncertainty reduction and management. *AAPG Bulletin, 95*, 267–297.

Bellion, Y. (1987). *Histoire geodynamique post-paieozdique de I'Afrique de I'Ouest d'apres I'Ctude de quelques bassins sedimentaires (Senegal, Taoudenni, Iullemmeden, Tchad)*. Ph.D. Thesis. Pub. Occasionnelle CIFEG 17.

Galloway, W. E. (1975). Process framework for describing the morphological and stratigraphic evolution of deltaic depositional systems. In M. E. Broussard (Ed.), *Deltas* (pp. 87–98). Houston Geological Society.

Guiraud, R., Bosworth, W., Thierry, J., & Delplanque, A. (2005). Phanerozoic geological evolution of Northern and Central Africa: An overview. *Journal of African Earth Sciences, 43*, 83–143.

Kogbe, C. (1981). Cretaceous and tertiary of the iullemmeden basin in Nigeria (West Africa). *Cretaceous Research, 2*, 129–186.

Lawal, M., & Hassan, M. H. A. (2023). Sedimentary and stratigraphic characteristics of the Maastrichtian Dukamaje formation of southern Iullemmeden Basin: Implications for the paleogeography of the upper cretaceous trans-Saharan seaway. *Journal of African Earth Sciences*, 104878.

Lawal, M., Hassan, M. H. A., Abdullah, W. H., & Otchere, D. A. (2023). Sedimentary facies and stratigraphy of the campanian-maastrichtian taloka formation, southeastern Iullemmeden Basin, Nigeria. *Journal of African Earth Sciences*, 104842.

Petters, S. W. (1979). Stratigraphic history of the south-central Saharan region. *Geological Society of America Bulletin, 90*, 753–760.

Ta, T. K. O., Nguyen, V. L., Tateishi, M., Kobayashi, I., Tanabe, S., & Saito, Y. (2002). Holocene delta evolution and sediment discharge of the Mekong River southern Vietnam. *Quaternary Science Reviews, 21*, 1807–1819.

Whiteman, A. (1982). *Nigeria: Its petroleum geology, resources and potential* (p. 381). Graham and Trotman.

Bio-Chemo-Petrographical Characteristics of the Kalambaina Formation as a Major Potential Carbonate Reservoir in the Sokoto Basin of Northwestern Nigeria

Nuhu Obaje, Adullahi Bomai, Sunday Moses, Muhammad Ali, Suleiman Adamu, Alfred Essien, Umar Lamorde, Umar Umar, Toochokwu Ozoji and Lukman Adamu

Abstract

Previous and ongoing studies have postulated the existence of a Cretaceous-Paleocene petroleum system in the Sokoto Basin of Northwestern Nigeria. The Nigerian National Petroleum Corporation is presently carrying out exploration projects in the Sokoto Basin. The Paleocene shallow to deep marine carbonate facies of the Kalambaina Formation is expected to be a major carbonate reservoir in the petroleum system build-up similar to major carbonate reservoirs in the Tethyan paleogeographical realms of North Africa and the Middle East to which the Sokoto Basin is intricately affiliated. Based on field observations, the carbonate rocks of the Kalambaina Formation have been classified into two facies, namely, the **lower permeable facies** and the **upper impervious facies**. The **lower permeable facies** is characterized by high diversity benthonic foraminifera, echinoids, pelecypods, gastropods, high CaO, granular in hand specimen and classified petrographically as intra-biosparite on the Folk's classification and packstone/grainstone on the Dunham's classification. The **upper impervious facies** is characterized by low diversity benthonic foraminifera, massive mudstone, high MgO content, lacks macrofossil content and petrographically classified as biomicrite on the Folk's classification and mudstone/wackestone on the Dunham's classification, whereby **the lower permeable facies** with up to 100 m stratigraphical thickness is expected to constitute the major good quality carbonate reservoir in the deep subsurface of the Sokoto Basin for possible hydrocarbon storage and fluid flow.

Keywords

Carbonate reservoir · Biomicrite · Wackestone · Packstone · Kalambaina formation

1 Introduction

Some previous studies by Adamu et al. (2020); Kogbe (1979, 1981); Obaje et al., (2013, 2019) and Ozoji and Obaje (2019) have hinted on the possibility of some petroleum systems in the Sokoto Basin of Northwestern Nigeria. The carbonaceous shales in the pre-Maastrichtian Gundumi/Illo Formation are expected to constitute the source rock in the first petroleum system. The sandstones of the Taloka Formation and carbonates of the Dukamaje Formation would constitute the reservoirs while shales of the Dukamaje Formation present the seal lithology. A recent field study (July 2020) by the authors uncovered some large quantity coal deposits within the Illo Formation exposed in Daranna village near Kaoje. The organic matter contents and qualities of these coals are still being studied. The

N. Obaje (✉)
Nigerian National Petroleum Corporation Chair in Basinal Studies, Ibrahim Badamasi Babangida University, Lapai, Nigeria
e-mail: nobaje@ibbu.edu.ng

A. Bomai · S. Moses · M. Ali · S. Adamu
Frontier Exploration Services, Nigerian National Petroleum Corporation, Abuja, Nigeria

A. Essien · U. Lamorde
Research and Development Division, Nigerian National Petroleum Corporation, P/Harcourt, Nigeria

U. Umar · T. Ozoji
Department of Geology, Ibrahim Badamasi Babangida University, Lapai, Nigeria

T. Ozoji
Oulu Mining School, University of Oulu, Oulu, Finland

L. Adamu
Department of Earth Sciences, Kogi State University, Anyigba, Nigeria

F. Lucci et al. (eds.), *Selected Studies in Geomorphology, Sedimentology, and Geochemistry*, Advances in Science, Technology & Innovation, https://doi.org/10.1007/978-3-031-43744-1_32

second potential petroleum system may be derived from the shaley and carbonate source rocks of the Dukamaje Formation, the clastic sandstones of the Wurno Formation as reservoirs with the laterally extensive Dange Formation shales as cap/seal lithology for the Maastrichtian-Paleocene petroleum system. For both petroleum systems, the geographically and stratigraphically extensively deposited limestone/carbonate of the Kalambaina Formation presents mega reservoir opportunities in which migrations from the older systems through fractures, cracks, dissolution tracks and structurally controlled pathways might have led to accumulations in the carbonate and constrained by stratigraphic traps.

2 Geological Background

The Sokoto Basin is a portion of a larger Iullemmden Basin whose geology is as shown of Fig. 1. In the portion referred to as the Sokoto Basin, the pre-Maastrichtian Continental Intercallaire made up of the Illo and Gundumi Formations are overlain unconformably by the Maastrichtian Rima Group, consisting of mudstones and friable sandstones (Taloka and Wurno Formations), separated by the fossiliferous, calcareous and shaley Dukamaje Formation. The Dange and Gamba Formations (mainly shales) separated by the calcareous Kalambaina Formation constitute the Paleocene Sokoto Group. The overlying continental Gwandu Formation forms the Eocene Continental Terminal. These sediments dip gently and thicken gradually towards the northwest and reached maximum thicknesses toward the border with Niger Republic (Obaje et al., 2019).

3 Results and Interpretations

Field studies carried out through traverses within the basin concentrated on the quarry of the Sokoto Cement Company where about 30 m of the Kalambaina Formation is exposed (Fig. 2) and on the Wurno Hill involving the Kalambaina Formation along with the Dange and Gamba Formations, the three constituting the Sokoto Group (entirely marine and Paleocene in age) lie on the widespread Wurno Formation (Fig. 3) of Maastrchtian age and with which they are separated by an unconformity of the Cretaceous-Tertiary boundary. Field studies reveal two types of carbonate facies within the Kalambaina Formation, namely a **lower permeable facies** (Fig. 2) which is generally whitish and granular. Above the lower permeable facies lies the **upper impervious facies** which is yellowish and massive. The **lower permeable facies** is characterized by high diversity benthonic calcareous and aglutinated foraminifera,

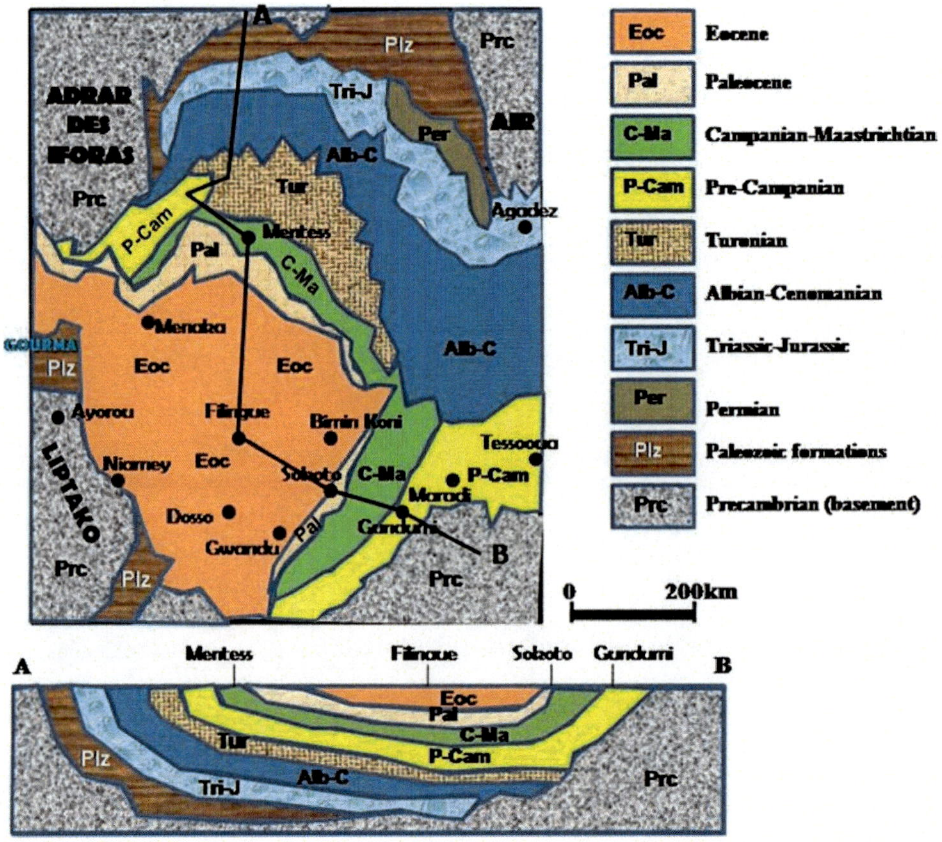

Fig. 1 Geological framework of the Iullemmeden Basin; showing that the formations become younger towards the south (modified from Kogbe, 1979) (Cross section drawn by present authors)

Fig. 2 Field studies reveal two carbonate reservoir facies within the Kalambaina Formation mappable at the quarry of the Sokoto cement company: a lower permeable/porous facies and an upper impervious facies

Fig. 3 A Wurno Formation section at the Wurno hill lies unconformably below the Sokoto Group to which the Kalambaina formation belongs

echinoids, pelecypods, gastropods, high CaO, and classified as intra-biosparite on the Folk classification and packstone/grainstone on the Dunham classification (Fig. 4). The foraminiferal microfossil assemblage of this facies is dominated by high-diversity rotaliids. The **upper impervious facies** is characterized by low-diversity benthonic foraminifera, massively indurated calcareous mud with high MgO content (dolomitic) and classified as biomicrite on the Folk classification and mudstone/wackestone on the Dunham classification (Fig. 5), whereby the lower permeable facies with up to 100 m stratigraphical thickness in the subsurface (Cement Company Quarry Manager-personal communications) is expected to constitute the major carbonate reservoir in the deep subsurface in the Sokoto Basin. Macrofossils lack in the upper impervious facies while the low population low diversity benthonic foraminiferal assemblages comprise mainly the agglutinated forms of *Ammobaculites* sp., *Haplophragmoides* sp., and a few *Rotalia* sp.

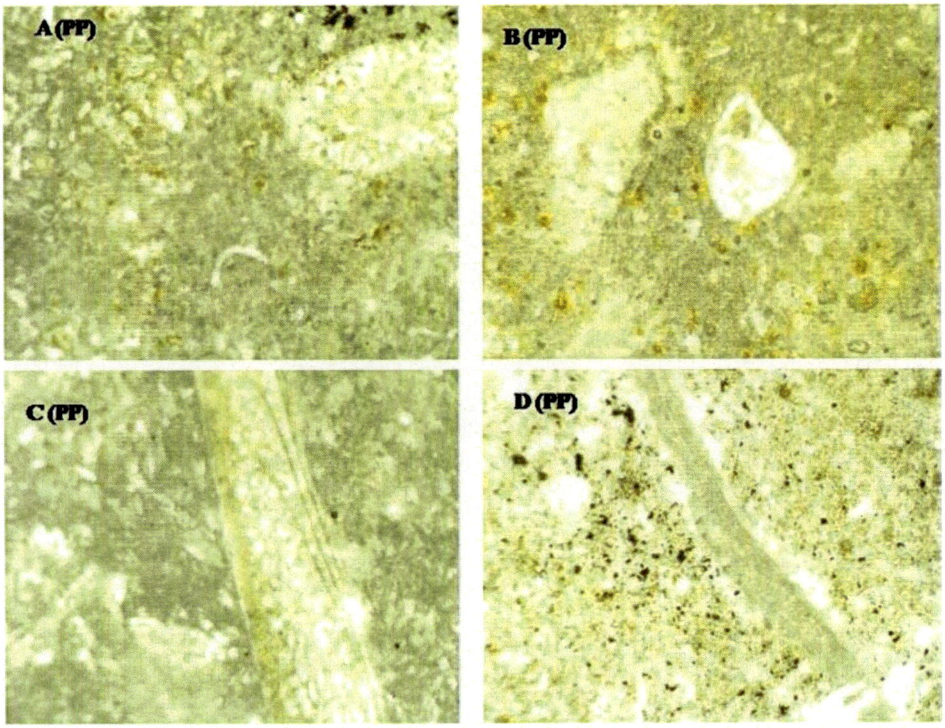

Fig. 4 Thin section petrography of the lower permeable facies of the Kalambaina formation classifies the carbonate as intra-biosparite (Folk classification) or packstone (Dunham classification). PP = plane polar

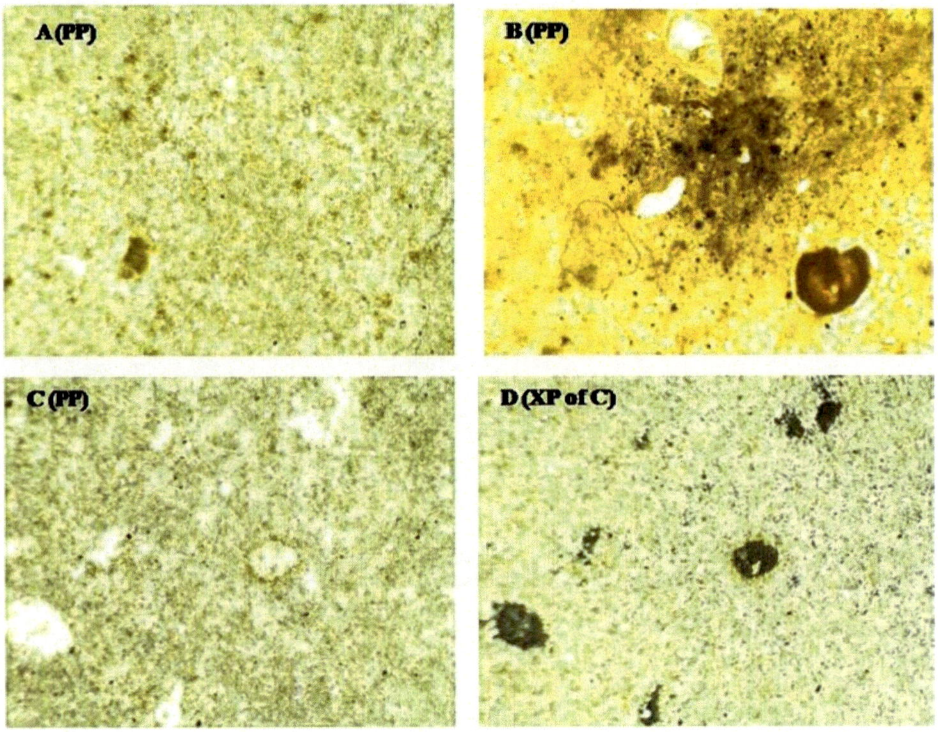

Fig. 5 Thin section petrography of the upper impervious facies of the Kalambaina formation classifies the carbonate as biomicrite (Folk classification) or mudstone/wackestone (Dunham classification). XP = cross polar

4 Conclusion

It is expected that the bio-chemo-petrographical characteristics of the Kalambaina Formation carbonates constitute important guides to identifying the carbonate reservoir intervals in exploratory and appraisal wells that may come up for reserve evaluation. Based on the cross-section constructed by this study authors from the geological configuration of the Iullemmeden Basin, it is possible that older sequences than the pre-Maastrichtian Gundumi/ Illo Formations may exist deep down in the Sokoto Basin beyond our current knowledge, particularly the possibility of Turonian carbonate deposit deep in the subsurface (Fig. 1) thereby increasing the volume of potential carbonate reservoirs in the basin.

Acknowledgements This study was funded by the Nigerian National Petroleum Corporation (NNPC) endowed Professorial Chair in Basinal Studies at the Ibrahim Badamasi Babangida University Lapai (IBBUL), Niger State, Nigeria. The managements of the NNPC and IBBUL are gratefully appreciated.

References

Adamu, L. M., Obaje, N. G., Okafor, P. N., & Umar, U. M. (2020). Integrated geochemical and multivariate statistical examination of source rocks in the Sokoto and Anambra Basins, Nigeria: implications for hydrocarbon prospectivity. *Petroleum Science and Engineering, 4*(2), 39–50.

Dunham, R. J. (1962). Classification of carbonate rocks according to depositional texture. In W. E. Ham (Ed.), *Classification of carbonate rocks* (Vol. 1, pp. 108–121). American Association of Petroleum, Geologists Memoir.

Folk, R. L. (1959). Practical petrographic classification of limestones. *American Association of Petroleum Geologists Bulletin, 43*, 1–38.

Kogbe, C. A. (1979). Geology of the southeastern (Sokoto) sector of the Iullemmeden Basin. *Bulletin Department of Geology, Ahmadu Bello University, Zaria, Nigeria, 2*(1), 420.

Kogbe, C. A. (1981). Cretaceous and tertiary of the Iullemmeden Basin of Nigeria (West Africa). *Cretaceous Research, 2*, 129–186.

Obaje, N. G., Aduku, M., & Yusuf, I. (2013). The Sokoto Basin of Northwestern Nigeria: a preliminary assessment of the hydrocarbon prospectivity. *Petroleum Technology Development Journal, 3*(2), 71–86.

Obaje, N. G., Faruq, U. Z., Bomai, A., Moses, S. D., Ali, M., Adamu, S., Essien, A., Lamorde, U., Umar, U. M., Ozoji, T., Okonkwo, P., Adamu, L., & Idris-Nda, A. (2019). A short note on the petroleum potential of the Sokoto Basin in North-western Nigeria. *Petroleum Science and Engineering, 4*(1), 34–38.

Ozoji, T. M., & Obaje, N. G. (2019). Paleobiology, paleoenvironment and petroleum geochemistry of the Taloka and Dukamaje formations of the Sokoto Basin, North-western Nigeria. In *Nigerian Association of Petroleum explorationists 37th annual international conference and exhibition, Lagos, Nigeria.*

Microfacies and Depositional Environment of Senakin Coal, Tanjung Formation, South Borneo, Indonesia

Ferian Anggara, Azmii Nanda Fadhillah, Aulia Agus Patria, Donatus Hendra Amijaya and Himawan Tri Bayu Murti Petrus

Abstract

The Tanjung Formation in South Borneo Province, Indonesia, is well-known as an extensive Eocene coal-bearing formation and is economically significant. Coal samples were collected from an open pit mine in Senakin area to study its characteristics. A combination of detailed petrographic composition and geochemical analysis has been used to reconstruct microfacies and depositional environment of the Senakin coal precursor. According to petrographic analysis, the Senakin coal is dominated by vitrinite (62–72 vol%), with less abundant liptinite (20–28 vol%) and inertinite (3.3–6.5 vol%). Mineral matter is rare (2.3–7.8% vol%) of which pyrite is the most abundant mineral. Resembling the modern tropical peats in Indonesia, this section can be derived from hemic to fibric peat types in which the sapric layer is observed at the upper part of the peat. The vertical sequence of maceral abundance in the coal seam shows that development starts from a topogenous to ombrogenous mire and back to a topogenous mire. In line with its variation of coal facies, high Tissue Preservation Index (TPI) and Gelification Index (GI) values suggest that coal samples were derived from woody plants and lower amounts of herbaceous plants. The telmatic setting was influenced by marine influx, which can be distinguished by the presence of framboidal-syngenetic pyrite.

F. Anggara (✉) · A. N. Fadhillah · A. A. Patria · D. H. Amijaya
Department of Geological Engineering, Universitas Gadjah Mada, Yogyakarta, Indonesia
e-mail: ferian@ugm.ac.id

F. Anggara · D. H. Amijaya · H. T. B. M. Petrus
Unconventional Geo-Resources Research Group, Faculty of Engineering, Universitas Gadjah Mada, Yogyakarta, Indonesia

H. T. B. M. Petrus
Department of Chemical Engineering (Sustainable Mineral Processing Research Group), Universitas Gadjah Mada, Yogyakarta, Indonesia

Keywords

Coal petrography · Microfacies · Senakin coal · Indonesia

1 Introduction

According to the ASTM classification, Eocene coal in Senakin, Indonesia, is mainly of high volatile bituminous C and sub-bituminous A rank (Moore & Ferm, 1992). Senakin coalfield provides a suitable area to study coal characteristics and properties because of its accessibility and that it is an economically significant deposit. Although, several studies have been carried out since the coalfield was first mined in 1982 (Febriadi, 2010; Moore & Ferm, 1992; Moore et al., 2020), there has been no comprehensive study using petrographical composition and geochemical analysis to reconstruct the microfacies depositional environment of the Senakin coal.

2 Sampling and Analytical Methods

Field observation and coal sampling were gathered from an active open pit mine in Sebuli area. Thirteen coal samples were collected from three sections by using the ply-by-ply sampling method. Each ply sample was identified on the basis of coal lithotype and macroscopic appearance referring to Diessel (1992) classification system. The samples were prepared for polished sections and proximate analysis. Coal petrography was done according to ASTM D2799-05a (2005) and maceral classification referring to International Committee for Coal and Organic Petrology (ICCP) (1998) for vitrinite, (International Committee for Coal and Organic Petrology (ICCP), 2001) for inertinite, and (Pickel et al., 1994) for liptinite. Proximate analyses for all coal samples were conducted by following ASTM Standards (ASTM, 2005) to determine the ash yield.

F. Lucci et al. (eds.), *Selected Studies in Geomorphology, Sedimentology, and Geochemistry*,
Advances in Science, Technology & Innovation, https://doi.org/10.1007/978-3-031-43744-1_33

3 Results

3.1 Macroscopic Characteristics

Coal in the investigated seam shows a relatively homogenous lithotype of a varying thickness. The main coal seam is divided into three sections (upper, middle, and lower main coal seam). Coal plies in each seam part were separated with volcaniclastic parting bands of varying thicknesses. Coal samples with relatively high sulfur content were found at the upper part of the seam. Below the basal part of the seam, thick layers of tuffaceous mudstone were found. These volcanogenic beds could be the source of mineral matter in the coals.

3.2 Maceral Composition

The petrographic analysis has shown that investigated samples are dominated by vitrinite (62–72 vol%), followed by liptinite (20–28 vol%) and inertinite (3.3–6.5 vol%). Telovitrinite which dominates the vitrinite maceral group has been found in all the analyzed samples (>50 vol% of the total coal components). Detrovitrinite is observed in much smaller amount, varying from 1.9 to 12.5 vol%, while gelovitrinite is found from 0.7 to 6.3 vol%. Liptinite percentages are high (20–28 vol%). The common liptinite macerals in investigated samples are sporinite, cutinite, resinite and alginite. Small amounts of exsudatinite, liptodetrinite and suberinite are also observed in all analyzed samples.

Inertinite occurred in the smallest amount compared to the other two maceral groups. Funginite dominates the inertinite group in all samples. Inertodetrinite, semifusinite and fusinite were also found in smaller amounts. The upper and lower parts of the seam are distinguished by higher abundance of inertinite macerals.

3.3 Mineral Matter and Ash Yield Contents

Based on the petrographic analysis, mineral matter percentages vary from 2.3 to 7.8 vol%. Pyrite is found as the most common mineral in all samples (1.4–6.1 vol%). Clay minerals and quartz are also observed in smaller amounts. Pyrites are observed as syngenetic and epigenetic types. Syngenetic pyrites occur in framboidal structures, euhedral crystals, and concretions, while epigenetic pyrites are present as cell-infilling, discrete pyrites and cross the macerals, especially associated with exsudatinite.

Average value for ash yield in Senakin bright coal lithotype is 8.2 vol% (Febriadi, 2010). According to proximate analysis, it is obtained that average ash yield value for coal samples are 29 vol%, 20 vol%, and 20 vol%, respectively.

3.4 Coal Microfacies

Coal microfacies were obtained by plotting the percentages of three maceral groups in the ternary diagram (mineral-matter-free, mmf) (Fig. 1). Three classes formed by plotting patterns were used as basic information to deduce paleoenvironments during the peatification stage (Table 1). A more detailed classification using maceral composition association was done to accomplish paleoenvironment reconstruction in the Senakin coal. This classification can be used to give a more integrated explanation of the evolution of the seam in the peatification stage. Based on maceral assemblages in vol.%, Senakin coal can be classified into four groups (Table 1).

4 Discussion

Based on a detailed maceral identification from base to top section, the volume of structured vitrinite decreases and degraded vitrinite increases, and vice versa. Gelified vitrinite is observed mainly in the middle part of this section. These trends are marked as an upward evolution from topogenous to ombrogenous mire. The base of the peat section might be derived from sapric and fine hemic peat, the middle section from hemic and fibric peat, and the top section from by sapric peat. The occurrence of mineral matter especially pyrite, indicates that the mire was influenced by marine influx during early stage diagenetic processes

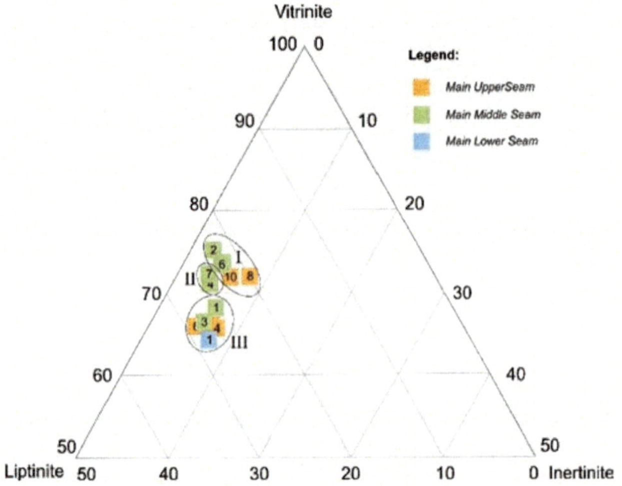

Fig. 1 Coal facies of Senakin coal based on their maceral assemblages (mmf basis). Most of all seams are classified in Facies III and I, Facies II is composed only of main middle seam

Table 1 Parameters used for the coal microfacies classification and their corresponding maceral assemblages

Microfacies (%, mmf)	Maceral Assemblages (%, vol)
Facies (I): vitrinite content>72 (%, mmf), liptinite content around 20–30 (%, mmf) and inertinite<10 (%, mmf)	Telovitrinite-rich group: assemblages of>62 (%, vol) telovitrinite macerals,<22 (%, vol) liptinite and<5 (%, vol) inertinite
	Detrovitrinite-rich group: assemblages of>6 (%, vol) detrovitrinite macerals, ≤20 (%, vol) liptinite and>5 (%, vol) inertinite
Facies (II): vitrinite content ≤72 (%, mmf), liptinite content>20 (%, mmf) and inertinite<5 (%, mmf)	Telovitrinite-liptinite-rich group: assemblages of<63 (%, vol) telovitrinite macerals,>22 (%, vol) liptinite and<5 (%, vol) inertinite
Facies (III): vitrinite content<70 (%, mmf), liptinite content>25 (%, mmf) and inertinite>5 (%, mmf)	Detrovitrinite-liptinite-rich group: assemblages of>5 (%, vol) detrovitrinite macerals,>25 (%, vol) liptinite and<10 (%, vol) inertinite

(Fig. 2). In line with maceral and mineral characteristics, the TPI versus GI (Amijaya & Littke, 2005) plot also suggests the Senakin depositional environment in a telmatic area, which was influenced by marine influx. Peat was deposited in wet-forest swamp condition with major input of woody plants and less herbaceous plants.

5 Conclusion

The studied coals in the Senakin area are rich in vitrinite and liptinite and contain less inertinite and mineral matter. The maceral assemblages can be correlated to the pre-maceral in Indonesian modern tropical peats. The ideal model of development from topogenous to ombrogenous peats is indicated by the series of detrovitrinite-liptinite rich group at the base to telovitrinite-liptinite rich group at the middle and telovitrinite rich and detrovitrinite rich group at the top

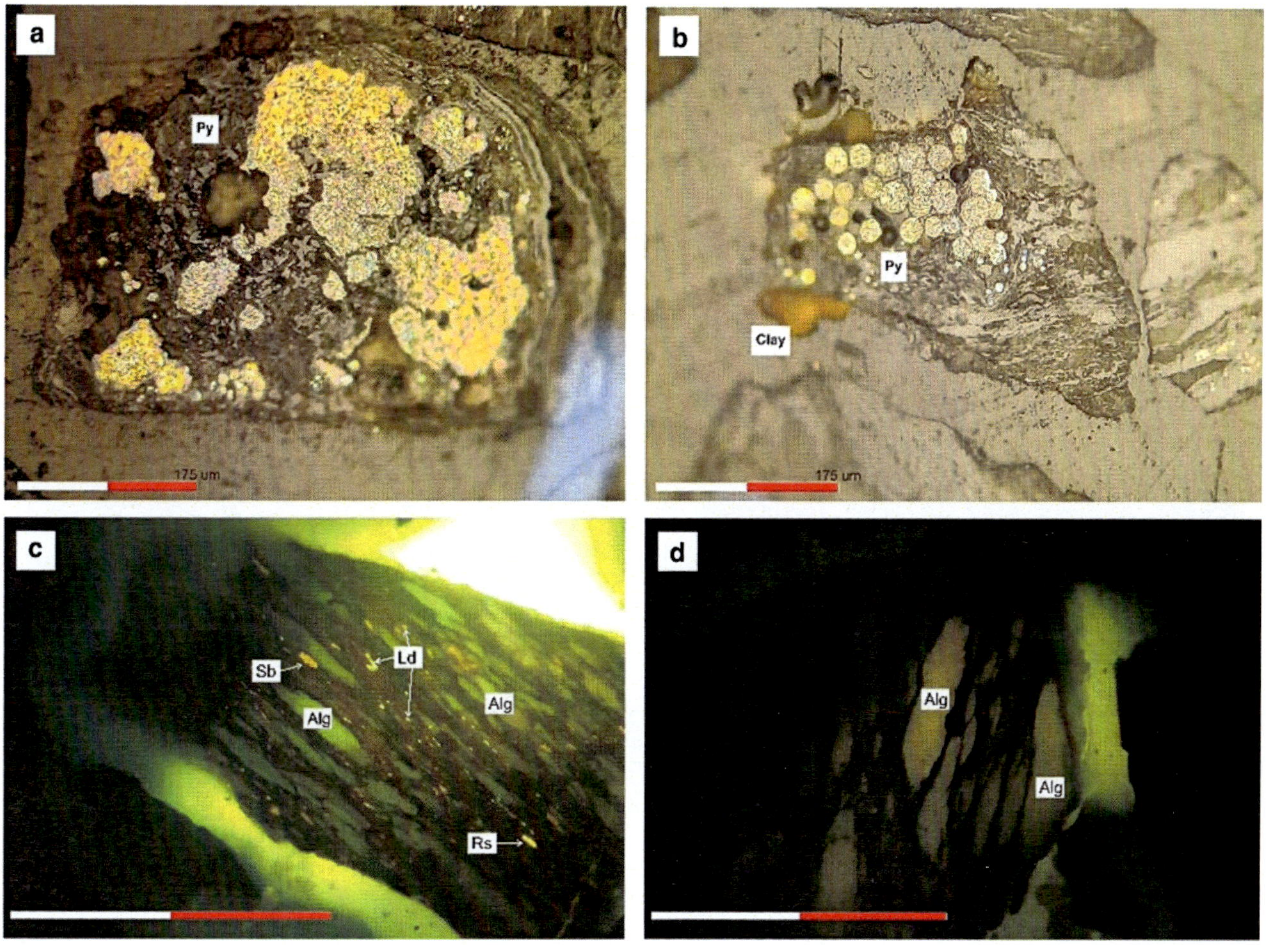

Fig. 2 Photomicrograph of Senakin coal. **a** and **b** Massive and framboidal pyrite in vitrinite macerals, indicate as syngenetic pyrite during peatification, **c** and **d** liptinite macerals in fluoroscence shows the abundant of alginite, resinite, sporinite and liptodetrinite

of the section. The use of a TPI-GI diagram indicates the predominance of woody plants and less herbaceous plants. The paleodepositional environment is represented by wet forest swamp conditions. The abundance of syngenetic pyrite indicates marine influence during peatification.

References

Amijaya, H., & Littke, R. (2005). Microfacies and depositional environment of Tertiary Tanjung Enim low rank coal, South Sumatra basin, Indonesia. *International Journal of Coal Geology, 61*, 197–221.

ASTM D3174-04. (2005). Test method for ash in the analysis sample of coal and coke: Gaseous fuels; Coal and coke. 05, 06.

ASTM D2799-05a. (2005). *Test method for microscopical determination of the maceral composition of coal.*

Diessel, C. F. K. (1992). *Coal-bearing depositional systems* (p. 721). Springer.

Febriadi, E. (2010). PT Arutmin Indonesia: Discovery of South Kalimantan coal. In *Proceedings, MGEI—IAGI, Kalimantan coal and mineral resources* (pp 27–48).

International Committee for Coal and Organic Petrology (ICCP). (1998). The new vitrinite classification (ICCP System 1994). Elsevier.

International Committee for Coal and Organic Petrology (ICCP). (2001). The new inertinite classification (ICCP System 1994). *Fuel, 80*, 459–471.

Moore, T. A., & dan Ferm, J. C. (1992). Composition and grain size of an Eocene coal bed in southeastern Kalimantan, Indonesia. *International Journal of Coal Geology, 21*, 1–30.

Moore, T. A., Friederich, M. C., Anggara, F., Amijaya, D. H., & Trofimovs, J. (2020). Volcanic and volcanogenic sediments in the coal-bearing Tanjung Formation (Late Eocene), Senakin Peninsula, South Kalimantan (Borneo), Indonesia. *Indonesian Journal on Geoscience, 7*, 65–85.

Pickel, W., Kus, J., Flores, D., Kalaitzidis, S., Christians, K., Cardott, B. J., Misz-Kennan, M., Rodrigues, S., Hentschel, A., Hamor-Vido, M., Crosdale, P., & Wagner, N. (2017). ICCP. Classification of liptinite—ICCP system 1994. *International Journal of Coal Geology, 169*, 40–61.

The Paleobiodiversity of the Famennian Ammonoid Genera on the Northern Margin of Gondwana: A Case from the Western Sahara of Algeria

Abdelkader Abbache, Aicha Fetati and Mansour Zaagane

Abstract

The Devonian is a cardinal stratigraphic period of the Paleozoic era due to major faunal, environmental, climatic and atmospheric changes. It contains one of the major biological crises during the Late Devonian (Frasnian-Famennian boundary). This coincides with a global eustatic fall and consequently the disappearance of large faunal groups. After the Frasnian, the dispersal of the ammonoid Order Goniatita and the appearance of the new Order Clymenida originated after these variations, and both are central to define the subdivisions and biozonation of the Famennian. From north to south, in NW Algerian Sahara and on the northern Gondwanan margin, three sedimentary basins were studied: the Bechar basin in the north (Ben-Zireg); the Ougarta basin (Saoura) in the south, and the Timimoun basin (Gourara) further south. During the Late Devonian and especially the Famennian, these regions had epicontinental sea as a consequence of paleocontinental collision and the extension of carbonate platforms. This led to the development of "griotte" facies, as a mixed sedimentation (clay-sandstone-limestone) controlled by tectonic activity (subsidence and resistance) and deposited in different sedimentary environments ranging from a shallow platform to a deep subsiding basin. In these basins, the generic ammonoid (Goniatitida, Clymeniida) biodiversity is interesting especially in the red carbonate facies of griotte types, highly rich in ammonoid macrofauna. Here, 13 families with 28 genera were identified, distributed unequally among families. This pelagic fauna enabled us to examine biodiversity at the generic level of this faunal group. The objective was to show fluctuations in the indices of diversity between the different sectors. The generic diversity and abundance index show fluctuating values among the respective sectors, with a higher relative abundance in the Saoura sector (Ougarta).

Keywords

Northwestern Sahara · Algeria · Biodiversity · Famennian · Goniatitis · Clymenies

1 Introduction

Ammonoids (Mollusca, Cephalopoda) are macrofauna with a very important biostratigraphic value (House, 1996; Petter, 1960). These pelagic organisms, together with conodonts are the main fossil groups to frame Late Devonian biozonation. Their adaptive shell morphologyis a key for inferring ecological information and general patterns of ecosystem diversification during the Devonian (Klug et al., 2010). Indeed, the Famennian corresponds to a major period in the evolution of ammonoids. This interval is characterized by carbonate sediments relatively rich in Goniatitida and Clymeniida.

The study material comes from three areas, from north to south: Ben-Zireg (Bechar basin), Saoura (Ougarta basin) and Gourara (Timimoun basin) (Fig. 1a) located in the mountain ranges of the northwestern Sahara. These mountains are made up of anticlines with Cambro-Ordovician cores and synclines with Siluro-Devonian cores (Fabre, 2005). The Upper Devonian strata of this region has been the subject of numerous mainly stratigraphic and paleontological studies (Abbache et al., 2019; Allaire et al., 2020). The abundance of ammonoids allowed (Petter, 1960) us to frame the modified standard zonation established in Germany and subsequently modified (Korn & Klug, 2015).

A. Abbache (✉) · A. Fetati · M. Zaagane
Mascara University, Mascara, Algeria
e-mail: kader.abbache@gmail.com

A. Abbache
Paleontology, Stratigraphy and Paleoenvironment Loboratory, Mohamed Ben Ahmed University (Oran II), Bir El Djir, Oran, Algeria

F. Lucci et al. (eds.), *Selected Studies in Geomorphology, Sedimentology, and Geochemistry*, Advances in Science, Technology & Innovation, https://doi.org/10.1007/978-3-031-43744-1_34

This study examined the biodiversity of ammonoids in the Famennian in the three areas of the NW Algerian Sahara. We aimed to document the fluctuations in biodiversity and evidence the factors controlling ammonoid distributions during the Famennian.

2 Materials and Methods

The three study sections display the following Famennian stratigraphic units: (i) The Griottes Limestone Formation of Ben-Zireg which strongly folded rocks from Famennian II to Famennian VI corresponding to a reddish series with “griottes” type carbonate, rich in ammonoid fauna (Fig. 1b). (ii) In the Saoura (Marhouma and Ouarourout sections, Ougarta Basin), a thick series with mixed facies (clayey, sandstone, carbonate) is represented successively in the Clays Formation of Marhouma and the Sandstones Formation of Ouarourout (Fig. 1c), particularly the median part of griottes type carbonates (Famennian III, IV, V and base of VI) of the Marhouma Clay Formation with a rich ammonoid fauna. (iii) In the Charouine section (Gourara area, Timimoun Basin), the Kahla Clay Formation is represented in a synclinal structure by clays with rare carbonate levels, more or less fossiliferous with ammonoid fauna (Fig. 1d).

The Famennian deposits of the study areas are characterized by an abundant ammonoids fauna of high biostratigraphic value. This diversified fauna group has been the core of many works (Allaire et al., 2020; Petter, 1960) documenting affinities between North Africa and Europe. Systematic sampling allowed a more precise analysis of their diversity during the Famennian. The identified ammonoids (Allaire et al., 2020) are represented by two orders (Goniatitida and Clymeniida) (Fig. 1b–d).

In order to assess the taxonomic diversity of Famennian ammonoids, several paleoecological indices were used (Hammer & Harper, 2006). The taxonomic richness index which corresponds to the total number of taxa, Dtot, was assessed at the generic level. The Shannon–Weaver diversity index, H (Shannon & Weaver, 1949), takes into account both the number of individuals and the number of taxa and is calculated as $H = -.\sum(ni/n)\ln(ni/n)$, with ni the number of individuals of taxa i. Margalef’s wealth index (Margalef, 1958) is calculated as $M = (S - 1)/\ln(n)$, with S the number of genera and n the number of individuals. The dominance index $D = \sum(ni/n)2$ ranges from 0 (fairly represented taxa) to 1 (dominant taxa). The equitability index $E = (-\sum(ni/n)\ln(ni/n))/\ln S$) measures the distribution of individuals within genders, regardless of generic wealth and considers the uniformity of the distribution of individuals among the present taxa. Based on relative abundance, measures of dominance D show high values for less represented assemblages with some common elements and low values when taxa are more or less regularly represented.

3 Results

Thirteen ammonoid families with 28 genera are present in the Famennian formations of the study areas (Figs. 2a and 3). They are represented by 11 genera (with 132 specimens) in the Griottes Limestone Formation of Ben-Zireg, 21 genera (with 230 specimens) in the Marhouma Clay Formation (Saoura area), and 15 genera (with 295 specimens) in the Kahla Clay Formation (Gourara area). The generic abundance is unequal over the considered areas, but particularly noteworthy in the Saoura area.

The rarefaction curves indicate that the Ben-Zireg (11 genera) and Gourara areas (15 genera) are less diverse than the Saoura area (21 genera). However, these three regions do not appear under-sampled (Fig. 2b). The genera that exist only in the Saoura sector are represented by *Procymaclymenia*, *Genuclymenia*, *Discoclymenia*, *Trigonoclymenia*, *Medioclymenia*. In the Saoura area (Marhouma and Ouarourout), the high values are consistent with the indices of diversity (H = 2.37) and wealth (M = 2.75). In the Gourara, values of the diversity index (H = 1.80) and richness index (M = 2.46) are higher lower, indicating a fairly wide range of taxa. In the Ben-Zireg area, the diversity index (H = 1.85) and the wealth index (M = 2.04) are equivalent to those of Gourara, indicating a relatively large range of taxa and showing a moderately high diversity. The H diversity indexes show higher values in the Saoura area (2.37%) than in Gourara (1.80%) and Ben-Zireg (1.85%) area (Fig. 2c). These values are supported by rarefaction analyses (Fig. 2b).

4 Discussion

In detail, the diversity at the generic level is higher in the Saoura area (Ougarta) compared to Ben-Zireg (north) and Gourara (south). Such a differentiation is controlled by environmental conditions of deposition which are more favorable in the Saoura area (photic and aerated in proximal to middle ramp) and less favorable in the other areas (distal ramp to basin) (Fig. 1b–d). The comparison at the generic level between these three areas shows some similarities and an arbitrary spatial distribution with a slight dominance at the generic level in the Saoura area.

The climatic (temperature increase) and sedimentological (carbonate deposits) conditions are controlled by the location of the study areas in the southern hemisphere between 40 and 50° latitudes during the Frasnian and between 30 and 40° latitudes during the Famennian. These

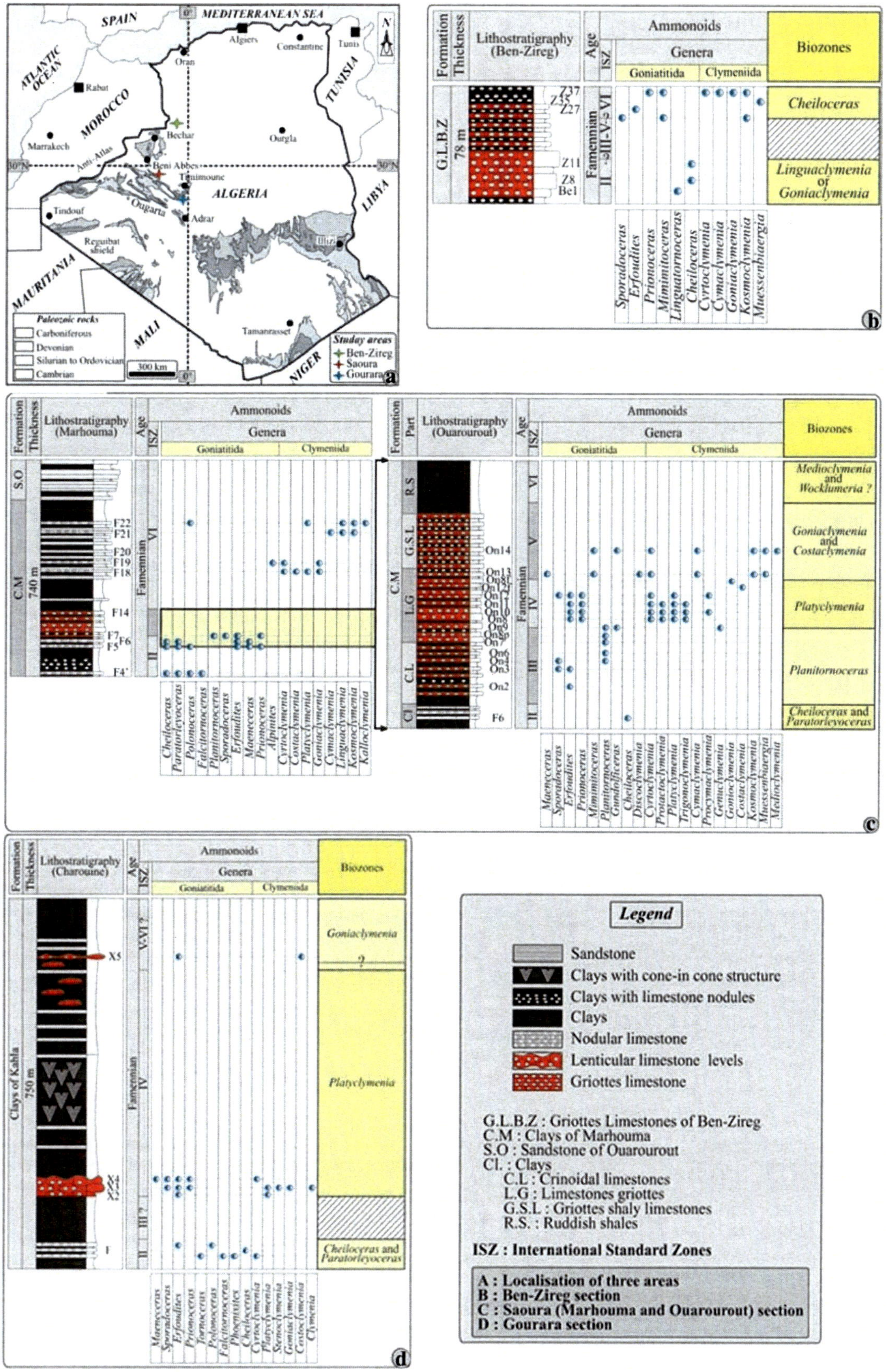

Fig. 1 **a** Geological map showing the Paleozoic exposures and particularly upper Devonian outcrops; **b–d** Lithostratigraphic succession and ammonoid distribution in Ben-Zireg, Saoura (Marhouma and Ouarourout) and Gourara (Charouine) respectively

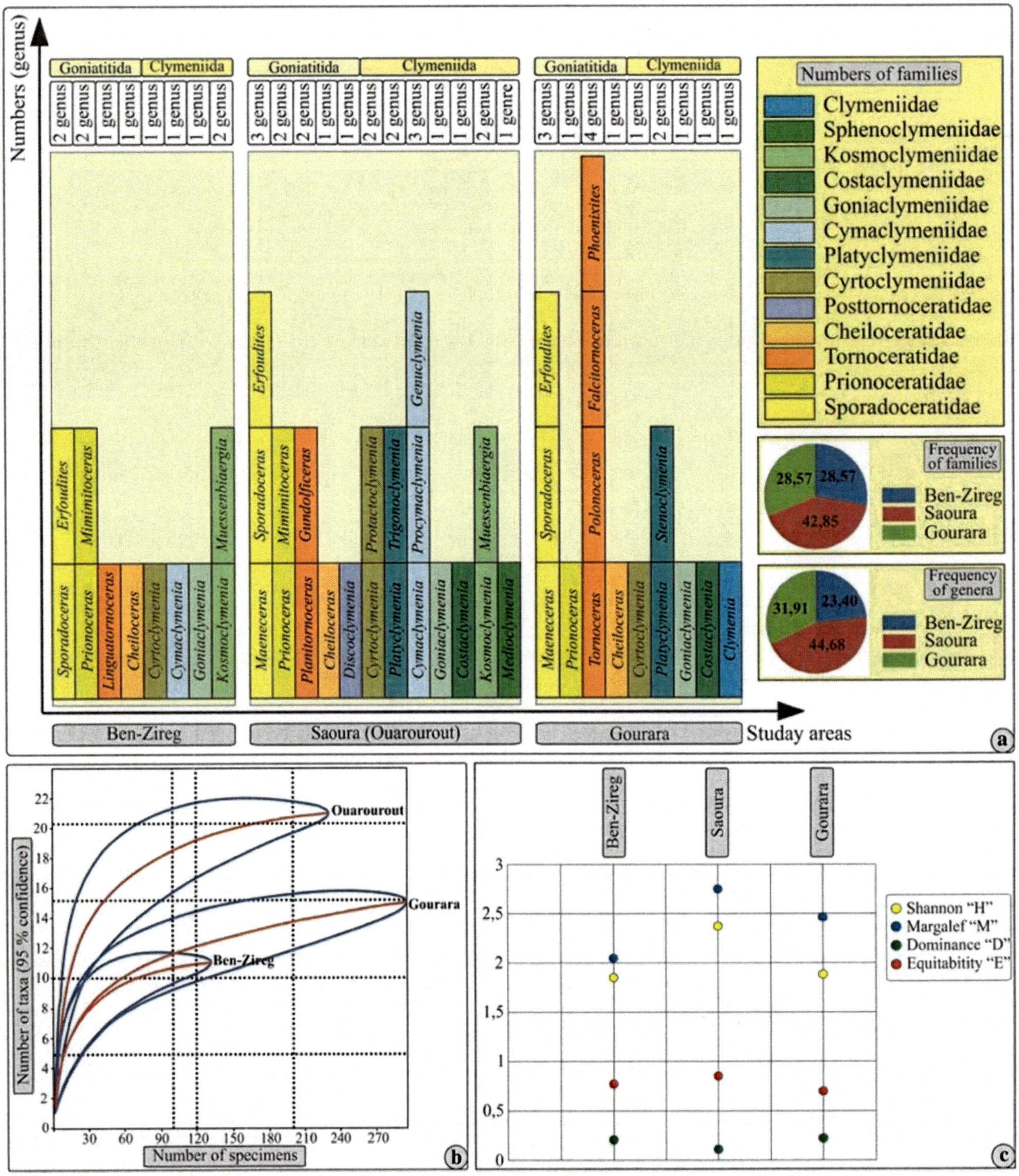

Fig. 2 **a** Number of families and genera in the three study areas during the Famennian; **b** Rarefaction curve at the generic level for each study area; **c** Comparison of paleoecological indices

conditions testify for moderate and relatively warm climate close to the tropic of Capricorn (latitude 30°). On a regional scale, the Frasnian marine transgression spread across the entire Western Saharan platform during the Late Devonian. It deepened the sedimentary environments (with the onset of seaways, sometimes well localized) and micritic carbonate sedimentation (condensed griottes facies) prone to the good conservation of pelagic ammonoid macrofauna associated with benthic forms and to maintenance of high diversity at the family level until the late Famennian (Fabre,

Fig. 3 Goniatititida and Clymeniida genera. **a** *Planitornoceras* (F6-1), **b** *Erfoudites* (F6-28), **c** *Sporadoceras* (F6-20), **d** *Genuclymenia* (On9-17), **e** *Platyclymenia* (On8-38), **f** *Trigonoclymenia* (On8-22), **g** *Kosmoclymenia* (On13-10), **h** *Goniaclymenia* (F19-10), **i** *Clymenia* (X3-17). The black vertical bar represents 1 cm

2005). In detail, diversity is higher in the Saoura area, particularly in Famennian biozones II, III, IV, V and early Famennian zone VI.

5 Conclusion

In the NW Algerian Sahara, in three regions (Ben-Zireg, Saoura and Gourara respectively) representing different basins, Famennian ammonoids belong to two orders (Goniatitida and Clymeniida), 13 families and 28 genera, represented sometimes by only a few specimens. The richness in ammonoid fauna is seemingly controlled by paleoecological/climatic and geographical conditions, and more favorable environmental conditions with fine micritic sedimentation in an oxygenated ramp-like environment. Between the three study regions, the indices of biodiversity and richness in ammonoids (family and genera) are appreciably more important in the Saoura (Ougarta) region, a proximal to median platform area, and less important in the Ben-Zireg region, a region represented by a (distal platform and the region of Gourara represented by a platform distal to a basin. Paleoecological indices are consistent with this faunal distribution.

References

Abbache, A., Ouali Mehadji, A., Cronier, C., Allaire, N., & Monnet, C. (2019). The upper Devonian of the North–Western Sahara (Algeria): Facies, environments and geodynamic significance of the griottes limestones. *Carnets Géologica, 19*, 71–96.

Allaire, N., Abbache, A., Crônier, C., & Monnet, C. (2020). Famennian (Late Devonian) ammonoids from the Ouarourout section (Saoura Valley, Algeria). *Annales de Paléontologie, 106*, 102363. https://doi.org/10.1016/j.annpal.2019.08.002

Fabre, J. (2005). *Geology of western and central Sahara* (vol. 108, p. 572). Tervuren African geoscience collection, Royal Museum for Central Africa.

Hammer, Ø., & Harper, D. A. T. (2006). *Paleontological data analysis* (p. 368). Wiley-Blackwell.

House, M. R. (1996). The middle Devonian Kačák Event. In *Proceedings of the Ussher society* (Vol. 9, pp. 79–84).

Klug, C., Kröger, B., Kiessling, W., Mullins, G. L., Servais, T., Fryda, J., Korn, D., & Turner, S. (2010). The Devonian nekton revolution. *Lethaia, 43*, 465–477.

Korn, D., & Klug, C. (2015). Paleozoic ammonoid biostratigraphy. In: C. Klug, D. Korn, K. De Baets, I. Kruta, & R. H.Mapes, (eds.), *Ammonoid paleobiology: From macroevolution to paleogeography* (Vol. 44, pp. 299–328). Springer, Topics in Geobiology.

Margalef, R. (1958). Information theory in ecology. *General Systematics, 3*, 36–71.

Petter, G. (1960). Clymenia of the Sahara. Publications of the service of the Geological Map of Algeria (new series). *Paleontology memory* (Vol. 6, pp. 1–58).

Shannon, C. E., & Weaver, W. (1949). *The mathematical theory of communication* (p. 117). University of Illinois Press.

Micropaleontology, Sedimentology and Biostratigraphy of the Aptian/Albian Boundary in the Ouadi Aïn Section (Jabal Ammar, Northeastern Tunisia)

Abdallah Elkhazri and Mabrouk Boughdiri

Abstract

In the Ouadi Aïn section (Jabal Ammar, northeastern Tunisia), the Upper Aptian strata are represented by a marly unit (40 mthick). These marls contain a foraminiferal assemblage and are associated with rare radiolarians. The presence of the planktonic index species *Hedbergella trocoidea* suggests an upper Aptian age for this interval. The Albian deposits (100 m thick) lies unconformably on the upper Aptian and are characterized by alternations of marl and limestone at its base, whereas its upper part consists of limestone beds. The occurrence of the planktonic index species *Ticinella primula* indicates a middle Albian age for the base of this unit. In Level 46, the planktonic foraminifer *Rotalipora appenninica* has its first occurrence and thus suggests a late Albian age for the upper part of the section. This taxon becomes frequent and is associated with Textulariidae. Micropaleontological analysis and the stratigraphic distribution of planktonic and benthonic foraminiferal species allowed us to locate the respective boundaries of the biozones in the Ouadi Aïn section. The analysis of the microfauna of the upper Aptian-Albian of the Ouadi Aïn section shows noticeable assemblage distributions in the three units. In the first unit (upper Aptian), benthonic foraminifers are frequent and diverse in both genera and species. In the second unit (middle Albian), the benthonic foraminifers are less common but more diverse. Ostracods and radiolarians are also present. In the last unit (upper Albian), the benthonic foraminifers are absent except in sample 53, whereas the planktonic foraminifers are more frequent and show two peaks of relative abundance. Radiolarians are frequent and represented by spumellarians. The data were compared with results from Tunisia and other localities from the Tethysian margin.

Keywords

Aptian–Albian boundary · Albian · Foraminifera

1 Introduction

The Cretaceous succession in northeast Tunisia consists mainly of marlstone and marly limestone ranging from the Berriasian to the Maastrichtian. Previous studies addressing the Lower Cretaceous period have mainly dealt with stratigraphy and geological mapping (Castany, 1951; Pini, 1971; Solignac, 1967). In Tunisia, the Aptian-Albian interval is still debated. Several studies have focused on the Aptian-Albian, especially its biostratigraphy, micropaleontology and geochemistry (Ben Fadhel et al., 2010, 2011; Burollet, 1956; Chida-Ben Haj Ali 2005; Chihaoui et al., 2010; Dali-Ressot, 1987; Elkhazri, 2012, 2019; Elkhazri et al., 2005, 2009, 2013, 2015a, 2015b; Memmi, 1999; Robaszynski et al., 2007; Saïdi & Belayouni, 1994; Salaj, 1980; Talbi, 1991; Tandia, 2001). The present study focused on the micropaleontology and biostratigraphy of the Aptian-Albian boundary and the Albian deposits in Jabal Ammar, within the Ouadi Aïn section of northeast Tunisia. The microfossil study documents various genera of benthonic and planktonic foraminifers through this section.

The study area of Jabal Ammar is located in northeast Tunisia (Fig. 1). The Mesozoic succession is exposed in the core of a dome with Triassic rocks bounded by tectonic contacts. Tectonics and Triassic diapirism have influenced Lower Cretaceous deposition. The succession investigated at Ouadi Aïn crops out continuously 2 km west of the village of Jebbes, south of the Triassic core and Oued Bekkar.

A. Elkhazri (✉) · M. Boughdiri
Faculty of Sciences of Bizerte, Department of Earth Sciences, University of Carthage, 7021 Bizerte, Zarzouna, Tunisia
e-mail: elkhazriabdallah@yahoo.fr

F. Lucci et al. (eds.), *Selected Studies in Geomorphology, Sedimentology, and Geochemistry*, Advances in Science, Technology & Innovation, https://doi.org/10.1007/978-3-031-43744-1_35

2 Materials and Methods

Both field observations and laboratory analyses were used. The section was logged bed by bed in the field and marly levels and indurated beds were systematically sampled. In the laboratory, samples were (1) washed for residues by soaking the samples with water to disaggregate them and retrieve microfossils, screening the samples on a sieve column, drying in an oven at 60 °C, and hand-picking microfossils under a microscope with a mounted needle. (2) Species were determined according to previous studies for benthonic foraminifers (Elkhazri et al., 2005), planktonic foraminifers (Ben Fadhel et al., 2010, 2011; Burollet, 1956; Chida-Ben Haj Ali 2005) and carbonate microfacial analysis (Dunham, 1962).

3 Results and Discussion

3.1 Results

The section exposes three units that are described from base to top. **Unit 1** is mainly marly (40 m thick), with alternating black carbonated levels (2 m thick) with clays (4 m) in the middle of the unit. The base of the unit (samples 1–5) contains abundant planktonic foraminifers with characteristic species *Hedbergella trocoidea*, *Praehedbergella aptiana*, *P. primare* and *P. luterbacheri*. The benthonic foraminifers are diversified in genera and species: *Ammodiscus*, *Dentalina* sp., *Dentalina linearis*, *D. legumen*, *D. guttifera*, *D. distincta*, *D. communis*, *Lenticulina guttata*, *Gavelinella barremiana*, *Vaginulinopsis excentrica*, *Valvulineria* sp. and *Saracenaria* sp. The microfacies have a mudstone-wackestone texture with benthonic foraminifera. The upper part of this unit (samples 6–12) is formed by marls contains frequent planktonic and benthonic foraminifers of the preceding planktonic species with *Praehedbergella sigali* and *Ticinella eubejaouensis*. Likewise, the previous benthonic species with *Dentalina gracilis*, *Lenticulina ouachensis*, *L. ouachensis multicella*, *Lenticulina* sp., *Lenticulina bartensteini*, *Praedohia hechti*, *P. trochus*, *Falsogaudryinella tealbyensis*, *Pseudonodosaria* sp., *P. mutabilis*, *Saracenaria frankei*, *Dentalina oligostegia*, *Lagena apiculata emaciata*, *Gavelinella* sp., rare radiolarians of spherical shape (spumellarians) and many ostracods. **Unit 2** contains alternating marls (80 m thick) with black limestone beds (1–7 m thick) (samples 13–45). This unit records the first appearance of *Ticinella primula* (sample 13). This first occurrence datum comes just above the erosion surface at level 13. The marly levels contain planktonic assemblages mostly consisting of *Ticinella primula*, *Ticinella* sp., *T. roberti* and *Hedbergella planispira*. Benthonic foraminifers are less common than in unit 1 and mainly represented by *Ammodiscus*, *Dentalina linearis*, *D. legumen*, *Lenticulina guttata*, *Gavelinella* sp., *G. barremiana*, *Dentalina distincta*, *Lenticulina ouachensis*, *Lenticulina* sp., *L. gibber*, *L. bartensteini*, *Praedohia hechti*, *P. trochus*, *Praedohia* sp., *Textularia* sp., *Pseudonodosaria mutabilis*, *P. humilis*, *Saracenaria frankei*, *Lagena apiculata emaciata*, *Marginulinopsis schloembachi*, *Epistomina cretosa* and *Nodosaria paupercula*. The ostracods are still present but less frequent than in unit 1, while radiolarians are very rare at the base of this unit and become abundant from sample 38 represented mainly by the spumellarians (*Cryptamorphella conara*). The microfacies has a mudstone-wackestone texture containing radiolarians of different shapes (Figs. 2 and 3). The lower part (18 m) of the 28 m-thick **unit 3** consists of alternating marls with green limestones (Samples 46 and 56). The marls and limestones display a wackestone texture with numerous planktonic

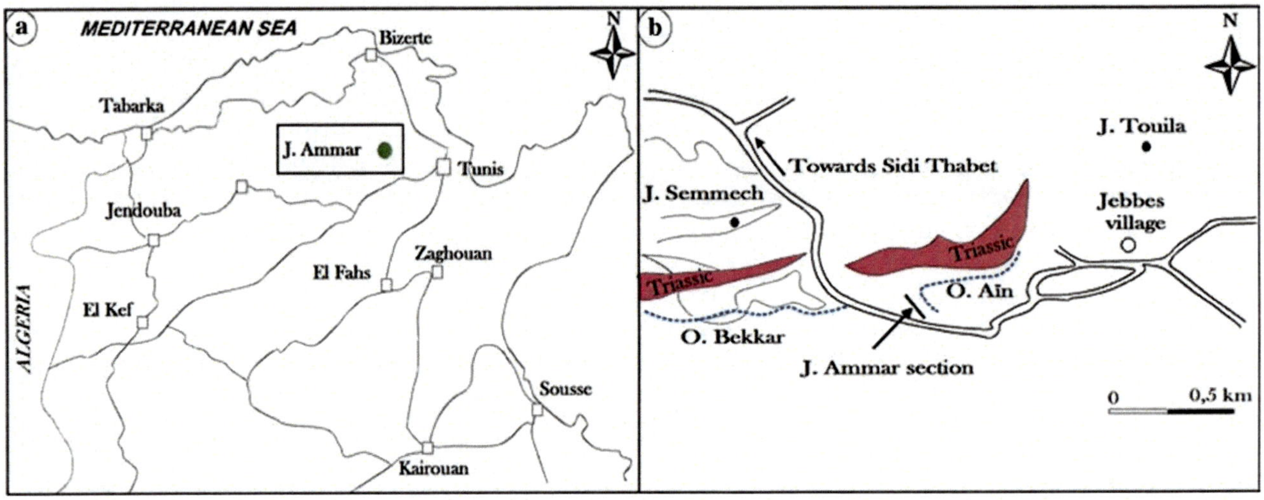

Fig. 1 a Location of the study area in northeast Tunisia, b detail of the investigated area

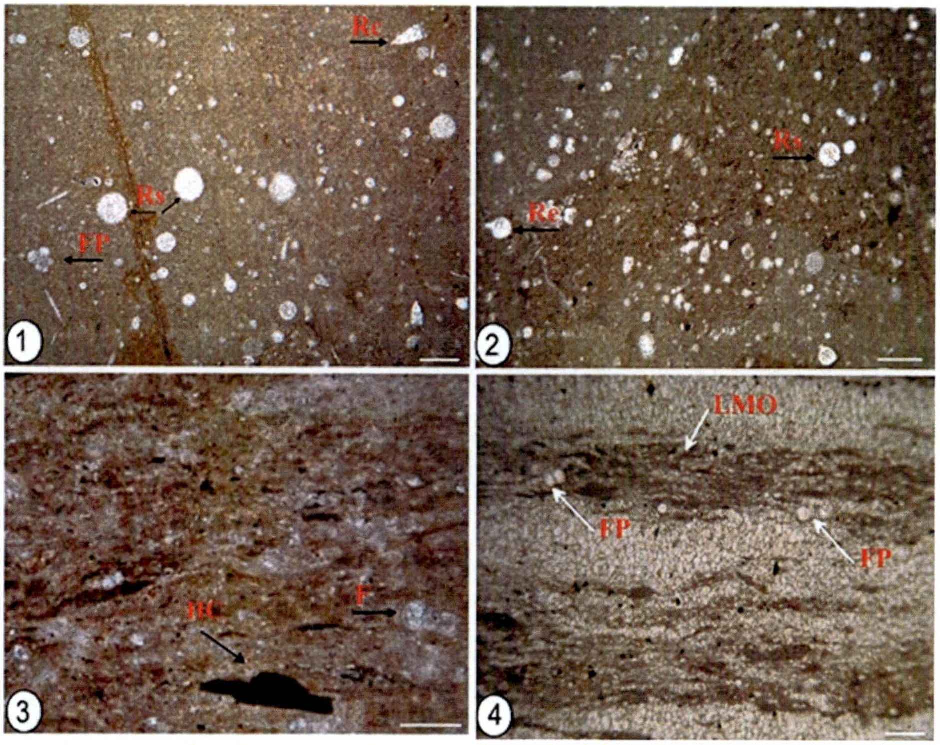

Fig. 2 Middle to upper Albian microfacies (1–2) Mudstone-wackestone containing sections of radiolarians of different shapes (Rs: spherical and Re: elongated), Planktonic foraminifers PF and laminae of organic matter LOM (1, sample 32; 2, sample 44). (3) Mudstone-wackestone including small foraminifera of globular chambers PF with discontinuous laminae containing HC (sample 54). (4) Mudstone with planktonic foraminifera PF with globular chambers, associated a laminae of organic matter LOM (sample 55). Scale bar represents 50 μm

foraminifers and few benthonic ones, except in sample 53: *Gavelinella barremiana*, *Lenticulina* sp., *Textularia* sp. and *Ammodiscus cretaceus*. Planktonic species include *Hedbergella planispira*, *Ticinella* sp, and *T. primula* all associated with spumellarian radiolarians. We note the first occurrence of *Rotalipora appenninica* in sample 46. The microfacies in samples 54–56 document a mudstone texture with planktonic foraminifers and organic matter (Figs. 2, 3, 4 and 5).

3.2 Discussion

The analysis of planktonic foraminiferal assemblages in unit 1 and the presence of index markers *Hedbergella trocoidea*, *Praehedbergella aptiana*, *P. primare*, and *P. luterbacheri* (Chida-Ben Haj Ali 2005; Elkhazri, 2012; Elkhazri et al., 2009, 2013, 2015a, 2015b; Memmi, 1999) all indicate a late Aptian age for the M'Cherga Formation (Burollet, 1956; Chida-Ben Haj Ali 2005; Elkhazri, 2012; Elkhazri et al., 2005, 2009, 2013, 2015a, 2015b; Tandia, 2001). We placed the lower Aptian–middle Albian boundary at level 13 based on the first occurrence of the marker *Ticinella primula*. In unit 2, the joint presence of planktonic foraminifera *Ticinella primula*, *T. roberti* and *Hedbergella planispira* (Bellier & Moullade, 2002; Ben Fadhel et al., 2010; Chida-Ben Haj Ali 2005; Elkhazri, 2019) suggests a middle Albian age. Unit 3 is assigned to the upper Albian based on the first appearance of *Rotalipora appenninica* associated with *Hedbergella planispira*, *Ticinella* sp and *T. primula*. We placed the middle Albian–late Albian boundary at level 46 based on the first occurrence of *Rotalipora appenninica* (Bellier & Moullade, 2002; Ben Fadhel et al., 2010; Chida-Ben Haj Ali 2005; Elkhazri, 2019; Moullade & Tronchetti, 2003). The middle to upper Albian succession belongs to the Fahdene Formation (Ben Fadhel et al., 2010, 2011; Burollet, 1956; Chida-Ben Haj Ali 2005; Chihaoui et al., 2010; Elkhazri, 2019).

The Albian strata unconformably overlie Aptian strata. A hiatus in the upper part of the upper Aptian is recognized in the zones *Planomalina cheniourensis* and *Ticinella eubejaouaensis*. At the same time, the Albian is not complete; we record the absence of a part of the lower Albian (*Hedbergella planispira* Zone) and upper Albian (*Biticinella breggiensis* and *Rotalipora ticinensis* Zones). This gap is probably related to Triassic diapirism which played an important role in this region (Elkhazri et al., 2013) and the Sidi Thabet fault and microfaults that influenced Lower

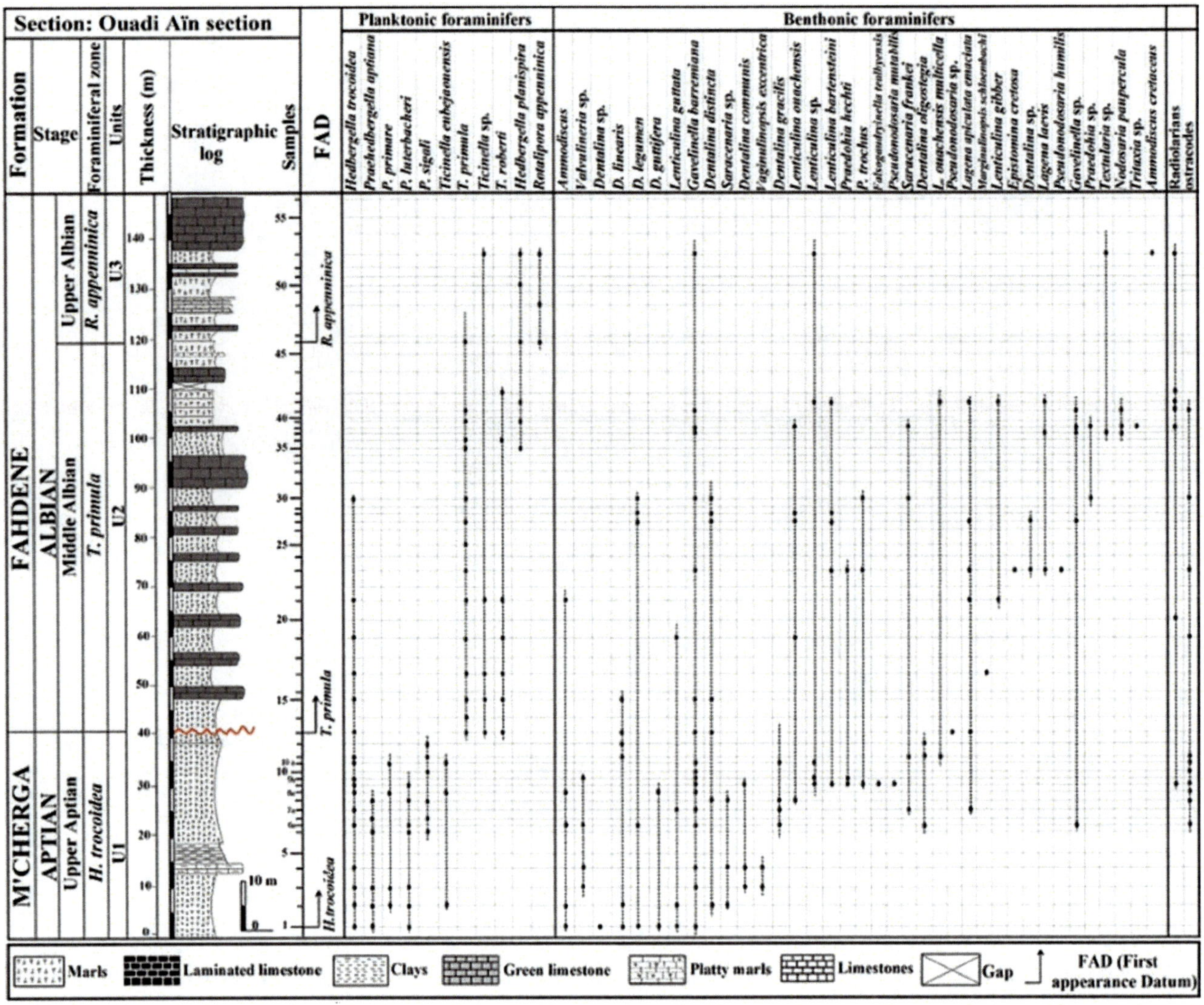

Fig. 3 Lithology and biostratigraphy (benthonic and planktonic foraminifera) distribution in the upper Aptian–Albian of Ouadi Aïn section (Jabal Ammar)

Cretaceous deposition (Elkhazri et al., 2013) and were highlighted in the Nebeur area (northern Tunisia) by Ben Fadhel et al. (2010); Ben Fadhel et al., 2011).This hiatus is similar to the one recorded by Elkhazri (2019) in Jebel Msella (northeast Tunisia) and underlined by Chihaoui (2009); Chihaoui et al., 2010; M'Rabet, 1981). In central Tunisia, it covers the late Aptian/lower to middle Albian interval. This work can be correlated with those of Bellier and Moullade (2002); Ben Fadhel et al., 2010; Ben Fadhel et al., 2011; Chida-Ben Haj Ali 2005; Elkhazri, 2019) for planktonic foraminiferal associations. The depositional environment is interpreted as a semi-confined basin under intensified upwelling with high nutrient supply, like in several upper Albian outcrops ascribed to the Tunisian trough.

A quantitative study of the microfauna in unit 1 (upper Aptian) shows fluctuations of benthonic foraminiferal assemblages both in frequency and in diversity with two peaks in samples 1 and 9 specific diversity does not exceed 12. Planktonic foraminifers are abundant at the base of this unit (intervals 1–3) with specific diversity of 18 to 23.In the upper part (samples 6–12), the planktonic component is frequent with specific diversity of 0 to 12. In unit 2, the benthonic foraminifer diversity fluctuates between 3 and 12. The planktonic specific diversity does not exceed 10. In unit 3, benthonic foraminifers are absent, except in sample 53 and planktonic foraminifers are more frequent and show two abundance peaks in samples 46 and 53 (10 and 8 taxa respectively) (Fig. 4).

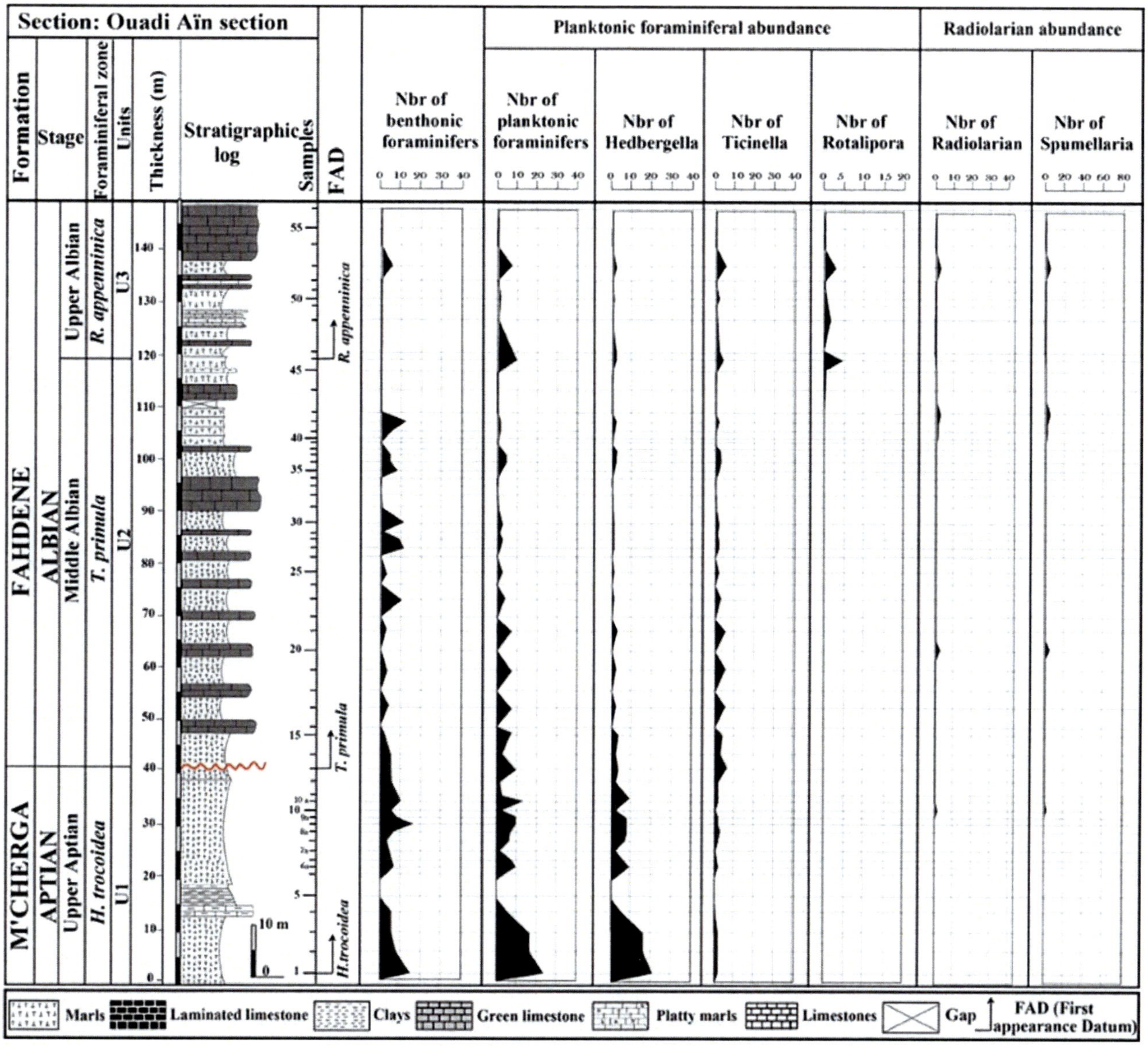

Fig. 4 Quantitative distribution of microfossils in the Upper Aptian-Albian of Ouadi Aïn section (Jabal Ammar)

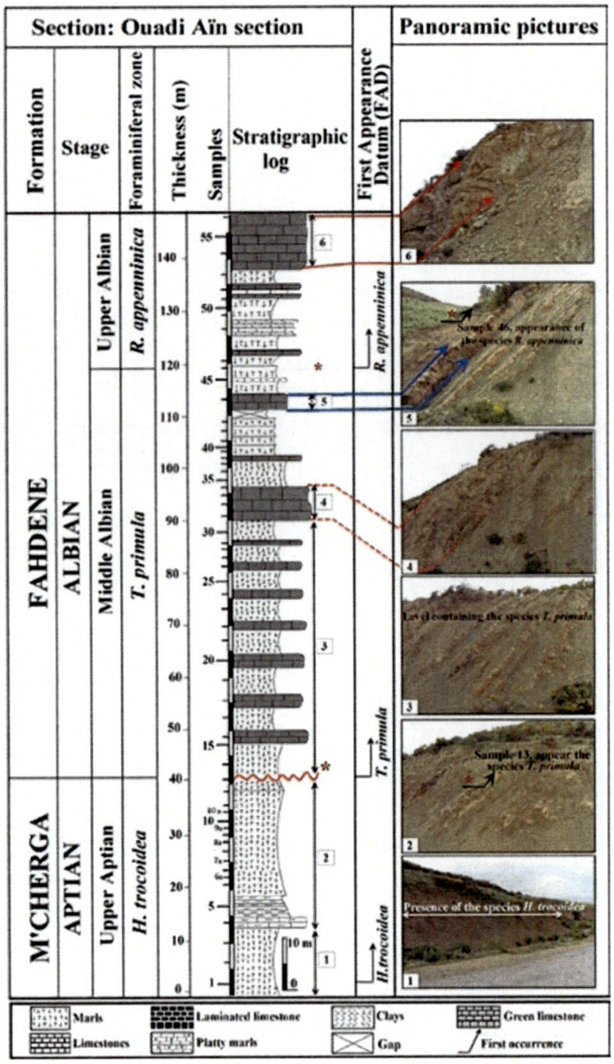

Fig. 5 Panoramic photos and their corresponding lithological level in the upper Aptian-Albian of the Ouadi Aïn section (Jabal Ammar)

4 Conclusion

The Albian lies unconformably on the upper Aptian in the study area. A gap in the upper part of the upper Aptian was recognized. Likewise, the Albian is not complete and is probably caused by a fault. Among the studied materials, we recognized 4 genera that belong to 10 species of planktonic foraminifers, and 18 genera that belong to 26 species of benthonic foraminifers, respectively.

Acknowledgements We would like to thank Dr. Bruno Bernard, Dr. Maurice Ferré, Dr. Orabi Orabi and the other reviewers for their constructive and helpful comments that improved the manuscript.

References

Bellier, J. P., & Moullade, M. (2002). Lower cretaceous planktonic foraminiferal biostratigraphy of the western north Atlantic (ODP LEG 171B), and taxonomic clarification of key index species. *Revue De Micropaléontologie, 45*, 9–26.

Ben Fadhel, M., Layeb, M., & Ben Youssef, M. (2010). Upper Albian planktic foraminifera and radiolarian biostratigraphy (Nebeur–northern Tunisia). *Comptes Rendus Palevol, 9*, 73–81.

Ben Fadhel, M., Layeb, M., Hedfi, A., & Ben Youssef, M. (2011). Albian oceanic anoxic events in northern Tunisia: Biostratigraphic and geochemical insights. *Cretaceous Research, 32*, 685–699.

Burollet, P. F. (1956). Contribution to the stratigraphic study of Central Tunisia (p. 350). Ph.D Thesis, University of Algiers and Annals of Mines and Geology of Tunis. [in French].

Castany, G. (1951). Geological study of the Eastern Tunisian Atlas. *Annals of Mining and Geology, 8*, 632p. [in French].

Chida-Ben Haj Ali, N. (2005). The planktonic foraminifera of the Cretaceous (Hauterivien to Lower *Turonian) of Tunisia: Systematic, biozonation and stratigraphic precisions (p. 348). Ph.D. Thesis, Faculty of Sciences of Tunis. [in French].*

Chihaoui, A. (2009). Albian transgression in the Tajerouine region of central Tunisia: Stratigraphy, sedimentology and synsedimentary tectonics (p. 267). Ph.D. Thesis, University of Grenoble/University of Sfax, Tunis. [in French].

Chihaoui, A., Jaillard, E., Latil, J. L., Zghal, I., Susperregui, A. S., Touir, J., & Ouali, J. (2010). Stratigraphy of the Hameima and lower Fahdene formations in the Tadjerouine area (Northern Tunisia). *Journal of African Earth Sciences, 58*, 387–399.

Dali-Ressot, M. D. (1987). Calcisphaerulidae of the Albian to Maastrichtian lands of central Tunisia (J. Bireno and J. Bou el Ahneche): Systematic interest, stratigraphic and palaeogeographic (p. 191). University of Tunis. [in French]

Danelian, T. (2008). Diversity and biotic changes of Archaeodictyomitrid Radiolaria from the Aptian/Albian transition (OAE1b) of southern Albania. *Micropaleontology, 54*, 3–12.

Dunham, R. J. (1962). Classification of carbonate rocks according to depositional texture. *American Association of Petroleum Geologists Memoir, 1*, 108–121.

Elkhazri, A., Ben Haj Ali, N., Abdallah, H., & Razgallah, S. (2005). Micropaleontology of the Barremo-Aptian deposits of Jabal Msella in north–eastern Tunisia. Global events during the quiet Aptian-Turonian Superchron, Grenoble 21–23 April, Alpine Geology, Special series conferences and excursions (Vol. 6, p. 99).

Elkhazri, A., Razgallah, S., Abdallah, H., & Ben Haj Ali, N. (2009). The Barremo-Aptian anoxic event « OAE 1a » in northeastern Tunisia: Interest of foraminifers. Revue de Paléobiologie (Vol. 28, pp. 93–130). [in French].

Elkhazri, A. (2012). Biostratigraphic study and geochemical signature of the Barremo-Aptian central-northern Tunisia. Ph.D. Thesis, Faculty of Sciences of Tunis, (p. 248). [in French].

Elkhazri, A. (2019). Micropaleontology and biostratigraphy of the upper Albian of Jabal Msella in north–eastern Tunisia. In *2nd conference of the Arabian journal of geosciences (CAJG)*. Sousse, Tunisia, 25–28 November.

Elkhazri, A., Abdallah, H., Razgallah, S., Moullade, M., & Kuhnt, W. (2013). Carbon-isotope and microfaunal stratigraphy bounding the lower Aptian oceanic anoxic subevent 1a in northeastern Tunisia. *Cretaceous Research, 39*, 133–148.

Elkhazri, A., Ben Abdesselam-Mahdaoui, S., Razgallah, S., & Abdallah, H. (2015b). The upper Jurassic-lower cretaceous at Jebel Ammar (Ariana Region). *Annales Du Service Géologique De Tunisie, 80*, 1–27. [in French].

Elkhazri, A., Razgallah, S., Abdallah, H., & Rabhi, M. (2015a). Barremian-Aptian of Oued Zarga, northeastern Tunisia: Micropaleontological and geochemical study. *Revue De Paléobiologie, 34*, 385–409. [in French].

Hanna, G. (1989). Radiolarian of the lower Campania of Cracovie (Poland). *Palaeontologica, 34*, 327–354. [in French].

Loeblich, A. R., & Tappan, H. (1984). Suprageneric classification of the Foraminiferida (Protozoa). *Micropaleontology, 30*, 1–70.

M'Rabet, A. (1981). Stratigraphy, sedimentation and carbonate diagenesis of the lower cretaceous series of Central Tunisia (p. 540). Ph.D. Thesis, Université Paris-Sud. [in French].

Memmi, L. (1999). The Aptian and the Albian of Tunisia. Biostratigraphy from ammonites. *Bulletin of the geological society of France* (Vol. 170, pp. 303–309). [in French].

Moullade, M., & Tronchetti, G. (2003). Upper Albian Schackoinas (planktonic foraminifera) from Blake nose (Atlantic margin off Florida, ODP Leg 171B) Revue de Micropaléontologie (Vol. 46, pp. 131–141). [in French].

Pini, S. (1971). Unique geological map of Ariana in 1/50.000. Direction of Mines and Energy. [in French].

Robaszynski, F., Amedro, F., Gonzalez-Donoso, J. M., & Linares, D. (2007). The bioevens of the Albian (Vraconnian)-Cenomanian boundary at the northern and southern margins of Tethys (S.E. of France and Central Tunisia). In: L. G. Bulot, S. Ferry, & D. Grosheny, (eds), *Relations between the northern and southern margins of Tethys in the cretaceous* (Vol. 2, pp. 3–15). [in French].

Saïdi, M., & Belayouni, H. (1994). Geological and geochemical study of Albovraconian rocks in the area of Central-Northern Tunisia. In *Proceeding of the 4th petroleum exploration conference* (Vol. 7, pp. 91–116). [in French].

Salaj, J. (1980). Microbiostratigraphy of the cretaceous and Paleogene of northern and eastern Tunisia (Tunisian Hypostratotypes) (p. 238). Geological Institute of Dionyz Stùr. [in French].

Solignac, M. (1967). Geological survey of northern Tunisia. Direction-General of Public Works—geological map of Tunisia. [in French].

Talbi, R. (1991). Geological and geochemical study of facies rich in organic matter of Albian age of the basin of Bir M'Cherga (NE of Tunisia): Determinism of their genesis and oil interest of the region (p. 223). Ph.D. Thesis, Faculty of Sciences of Tunis. [in French].

Tandia, I. S. (2001). Lithostratigraphic and sedimentological study of the lower cretaceous (Barremian-Albian) of central-northern Tunisia (Krib and Tajerouine regions) (p. 298). Ph.D Thesis, Faculty of Sciences of Tunis. [in French].

The Adaptive Strategy of Foraminifers During the Early Aptian OAE1a Anoxic Event: An Example from Northeastern Tunisia

Abdallah Elkhazri and Mabrouk Boughdiri

Abstract

In northeastern Tunisia, the anoxic event OAE1a of the early Aptian strongly affected the distribution of foraminifers. This study used temporal analysis of all foraminifers to assess oxygen deficiency and adaptive strategies during this event. A multidisciplinary approach was applied in this study using lithological data, microfacies analyses, micropaleontology data and organic matter content. Predating the OAE1a event, the Barremian sequence reveals alternating marly limestones with interbedded carbonates. These marls yield planktonic foraminifers with globular chambers represented mainly by *Praehedbergella sigali*, *P. primare*, *Gorbachikella depressa*, *G. kugleiri*, *G. anteroapertura*, *Blowiella duboisi* and *Microhedbergella* spp. In addition, the appearance of foraminifers with elongated chambers belonging to the genus *Praehedbergella* was noted. During the OAE1a event, the upper part of the Lower Aptian is characterized by marls, interbedded marl-limestones and grey-black flaky limestones with low organic matter contents (average TOC 1–3 wt%). Planktonic foraminifers with elongated chambers become abundant and comprise more than 80% of the planktonic population belonging to the genera *Leupoldina* and *Praehedbergella*. The microfacies show organic matter, oxidation traces, radiolarian sections of various shapes, and planktonic foraminifera sections with elongated chambers belonging to the genus *Leupoldina*. Postdating the OAE1a event, the upper Aptian is characterized by marly sedimentation, in association with rare limestone beds, which yield planktonic foraminifera with globular chambers of the genus *Globigerinelloides*. Within the same samples, abundant larger sized benthonic (Epistomina, *Lenticulina*) and agglutinated foraminifera were also noted. The planktonic foraminifers with elongated chambers disappeared and gave way to those forms with globular chambers which increased in diversity and abundance, inferring a return to an oxygenated environment on a regional scale. The benthonic foraminifers of the genera *Lenticulina*, *Dentalina*, *Nodosaria* and *Epistomina* increased in size. In the upper Aptian, the presence of the keeled form *Planomalina cheniourensis* accompanied by *Epistomina* spp. and agglutinated foraminifers confirm a return to normal oxygenated marine environments. This study highlights the impact of the OAE1a event on the distribution of planktonic foraminifers and their adaptive strategy to oxygen deficiency. The results are correlated with others in the same stratigraphic interval in Tunisia and other Tethyan realms.

Keywords

Anoxic event OAE1a · Planktonic foraminifera · Oxygen deficiency · Adaptive strategy · Northeastern Tunisia

1 Introduction

Cretaceous organic-rich deposits, representing distinct periods of global ocean anoxia, are geographically widely distributed. This widespread partition pattern and the concentration of organic-rich sediments around certain time intervals led to the concept of Oceanic Anoxic Events (OAEs) (Bralower et al., 1994; Schlanger & Jenkyns, 1976). However, modern high-resolution geochemical studies concentrated mainly on European outcrop sections. In the Tethyan area, many authors have focused their work on the Early Aptian event OAE1a (Arthur & Schlanger, 1979; Elkhazri et al., 2013; Föllmi, 2012; Heldt et al., 2008; Li et al., 2008; Menegatti et al., 1998; Moullade et al., 1998;

A. Elkhazri (✉) · M. Boughdiri
Faculty of Sciences of Bizerte, Department of Earth Sciences, University of Carthage, 7021 Bizerte, Zarzouna, Tunisia
e-mail: elkhazriabdallah@yahoo.fr

F. Lucci et al. (eds.), *Selected Studies in Geomorphology, Sedimentology, and Geochemistry*, Advances in Science, Technology & Innovation, https://doi.org/10.1007/978-3-031-43744-1_36

Mutterlose et al., 2009; Rückheim & Mutterlose, 2002). In Tunisia, Cretaceous organic-rich deposits consist of laminated and platy limestones of the M'Cherga (*pro parte*), Allam, Mouelha and Bahloul Formations, which include the OAE1 and OAE2 events. The OAE1a event is still poorly analyzed, even though Lower Aptian organic-rich deposits have been identified (Heldt et al., 2008). Only recently was the typical carbon isotope signature of OAE1a identified and the biostratigraphic and sedimentologic expression of the event fully described in the Jebel Serj (northcentral Tunisia) (Heldt et al., 2008) and in the Jebel Ammar (Elkhazri et al., 2013). The present work uses quantitative micropaleontological data (foraminifers, radiolarians), and geochemical data from the Jebels Ammar, Msella, Ressas and Oued Zarga sections (northeastern Tunisia). The aim of the paper is the study of the influence of the OAE1a event on the stratigraphic distribution of planktonic and benthonic foraminiferal species and their adaptive strategy on the southern margin of the Tethys.

2 Materials and Methods

Aptian rocks are found at Jebel Ammar, Msella, Ressas in northeast Tunisian, and Jebel Oued Zarga in northwest Tunisia (Fig. 1). The Mesozoic succession begins in the core of a dome with Triassic rocks which are bounded by tectonic contacts. In Jebel Ammar, the Sidi Thabet fault, some microfaults, and Triassic diapirism have influenced the Lower Cretaceous succession (notably that of the Aptian). In Jebel Msella and Jebel Ressas, the horst and graben structures have largely contributed to the deposition of sediments rich in organic matter. In total, 58 rock samples were collected from low–middle Aptian successions in the Jebels Ammar (northeastern Tunisia). Samples of the lower Aptian succession of alternating marly limestones, limestones and platy limestones were taken at intervals of 5–20 cm. Washed residues (100 g) and thin sections were used for the study of microfossil distribution and for the sedimentological and organic matter analyses. Marker species were used to determine the age of benthonic foraminifers (Moullade et al., 1998) and planktonic foraminifers (Moullade et al., 1998; Premoli Silva & Sliter, 1999; Rückheim & Mutterlose, 2002). Species abundance was estimated by its frequency (number species/g sediment). A minimum of 100 individual foraminifers was used for sample comparison. All species were examined to establish a complete list of foraminifers. Organic matter evaluation was based on Rock–Eval pyrolysis, using a Rock–Eval VI analyzer. This technique consists of a quantitative analysis of the hydrocarbonate and oxygenate components which were released during the pyrolysis of the organic matter included in 70 mg of crushed rock. These analyses were performed in the Laboratory of Organic Geochemistry of the Entreprise Tunisienne d'Activités Pétrolières, Tunisia.

3 Results

3.1 Lithology, Biostratigraphy and Quantitative Micropaleontological Data

Barremian-Lower part of the lower Aptian (Lower Bedoulian), predating OAE1a. The Barremian is composed of alternating green marly limestone and claystone with interbedded carbonates. These marls yield planktonic foraminifera with globular chambers represent the species *Gorbachikella depressa*, *G. kugleiri*, *G. anteroapertura*, *Praehedbergella sigali*, *P. aptiana*, *P. globulifera*, *P. roblesae*, *Globigerinelloides duboisi*, *G. gottisi*, *G. blowi* and tiny praehedbergellids. These planktonic foraminifera are associated with benthonic such as the genera *Lenticulina*, *Dentalina*, *Gavelinella*, *Ammodiscus*, *Praedorothia*, *Epistomina* and Textularids. In the topmost Barremian and the lower part of the lower Aptian, immediately below the anoxic event, there is an appearance of limestone beds with marl-limestone intercalations. There is also a foraminiferal composition change at this level with the appearance of foraminifers with elongated chambers such as *Praehedbergella sigali*, *P. aptiana*, *P. globulifera* and *P. roblesae* that have become frequent always associated but rarely with species belonging to the genera *Globigerinelloides* and *Blowiella*. The benthonic foraminifera are present and represented by the previous genera which fluctuate in values.

The upper part of the lower Aptian (upper Bedoulian), during OAE1a. This upper part is characterized by green to grey marls, interbedded marl-limestones and grey-black flaky limestones with low organic matter contents (low TOC average 1–3%). Planktonic foraminifers become abundant and increase up to 80–90% and comprise more than 80% of the plankton population belonging to genera with elongate chambers such as the species *Praehedbergella sigali*, *P. aptiana*, *P. globulifera*, *P. roblesae*, *Leupoldina cabri*, *L. pustulans*, *Pseudoschackoina saundersi* and *Pseudoschackoina* (Fig. 2). The microfacies show organic matter, radiolarian sections of various shapes and planktonic foraminifera sections with elongated chambers belonging to the genus *Leupoldina*. Benthonic foraminifers (*Gavelinella*, *Lenticulina*, *Textularia* and biserial agglutinated forms) are also present. This microfacies includes many sublithified lithoclasts, which are darker (enriched in organic matter). The globular-chambered forms are *Praehedbergella sigali*, *P. primare*, *Globigerinelloides duboisi* and *G. blowi* which are present but do not exceed 20% of the planktonics population (Fig. 2). Radiolarians

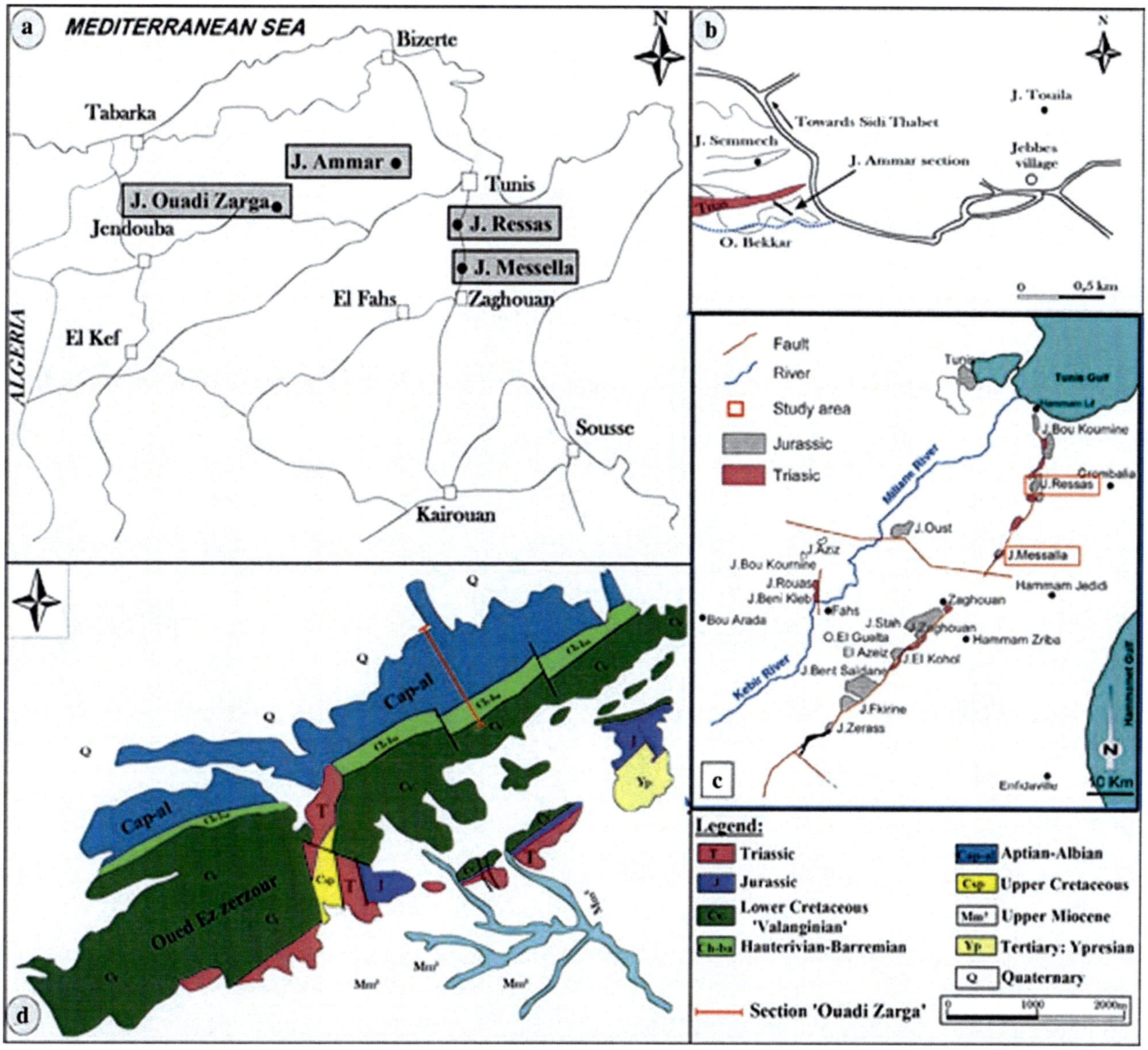

Fig. 1 **a** Location of the study area in northeastern and northwestern Tunisia, **b, c, d** detail of the investigated sectors

are the most common, diversified and represented by spherical forms (Spumellarian). Three substantial biological events are associated with this slight oxygen deficiency in the upper part of the Lower Aptian: (1) the first apparition of *Leupoldina cabri* and its abundance (Fig. 2), (2) the radiolarian bloom (Fig. 2) the presence of organic deposits (Fig. 3). The microfacies reveals a mudstone texture clear in some sections of planktonic foraminifera with elongate chambers, we can also guess the existence of some organic matter indices associated with sections of radiolarians.

Middle-upper Aptian, postdating OAE1a. The Middle to Upper Aptian is characterized by marly sedimentation, in association with rare limestone beds, which yield planktonic foraminifera with globular chambers of *Globigerinelloides* and *Hedbergella*. *L. cabri* disappears and the genus *Praehedbergella* becomes rarefied and gives way to *Hedbergella* and *Globigerinelloides*. Benthonic foraminifers are reduced in percentage (not exceeding 10% of total foraminifers). Within the same samples, abundant larger sized benthonic (*Epistomina, Lenticulina, Dentalina*) and agglutinated foraminifera are also noted. The planktonic foraminifers with elongated chambers disappear and give way to those forms with globular chambers which increase in diversity and abundance, inferring a return to an oxygenated environment on a regional scale. Benthonic foraminifers of *Lenticulina* and *Dentalina* increase in size. In the Upper Aptian, the presence of the keeled form *Planomalina cheniourensis* accompanied by *Epistomina* spp. and agglutinated foraminifers confirm a return to normal oxygenated marine environments.

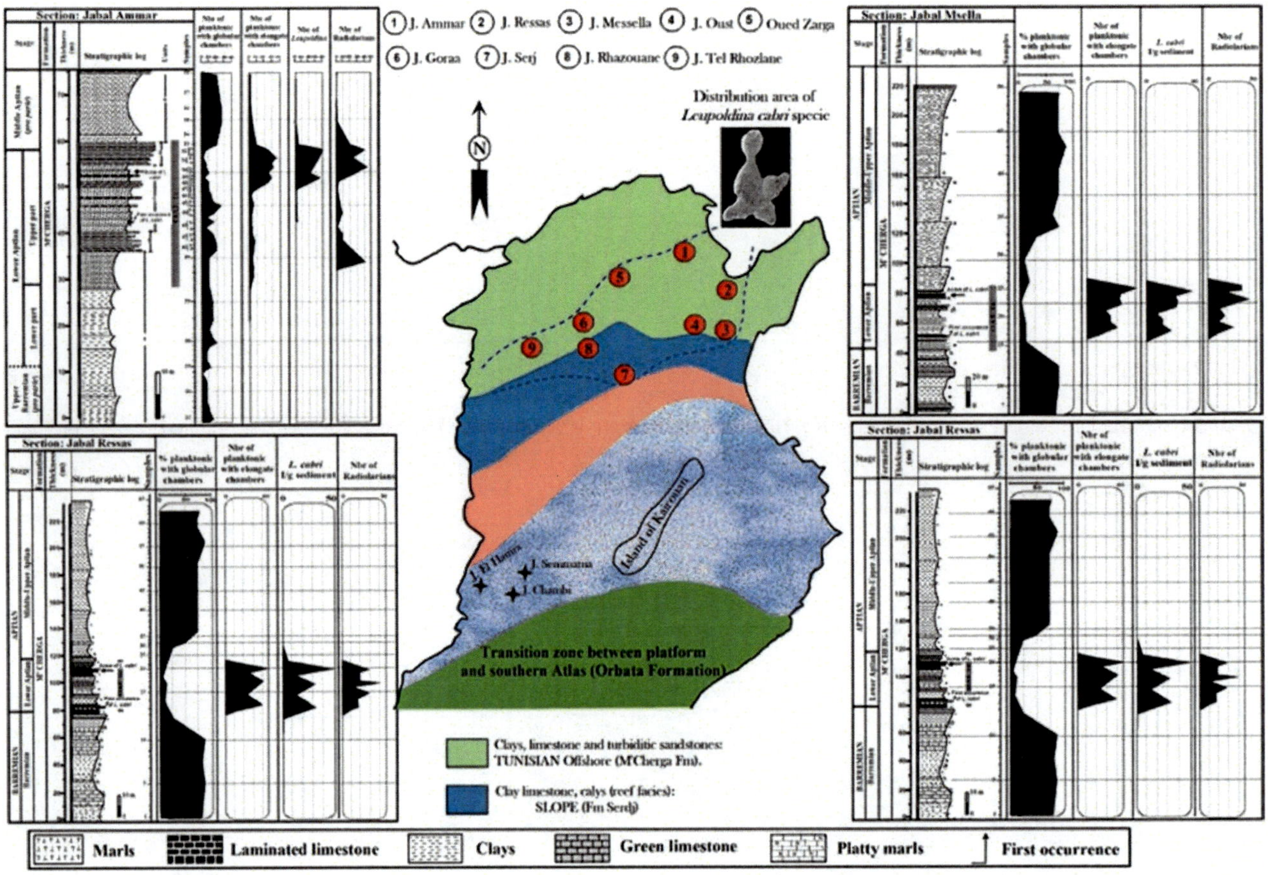

Fig. 2 Quantitative micropaleontological data from the Aptian of the Jebel Ammar, Messella, Ressas and Ouadi Zarga section

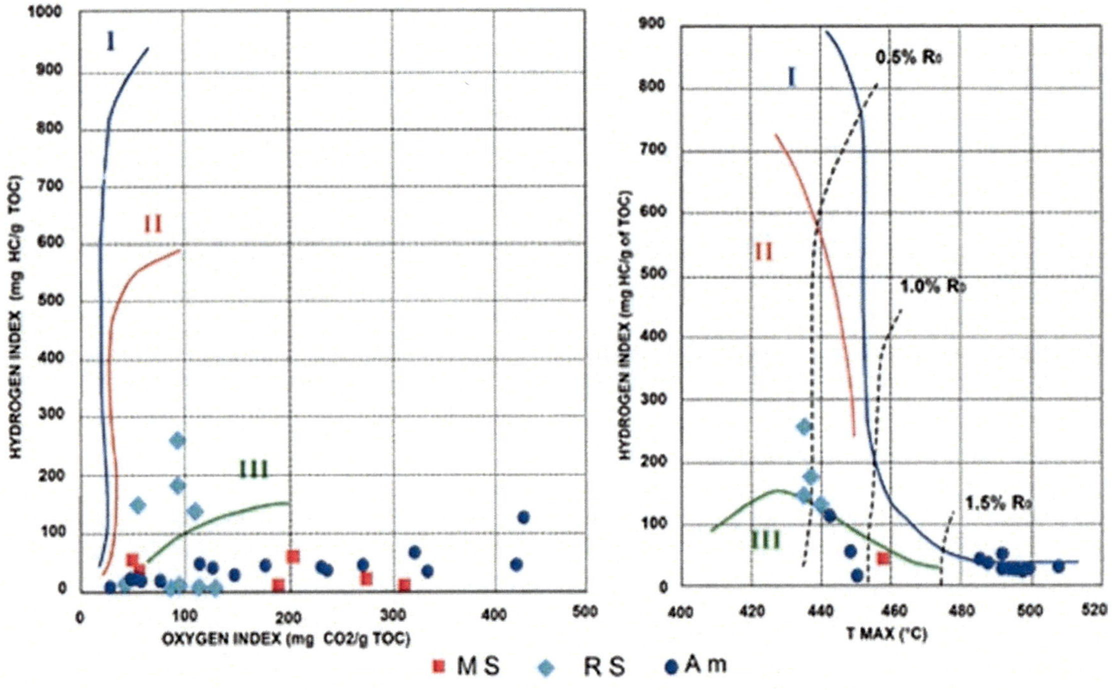

Fig. 3 IH-IO and IH-T_{max} diagrams summarising the different results at Jebel Messella, Jebel Ressas and Jebel Ammar

3.2 Organic Geochemistry

Organic matter characterizing the grey-black flaky limestones of the Lower Aptian is characterized by low TOC contents not exceeding 2% with the exception of Jebel Ammar where it reaches 4.59% (Fig. 3). These low levels can be attributed to low organic production rates, moderate anoxia or the burial of this organic matter. Based on HI/IO and HI/Tmax diagrams (Fig. 3) all the analyzed organic matter is distributed between lines II and III. However, it is more concentrated in lineage III. Geochemical analyses show that organic matter is dissimilated in the marl-limestone alternations whereas they concentrate with higher contents in the grey-black flaky limestones of the lower Aptian.

3.3 C^{12}/C^{13} Results

The $\delta^{13}C$ curve of the Jebel Ammar section shows excursions that are closely correlated with those described elsewhere (Menegatti et al., 1998; Moullade et al., 1998). Three isotopic excursions are clearly expressed (one negative anomaly and two are positive), correlated with the isotopic events observed at La Bédoule (SE France) (Moullade et al., 1998). On the other hand, the high-resolution interpretation of the $\delta^{13}C$ curve reveals six intervals (Fig. 4).

4 Discussion

4.1 *L. cabri* in the Upper Part of the Lower Aptian During OAE1a

Through all the sections studied in northeastern Tunisia, *L. cabri* appears within the grey-black flaky limestones at high percentages (Fig. 5) indicating the middle part of the Upper Bedoulian (Elkhazri et al., 2013, 2015). Several authors have noted the abundance of this species which sometimes increases with significant percentages in relation to other forms. This species seems to adapt better than other species to environmental stress conditions. The abundance of *L. cabri* coincides with the timing of OAE1a. According to Magniez-Jannin (1998), the abundance of Schackoines occurs in limestones and grey-black flaky limestones at the Barremian/Aptian boundary. *L. cabri* occurs towards the middle part of the Upper Bedoulian in Jebel Serdj (Heldt et al., 2008); it covers an interval of 60 to 235 m and constitutes a zone of total extension. These elongated forms proliferate with a tendency to multiplication of the number of chambers and bulbs towards the middle part of the Bedoulian (grey-black flaky limestones) where dysaerobic conditions are accentuated (Fig. 5). The latest levels characterizing the top of the lower Aptian show low percentages of Leupoldines (5%), testifying to reoxygenation of the environment and a return to normal conditions hence

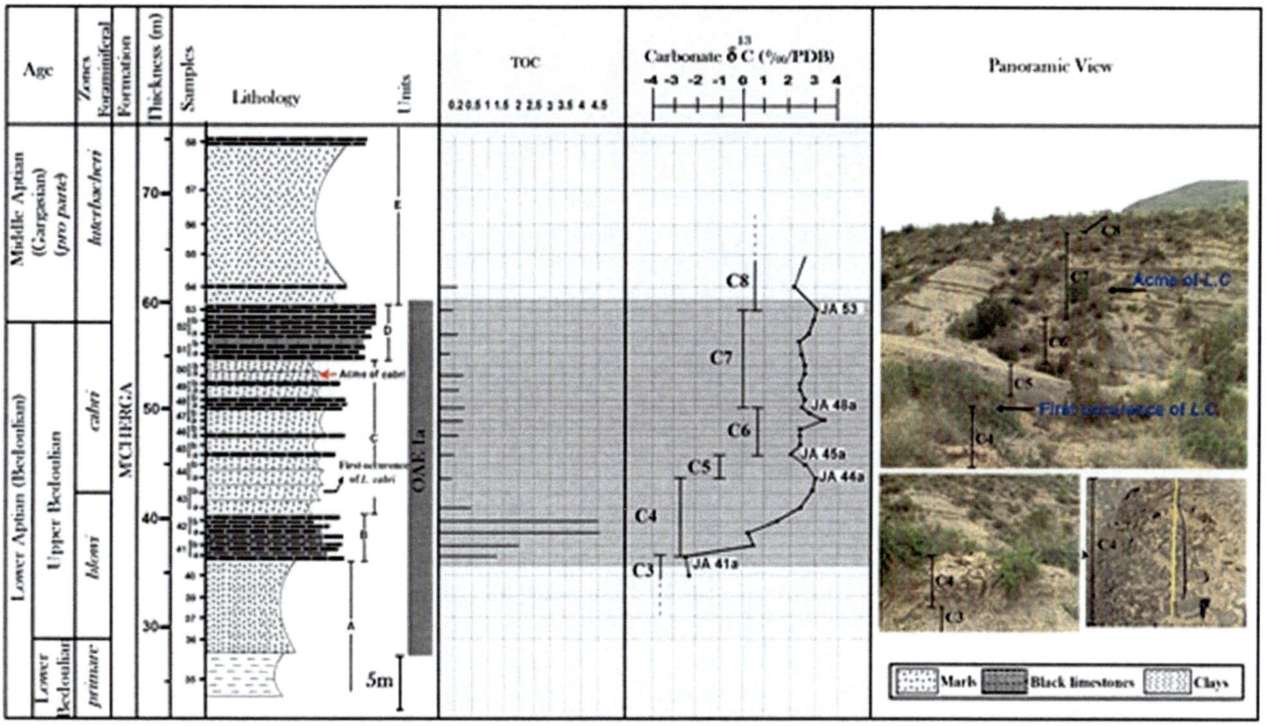

Fig. 4 Correlation of $\delta^{13}C$ curve with biostratigraphic and geochemical data of the Lower Aptian at Jebel Ammar

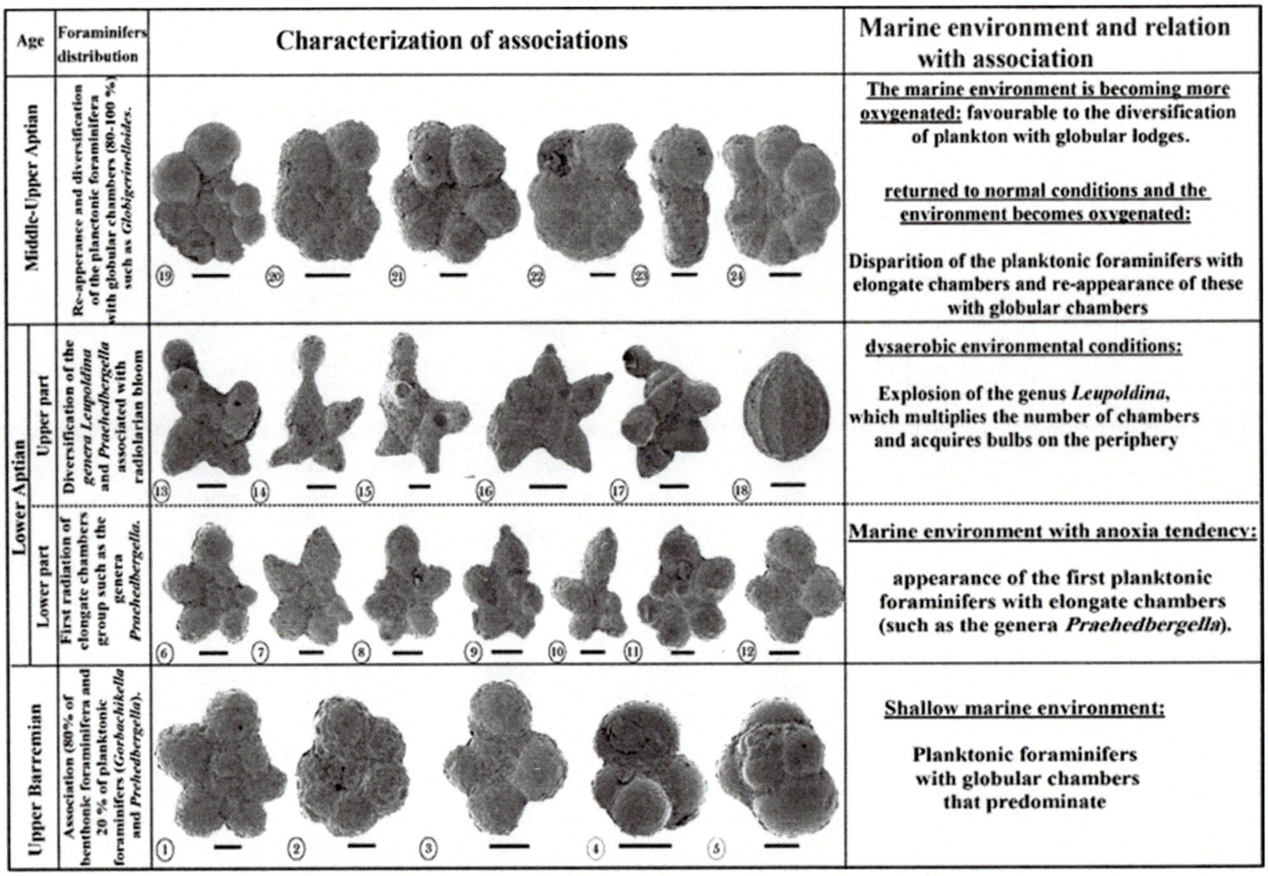

Fig. 5 Evolutionary scheme and evidence of dysoxia on the planktonic and benthonic foraminifera in the study area

a general decline of these elongated forms from the lower Gargasian.

4.2 Adaptive Strategy of Planktonic Foraminifera During OAE1a

Just like (Coccioni et al., 2007; Magniez-Jannin, 1998), we believe that the elongate chambers of Leupoldina reflects an adaptation to an oxygen-poor environment (Fig. 5). Indeed, these authors show that radial elongate chambers do not appear randomly but coincide with the deposition of organic matter. This environment is responsible for dysoxic conditions and the radial elongate chambers are related to environmental characteristics such as variations in physicochemical factors, ecological, dissolution of oxygen concentrations, temperature, salinity, nutrients and food supply (Coccioni et al., 2007). Several studies also show that the increase of the surface/volume ratio at the level of the Leupoldinidis chambers through the organization of the branches in light bulbs promotes and helps increase of the consumption of oxygen and therefore allows to better resist dysoxia (Elkhazri et al., 2013; Magniez-Jannin, 1998; Menegatti et al., 1998). Comparing the characteristics of anoxic event OAE1a in Tunisia with the data reported at the Tethys scale shows the appearance of *L. cabri* within the level «Livello Selli» (Bralower et al., 1994; Coccioni et al., 2007; Rückheim & Mutterlose, 2002). Some authors (Bralower et al., 1994; Coccioni et al., 2007; Premoli Silva & Sliter, 1999) established that these "Black shales" are characterized by a high abundance of radiolaria during OAE1a, which is confirmed in the studied sections of northeastern Tunisia.

5 Conclusions

This work highlights the presence of *Leupoldina cabri* highlighted for the first time in these areas, is evidence for dysaerobic surrounding conditions. TOC and $\delta^{13}C$ analyses of the Aptian succession confirm the existence of the ocean anoxic event OAE1a in this area, and analysis of planktonic foraminifers shows their adaptive strategy to oxygen deficiency. TOC levels highlighted in northeastern and northwestern Tunisia can be correlated with the «Livello Selli» levels in Italy, «Goguel» in the Vocontien Basin in southeastern France, and «Fischschiefer» in northwestern Germany (Elkhazri et al., 2013, 2015). For the first time in Tunisia, there is evidence for a relationship between the radiations of planktonic foraminifera and radially elongated chambers coincide with the anoxic environment. The adaptive strategy to this oxygen deficiency related to the elongation of the chambers has been confirmed in the northern margins of the Tethys.

Acknowledgements We would like to thank Dr. Nigel Ainsworth, Dr. Orabi Orabi, Dr. Malcolm Hart and the other reviewers for their constructive and helpful comments that improved the manuscript.

References

Arthur, M. A., & Schlanger, S. O. (1979). Cretaceous 'oceanic anoxic events' as causal factors in development of reefreservoired giant oil fields. *American Association of Petroleum Geologists Bulletin, 63*, 870–885.

Bralower, T. J., Arthur, M. A., Leckie, R. M., Sliter, W. V., Allard, D. J., & Schlanger, S. O. (1994). Timing and paleoceanography of oceanic dysoxia/anoxia in the late Barremian to early Aptian (Early Cretaceous). *Palaios, 9*, 335–369.

Coccioni, R., Silva, I. P., Marsili, A., & Verga, D. (2007). First radiation of Cretaceous planktonic foraminifera with radially elongate chambers at Angles (Southeastern France) and biostratigraphic implications. *Revue de Micropaléontologie, 50*, 215–224.

Elkhazri, A., Abdallah, H., Razgallah, S., Moullade, M., & Kuhnt, W. (2013). Carbon-isotope and microfaunal stratigraphy bounding the Lower Aptian Oceanic Anoxic Subevent 1a in Northeastern Tunisia. *Cretaceous Research, 39*, 133–148.

Elkhazri, A., Razgallah, S., Abdallah, H., & Rabhi, M. (2015). Barremian-Aptian of Oued Zarga, northeastern Tunisia: Micropaleontological and geochemical study. *Revue de Paléobiologie, 34*, 385–409 [in french].

Föllmi, K. B. (2012). Early cretaceous life, climate and anoxia. *Cretaceous Research, 35*, 230–257.

Heldt, M., Bachmann, M., & Lehmann, J. (2008). Microfacies, biostratigraphy, and geochemistry of the hemipelagic Barremian-Aptian in North Central Tunisia: Influence of the OAE 1a on the southern Tethys margin. *Palaeogeography, Palaeoclimatology, Palaeoecology, 261*, 246–260.

Li, Y.-X., Bralower, T. J., Montañez, I. P., Osleger, D. A., Arthur, M. A., Bice, D. M., Herbert, T. D., Erba, E., & Premoli Silva, I. (2008). Toward an orbital chronology for the early Aptian Oceanic Anoxic Event 1a (OAE 1a, ~120 Ma). *Earth and Planetary Science Letters, 271*, 88–100.

Magniez-Jannin, F. (1998). Chamber elongation in Early Cretaceous planktic foraminifera: An adaptive response to oxygen depleted sea water? *Comptes Rendus De L'académie Des Sciences De Paris, Sciences De La Terre Et Des Planètes, 326*, 207–213 [in french].

Menegatti, A. P., Weissert, H., Brown, R. S., Tyson, R. V., Farrimond, P., Strasser, A., & Caron, M. (1998). High resolution $\delta^{13}C$ stratigraphy through the early Aptian "Livello Selli" of the Alpine Tethys. *Paleoceanography, 13*, 530–545.

Moullade, M., Kuhnt, W., Bergen, J. A., Masse, J. A., & Tronchetti, G. (1998). Correlation of biostratigraphic and stable isotope events in the Aptian historical stratotype of La Bédoule (SE France). *Comptes Rendus De L'académie Des Sciences, 327*, 693–698.

Mutterlose, J., Pauly, S., & Steuber, T. (2009). Temperature controlled deposition of early Cretaceous (Barremian-early Aptian) black shales in an epicontinental sea. *Palaeogeography, Palaeoclimatology, Palaeoecology, 273*, 330–345.

Silva, I. P., & Sliter, W. V. (1999). Cretaceous paleoceanography; evidence from planktonic foraminiferal evolution. In E. Barrera, & C. C. Johnson (Eds.), *Evolution of the Cretaceous Ocean-Climate System. Geological Society of America, Special Paper* (Vol. 332, pp. 301–328).

Rückheim, S., & Mutterlose, J. (2002). The Early Aptian migration of planktonic foraminifera to NW Europe: The onset of the mid-Cretaceous plankton revolution in the Boreal Realm. *Cretaceous Research, 23*, 49–63.

Schlanger, S. O., & Jenkyns, H. C. (1976). Cretaceous oceanic anoxic events: Causes and consequences. *Geology En Mijnbow, 55*, 179–184.

Upper Jurassic (Oxfordian—Kimmeridgian) Benthic Foraminifera and Algae from the Blue Nile Basin, Ethiopia

Gilamichael Kidanemariam Domenico

Abstract

Micropaleontology and field studies were carried out on the top part of the Middle to Upper Jurassic Antalo Limestone Formation exposed near Gohatsion (Blue Nile Basin, central western Ethiopia) to precisely determine the ages and understand depositional conditions. This succession (ca. 30 m thick) consists of pack to wackestones, an alternation of micritic mudstones and oolitic grainstones, lime mudstones and stromatoporoid floatstone (containing *Cladocoropsis mirabilis* Felix), in stratigraphic order. The benthic foraminiferal assemblage is represented by *Alveosepta jaccardi*, *Pseudomarsonella maxima*, *Redmondoides lugeoni*, *Kurnubia palastiniensis Nautiloculina oolithica* and *Siphovalvulina variabilis, Glomospira* and *Placopsilina.* The identified dasycladaean algae include *Salpingoporella annulata, Salpingoporella pygmaea* and *Thyrsoporella pseudoperplexa. Salpingoporella annulata* ranges in age from the Bathonian to Hauterivian, whereas *Thyrsoporella pseudoperplexa* is indicative of the Upper Jurassic. Other identified species include *Garwoodia* (a cyanobacteria) and some scleractinian corals. Based on the foraminiferal-algal assemblage, and in particular the presence of the globally recognized upper Oxfordian-lower Kimmeridgian *Alveosepta jaccardi* Taxon Range Zone, a late Oxfordian to early Kimmeridgian age is proposed for the studied section.

Keywords

Jurassic · Ethiopia · Antalo formation · Foraminifera · Algae

G. K. Domenico (✉)
East Los Angeles College, 1301 Avenida Cesar Chavez, Monterey Park, CA, USA
e-mail: domenigk@laccd.edu

1 Introduction

During the Jurassic, the continental margins of Northeast Africa and Arabia were epicontinental shallow water platforms, where Middle to Upper Jurassic carbonate sediments of the Antalo Formation in Ethiopia (Bosellini, 1989) and coeval sequences in Arabia (Ziegler, 2001) were deposited. The Antalo Limestone sequence of the Blue Nile Basin (Fig. 1) is known for its microfossil content (Atnafu, 2003; Domenico, 2010; Kurita et al., 2009; Russo et al., 1994; Turi et al., 1980). The objective of the present work was to record the presence of a relatively rich assemblage of foraminifers and algae identified at the top sections of the Antalo Limestone in the Blue Nile Basin. This is a contribution to the knowledge about the distribution of taxa in the Jurassic carbonates of Ethiopia and their age. The method includes laboratory and field analyses. Two stratigraphic sections were chosen from the topmost part of the Antalo Formation (Fig. 2) to be studied and about 50 thin sections were microscopically examined.

2 Geologic Setting

The studied sections crop out in the Blue Nile Basin of north-central Ethiopia (Fig. 1). The area is characterized by Permian to Triassic rifting and subsequent marine flooding resulting in the deposition of the siliciclastic Anisian-Pliensbachian Adigrat Formation (Wolela, 2010), an approximately 30 m thick glauconitic mudstone, an approximately 400 m-thick gypsum–dolostone sequence of Toarcian age, the Gohatsion Formation (Assefa, 1980), and the Middle to Upper Jurassic Antalo Limestone (~500 m thick). The Gohatsion and Antalo formations represent transgression of the southwest end of the Tethys Ocean on east Africa during the Middle to Upper Jurassic (Bosellini, 1989; Domenico, 2010).

F. Lucci et al. (eds.), *Selected Studies in Geomorphology, Sedimentology, and Geochemistry*, Advances in Science, Technology & Innovation, https://doi.org/10.1007/978-3-031-43744-1_37

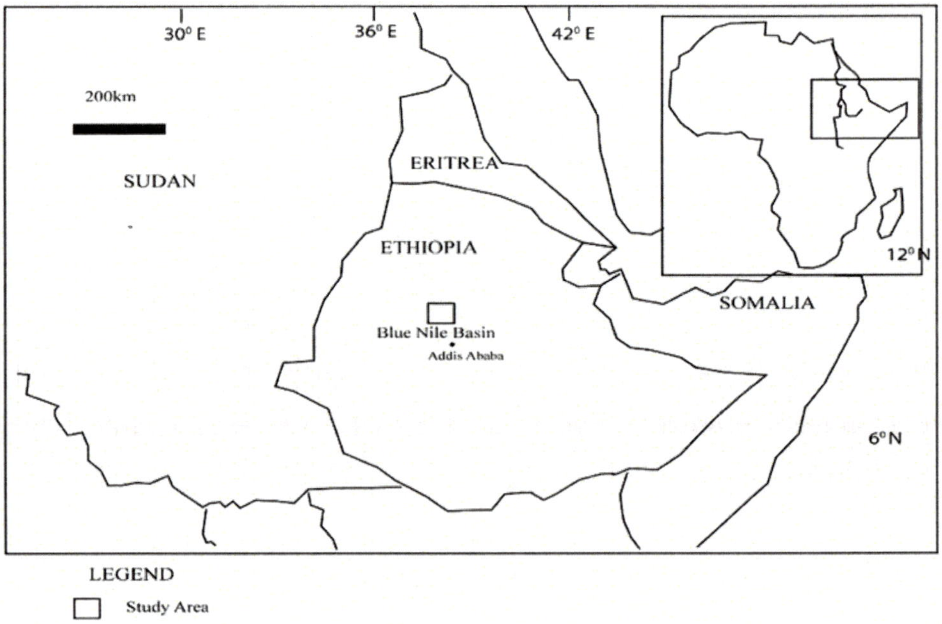

Fig. 1 Location map of the study area

3 Results

The top of the Antalo Limestone in the Blue Nile Basin (Fig. 2) consists of shallow marine carbonate beds characterized by foraminifera, algae, stromatoproids and cyanobacteria. The foraminifera observed include textularia and miliolina (Domenico, 2010). These are *Pseudomarsonella redmondoides, Alveosepta jaccardi, Kurnubia palastiniensis, Redmondoides lugeoni, Nautiloculina oolithica, Siphovalvulina variabilis, Spirocyclinidae, Glomospira* sp, *Lituolidae* sp., *Miliolidae and Valvulinida* sp. (Fig. 3). The identified calcareous algae belong to two groups, dasycladaean algae and siphonophytes (siphonous green algae). The former includes *Salpingoporella annulata, Thyrsoporella pseudoperplexa* and *Salpingoporella pygmaea*, while the latter includes *Garwoodia bardoosi* and *Garwoodia* sp. Macroinvertebrate fauna were also identified: bivalves, gastropods, hexacorals, and the stromatoporoid *Cladocoropsis mirabilis.*

4 Discussion

In most representative samples, the most abundant foraminifer is *Kurnubia palastiniensis*. According to Clark and Boudagher-Fadel (2002) it is Oxfordian–Kimmeridgian in age. Septfontaine (1988) and Velic (2007) dated *K. palastiniensis* as late Jurassic (Oxfordian–Tithonian). The most important age-diagnostic foraminifer is *Alveosepta jaccrdi*, which is dated as late Oxfordian to early Kimmeridgian by Septfontaine (1988), Tasli (2001) and Kimmeridgian by Velic (2007).

Among the dasycladaean algae, *Salpingoporella annulata* is possibly Callovian–Oxfordian. *Salpingoporella pygmaea* is certainly upper Kimmeridgian but it could range from Oxfordian to Aptian (Granier & Deloffre, 1993). *Thyrsoporella pseudoperplexa* ranges from Oxfordian to Kimmeridgian. The association of *Labrynthia mirabilis* with *K. palastiniensis* and *A. jaccardi* is considered as Late Oxfordian–Early Kimmeridgian in Morocco. In general, the

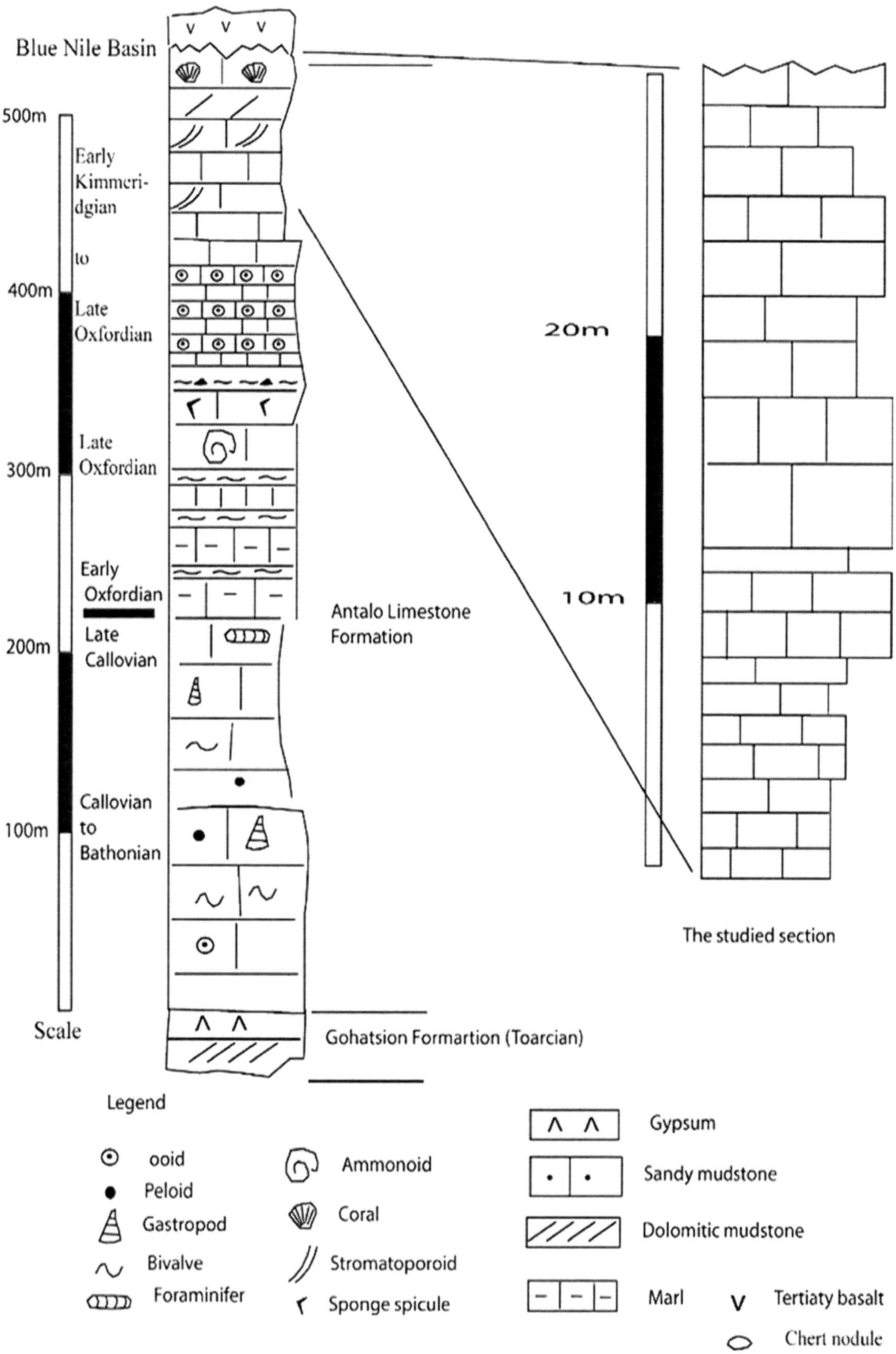

Fig. 2 Composite stratigraphy of the Antalo Limestone in the Blue Nile Basin, and stratigraphic position of the studied section

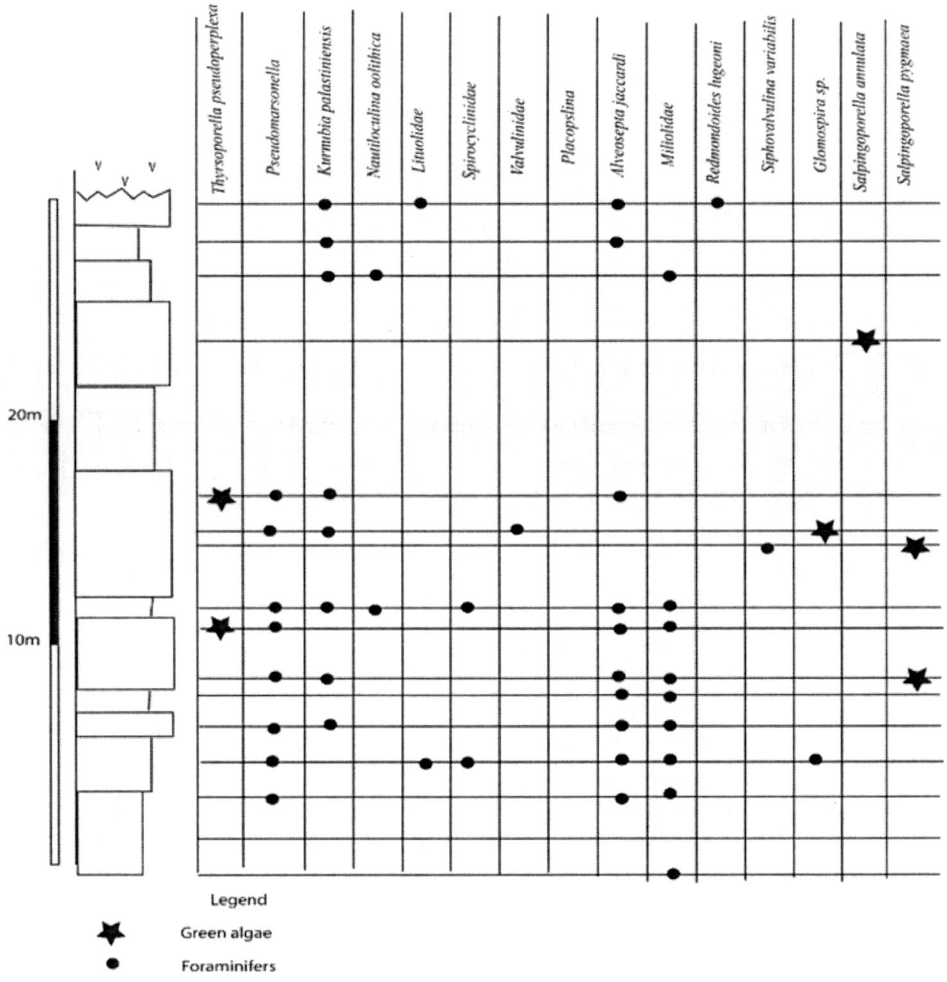

Fig. 3 Stratigraphic distribution of taxa indicative of an upper Oxfordian to lower Kimmeridgian age in the top part of the Antalo Limestone Formation, Blue Nile Basin

microfossil assemblage of the studied sections indicate a late Oxfordian to early Kimmeridgian age.

5 Conclusions

The top most sections of the Antalo Limestone Formation in the Blue Nile Basin of Ethiopia yield a rich association of foraminifera and algae, some of which are mentioned for the first time. A late Oxfordian–Early Kimmeridgian age is assessed for this sedimentary succession.

References

Assefa, G. (1980). Stratigraphy and sedimentation of the type Gohatsion Formation (Lias–Malm), Abbay River Basin, Ethiopia. *Sinet: Ethiopian Journal of Science, 3*, 87–110.

Atnafu, B. (2003). Facies and diagenetic development of Jurassic carbonates (Callovian, Oxfordian, Kimmeridgian) in the Abay River Basin (Upper Blue Nile, Ethiopia). Ph.D. thesis, Friedrich–Alexander Universitat Erlangen, Germany.

Bosellini, A. (1989). The continental margins of Somalia: Their structural evolution and sequence stratigraphy. Memoirie di Scienze Geologiche, *XLI*, 373–458.

Clark, G., & Boudagher-Fadel, M. (2002). The larger benthic foraminiferal assemblages and stratigraphy of the late Jurassic Bhanness complex, Central Lebanon. *Revue De Paleobiologie, 21*, 679–695.

Domenico, G. (2010). *Facies analysis of Middle to upper Jurassic carbonates in Eastern and North-Central Ethiopia*. Ph.D. thesis, Kyushu University, Japan.

Granier, B., & Deloffre, R. (1993). Inventaire critique des algues dasycladales fossiles: II° Partie-les algues dasycladales du Jurassique et du Crétacé. *Revue De Paléobiologie, 12*, 19–65.

Kurita, H., Domenico, G., & Sano, H. (2009). Timing of Jurassic carbonate deposition in an East African aulacogen; dinoflagellate cysts from the Dejen section, Blue Nile Basin, North Central Ethiopia. In *Annual Meeting of the Palaeontological Society of Japan, Chiba, Japan* (Vol. 66).

Russo, A., Getaneh, A., & Balemwal, A. (1994). Sedimentary evolution of the Abay River (Blue Nile) Basin, Ethiopia. *Neues Jahrbuch Für Geologie Und Paläontologie—Monatshefte, 5*, 291–308.

Septfontaine, M. (1988). Vers une classification evolutive des lituolides (foraminiferes) Jurassiques en milieu de plate-forme carbonatee. *Revue De Paléobiologie, Spec, 2*, 229–256.

Taslı, K. (2001). Benthic Foraminifera from the Upper Jurassic platform carbonates of the Aydıncık (İçel) area, Central Taurides, S Turkey. *Geologia Croatica, 54*, 1–13.

Turi, A., Bigi, L., & Radrizzani, C. (1980). Microfacies of the Antalo Limestone (Middle to Upper Jurassic) in some sections of East-Central Ethiopia. *Bollettino Societa Geologica Italiana, 99*, 437–454.

Velic, I. (2007). Stratigraphy and paleobiogeography of Mesozoic benthic foraminifera of the Karst Dinarides (SE Europe). *Geologia Croatica, 60*, 1–113.

Wolela, A. (2010). Diagenetic evolution of the Ansian-Pliensbachian Adigrat sandstone, Blue Nile Basin, Ethiopia. *Journal of African Earth Sciences, 56*, 29–42.

Ziegler, M. (2001). Late Permian to Holocene Paleofacies evolution of the Arabian Plate and its hydrocarbon occurrences. *GeoArabia, 6*, 445–504.

Praealveolina Quantitative Data of the Mishrif Formation Reservoir (Cenomanian–Early Turonian) in Selected Oil Fields, Southern Iraq

Qahtan A. M. Al Nuaimy

Abstract

The Mishrif Formation reservoir (Cenomanian-lower Turonian) of the southern Mesopotamian Basin is hosted within rudist-bearing deposits that are rich in larger benthonic foraminifera, including A and B forms of the genus *Praealveolina*. In this study, observations were made on both generations of external and internal parameters, and *Praealveolina* species recognition criteria were examined, based on a new material obtained from more than 200 individual equatorial and axial thin sections examined. The following species of *Praealveolina* were recognized: *Praealveolina simplex* Reichel, *P. osimoi* Zuffardi Comerci, *P. cretacea* Reichel, *P. lata* Reichel, and *P. tenuis* Reichel. Evolutionary changes were observed for the first time in Iraq in successive *Praealveolina*-bearing carbonate units of the Mishrif Formation, highlighting that toward the end of the Cenomanian, a significant increase in length and size of the test, enlargement of megalospheric nepiont proloculus, and progressive complication in secondary chamberlets, in addition to the number of apertural rows on the oral face in the last whorl of the polar region. Quantitative data obtained from *Praealveolina* assemblages allowed us to distinguish two lineages, the *P. simplex* lineage, showing the same evolutionary period without any complication in secondary chamberlets, and the *P. osimoi*—*P. cretacea*—*P. lata*—*P. tenuis* lineage showing an increase in size of the test and nepiont proloculus, in addition to the complication of secondary chamberlets in polar whorls toward the equator of the test during the early middle to late Cenomanian.

Keywords

Mishrif formation · Southern Iraq · *Praealveolina* · Morphometric data

1 Introduction

The genus *Praealveolina* (Cenomanian) was abundant in shallow, carbonate environments and thus are useful for high-resolution biostratigraphy. Many previous studies have examined the systematics, morphometric data and biostratigraphy of the genus *Praealveolina* worldwide: for example, Reichel (1936–1937), Neumann and Fourcade(1985), Schroeder and Neumann (1985), Cherchi and Schroeder (1989), Al-Nuaimy (1990), Caus et al., (1993, 1997 and 2002), Caus and Hottinger (1986) and Calonge et al. (2002). However, there is little knowledge concerning the stratigraphic distribution and phylogenetic relationships between the respective gen era genera and their species. Therefore, the present study addressed the first biometric data for the genus *Praealveolina* and brought new evidence for megalospheric and microspheric forms from southern Iraq (Fig. 1). The study aimed to recognize both generation forms of the genus *Praealveolina* and present their morphometric changes during the Cenomanian. The research is based on core materials from four sub-surface sections drilling the Cenomanian Mishrif Formation, in southern Iraq (Tuba-1, West Qurnah-3, Zubair-49, and Ahdab-1).

2 Materials and Methods

Axial and equatorial thin sections of *Praealveolina* specimens were prepared in the North Technical University Geology Laboratory. More than 200 new thin sections of the Alveolinid carbonate facies from the Mishrif Formation in southern Iraq, were studied and photographed. For

Q. A. M. Al Nuaimy (✉)
Technical College Kirkuk, Northern Technical University, Mosul, Iraq
e-mail: qahtaniraqi@ntu.edu.iq

F. Lucci et al. (eds.), *Selected Studies in Geomorphology, Sedimentology, and Geochemistry*,
Advances in Science, Technology & Innovation, https://doi.org/10.1007/978-3-031-43744-1_38

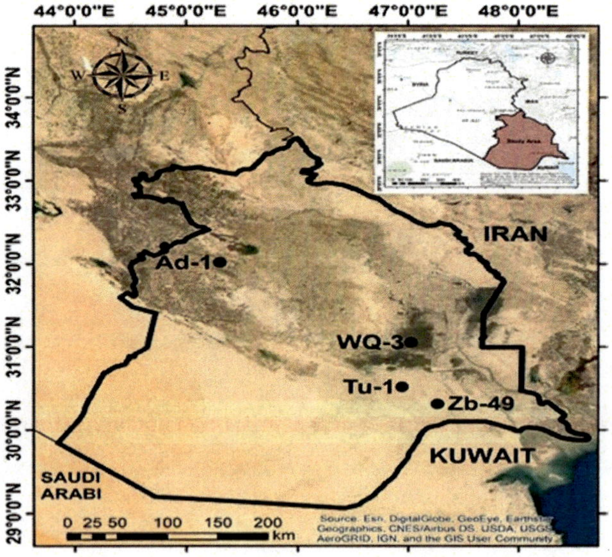

Fig. 1 Significant localities of studied sections in Southern Iraq

megalospheric and microspheric forms of *Praealveolina*, internal and external parameters were measured according to the identification criteria used in Calonge et al. (2002).

3 Results

3.1 Alveolinidae Units in the Mishrif Formation

The Alveolinidae-bearing carbonate units in the studied subsurface sections consist of dense dark in most of the upper units, white wackestone to packstone without planktonic foraminifera, grading downwards to brown detrital, porous, partly very shelly, foraminiferal wackstone to packstone with patches of rudist and planktonic foraminifera in the lower Alveolinidae unit. These facial changes in the Mishrif Formation between lower and upper units suggest a regressive signature. The Mishrif Formation is overlain conformably by the Khasib Formation (Early Turonian) in places where the Kifil Formation is absent, and is gradationally underlain by the Rumaila Formation (Early Cenomanian) (Fig. 2).

3.2 Morphometry of *Praealveolina* Species According to Datation

The morphometric characteristics of the genus *Praealveolina* were examined in the axial and equatorial planes of free specimens. We followed herein the morphometric approach defined in Calonge et al. (2002), grouping specimens into populations, the representatives of which are clearly distinguishable from those of other populations within the same sample. Taxonomic determinations were usually based on certain individuals within populations according to eight parameters (Table 1, Fig. 2).

A single phylogenetic group defined by *P. simplex—P. iberica—P. pennensis—P. dibilis—P. brevis—P. lata—P. cretacea—P. tenuis* has been assigned so far to the genus *Praealveolina* as Cenomanian larger foraminifera (Neumann & Foucade, 1985; Schroeder & Neumann, 1985; Calonge et al., 2002). However, we assumed that the aforementioned species belong to two separate phylogenetic lineages: the first one represented by *P. simplex*, and the second *P. osimoi. P. simplex* lineage populations that are distinguished by the absence of secondary chamberlets in both (A and B) generations, in current study, were reported from the same units with *P. tenuis* in West Qurnah-3 and Zubair-49 sections, and associated with *P. tenuis* and *P. cretacea* in Tuba-1 section, probably of late Cenomanian age.

Second lineage species (*P. osimoi—P. cretacea—P. lata—P. tenuis*) were identified in the Middle to Upper Cenomanian sequence of Alveolinid units in the studied parts. *P. osimoi* differentiated from the lower sequence of Mishrif Formation (Lower Cenomanian) in Ahdab-1, West Qurna-3, and Zubair-49. The stratigraphic distribution of *P. cretacea* ranges from the Middle to Lower Upper Cenomanian under the fossiliferous calcareous facies overlapping an algal-rudist facies in the Tuba-1 and West Qurnah-3 sections, while overlying a planktonic foraminiferal facies in the Zubair-49 section. The *P. cretacea* type level includes *Ovolveolina ovum* d'Oarbigny, *Cisalveolina fraasi* Gümbel, *Pseudorhapydionina dubia* De Castro, *Pseudolituonella reicheli* Marie, *Charentia cuvillieri* Neumann, *Cuneolina pavania parva* d'Orbigny, *Chrysalidina gradate* d'Orbigny nizzazatids, orbitolinids and miliolids.

The type level of *P. lata* was found across the whole lower Upper Cenomanian associated with the same foraminiferal assemblages mentioned above, in addition to *Ovolveolina lehneri* Reichel and *Multispirina iranensis* Reichel. In the Upper Cenomanian, major elongated, dominated faunas in Tuba-1, West Qurna-3 and Zubair-49 were observed to be well preserved (Fig. 3), and thought to belong to the species *P. tenuis*. These specimens differ from any other species of the genus by their elongated-fusiform shape and by multiplying the cellars in the polar area are at very early stages of development. *P. tenuis* is associated with *P. simplex* Reichel, *Dictyopsella libanica* Saint-Marc, *Vidalina radoicicae* Cherchi and Schroeder, *Nezzazata simplex* Omara, *Biconcava bentori* Hamaoui and Saint-Marc, *Cisalveolina fraasi* Gümbel, *P.seudorhapydionina dubia* De Castro, *P.seudolituonella reicheli* Marie, and *Chrysalidina gradate* d'Orbigny.

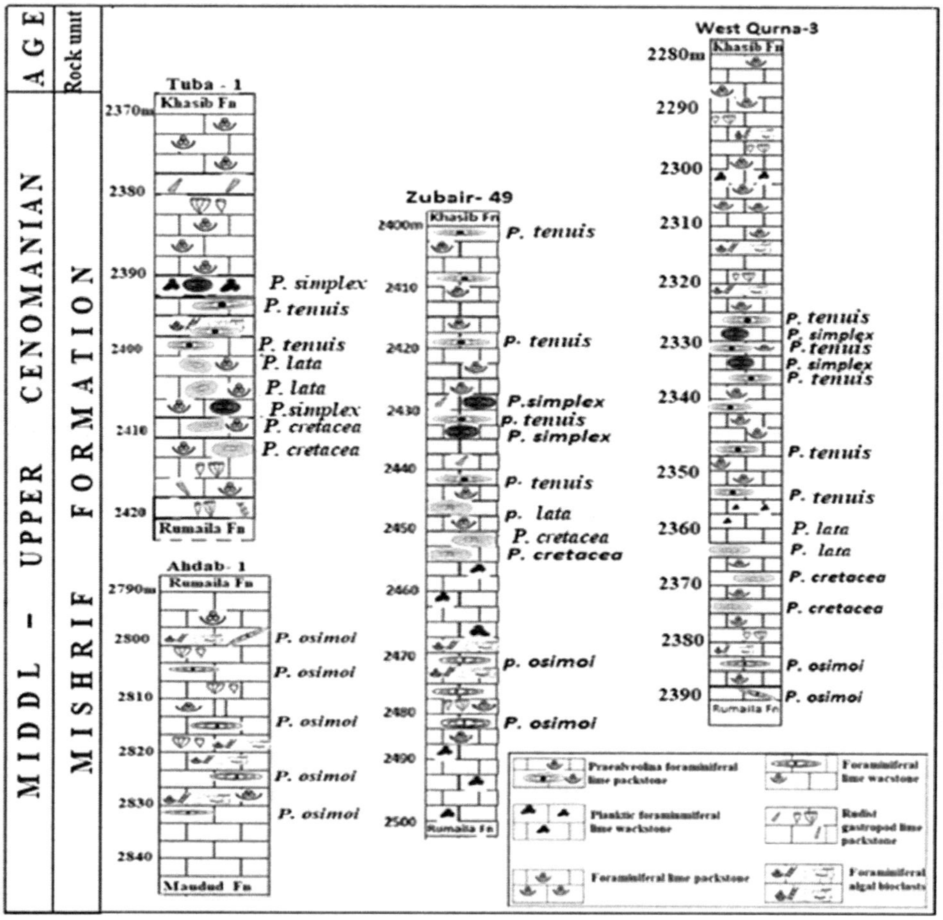

Fig. 2 Stratigraphy of successive *Praealveolina* species of the Middle–Upper Cenomanian (Mishrif Formation) in studied boreholes

Table 1 Morphometric parameters for *Praealveolina* species identification

AGE		Praealveolina species	Shell shape	Axial diameter (mm)	Index of elongation	No. of whorls A form	Apertures A form	Supplementary chamberletes A form	No. of chamberletes Forth whorl A form	Shape and Diameter of proloculus A form (µm)
CENOMANIAN	UPPER	simplex	Sub spherical to oval	A: 1–1.3 B: 2.1-2.2	A:1.15 -1.4 B: 1.5	11 - 12	One row	Absent	10- 14	Spherical 110 - 180
		tenuis	Fusiform	A: 6–12 B: 14 -22	A: 4 -6 B: 7 - 9	9 - 12	One row	Cellars α<40 4/6 rows	26 - 34	Oval 440 - 520
		lata	Oval to sub-cylindrical with truncate at two poles	A: 5.4–6 B: 12 -13	A: 2.5 -2.8 B: 5.5	8 - 10	Two rows	Cellars α<85 6/7 rows	24 -30	Spheric slightly oval 360 - 440
		cretacea	Oval	A: 4–5 B: 13 -15	A: 2.3 -2.9 B: 5	8 - 10	Two rows	Cellars α<90 6/7 rows	18 - 21	Spheric slightly oval 320 - 430
	MIDDLE	osimoi	Fusiform	A: 2.7–3.4 B: 3.7 -5	A: 3.3 – 4.5 B: 4 – 5.2	8 - 9	One row	Cellars α<25 6/7 rows	8 - 9	Globular 100 – 200

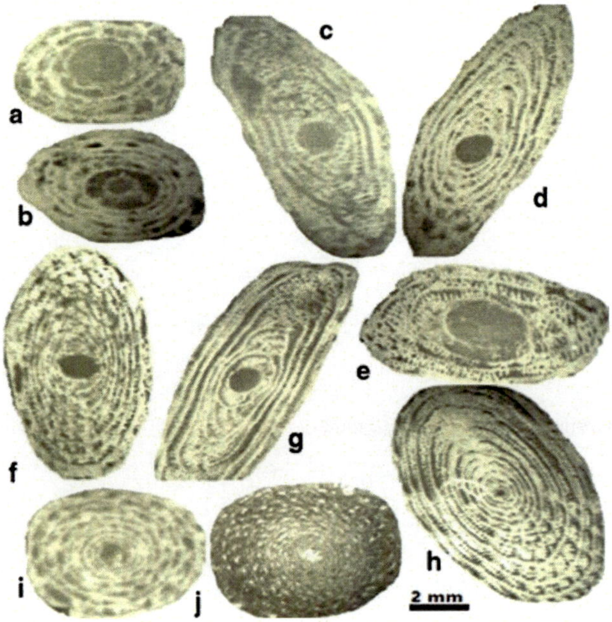

Fig. 3 **a–e** *Praealveolina tenuis* Reichel, Tuba-1. **a** form, A-B Equatorial section, **a**—depth 2.391 m. **b**—depth 2.390 m. **c**—Axial section depth 2.391 m. **d**—Subaxial section, depth 2.390 m. **e**—Axial section of young (A) form, depth 2.400 m. **f**, I-H *Praealveolina cretacea* Reichel, Zubair-49, **f**—Axial section, depth 2.460 m. **i**—Equatorial section, depth 2.455 m. **j**—Equatorial section, Tuba-1, depth 2.403 m. **g**, *Praealveolina lata* Reichel, Axial section, West Qurna-3, depth 2.364 m

4 Discussion

Comparison of both lineages indicates a remarkable difference in evolution rate. *P. simplex* shows no progressive difficulty of internal structure within two generations; however, in the second lineage between the Early Middle and Upper Cenomanian, the length of the test and the structural elements progressively increased. Deciding whether this size increase justifies the splitting of this lineage into several taxa requires a further study. However we can conclude that the two lineages formed independently of the primitive forms of miliolaceans.

5 Conclusions

New observations on the genus *Praealveolina* in successive alveolinid-bearing units allowed us to recognize the following species: *P. simplex* Reichel, *P. osimoi* Zuffardi Comerci, *P. cretacea* d'Archiac, *P. lata* Reichel, and *P. tenuis* Reichel. Quantitative data obtained from *Praealveolina* assemblages indicate two lineages, the *P. simplex* lineage, showing the same evolutionary period without any complication in secondary chamberlets, and the *P. osimoi–P. cretacea–P. lata–P. tenuis* lineage showing an increase in size of the test and nepiont proloculus, in addition to the complication of the secondary chamberlets in polar whorls toward the equator of the test during the early middle to late Cenomanian.

References

Al-Nuaimy, Q. A. (1990). *Study of large foraminifera in middle cretaceous (Albian–Cenomanian)*. Ph.D. thesis, University of Baghdad.

Calonge, A., Caus, E., Bernaus, J. M., & Agui Jar, M. (2002). *Praealveolina* (Foraminifera) species: A tool to date Cenomanian platform sediments. *Micropaleontology, 48*, 53–66.

Caus, E., & Hottinger, L. (1986). Particularidades de la fauna (foraminiferos) del Cretacico superior pirenaico. *Paleontol i Evol, 20*, 115–123.

Caus, E., Gomez-Garrido, A., Simo, A., & Soriano, K. (1993). Cenomanian-Turonian platform to basin integrated stratigraphy in the South Pyrenees (Spain). *Cretaceous Research, 14*, 531–555.

Caus, E., Teixell, A., & Bernaus, J. M. (1997). Depositional model of a Cenomanian-Turonian extensional basin (Sopeira Basin. NE Spain): Interplay between tectonics, eustasy and biological productivity. *Palaeogeography, Palaeoclimatology, Palaeoecology, 129*, 23–36.

Caus, E., Tambareau, Y., Colin, J-P., Aguilar, M., Bernaus, J-M., Gomez-Garrido, A., Brusset, S. (2002). Upper Cretaceous microfauna of the Cárdenas Formation, San Luis Potosí, NE Mexico. Biostratigraphical, palaeoecological, and palaeogeographical significance. *Revista Mexicana de Ciencias Geológicas, 19*, 137–144.

Cherchi, A., & Schroeder, R. (1989). Présence de deux lignes phylétiques à vitesse évolutive différente dans les Préalveolines du Cénomanien de l'Europe sud-occidentale. *Compte Rendu Académie Sciences Paris, Série II, 308*, 801–807.

Neumann, M., & Fourcade, É. (1985). Les grands foraminifères du Crétacé moyen de la région méditerranéenne. In *Geobios, Lyon, Memoire Special* (Vol. 7, p. 160).

Reichel, M. (1936–1937). Étude sur les alvèolines. *Mèm Soc Palèont Suisse, 57*, 95–147.

Schroeder, R. & Neumann, M. (Eds.). (1985). Les Grands. Foraminiferes du Cretace moyen de la region mediter-raneenne. *Memoire Special 7, Geobios* 161, 68.

Late Cretaceous Crocodyliform Nest Areas of the Parana Basin, Southeastern Brazil

Paulo Roberto de Figueiredo Souto and Willian Nava

Abstract

This study brings new information about the behavior and adaptation on the crocodyliforms through the analysis of morphology and preservation of coprolites, eggs and clutches preserved in the Adamantina Formation, Parana Basin, Brazil. Although different fossil vertebrates are found in these sediments, crocodilians are the most dominant taxa with 15 specimens described. Now, we present a recent discovery of associated eggs, clutches and coprolites with strong evidence of nesting areas of crocodyliforms. Here, we extend understanding of the behavior of these reptilian groups that lived during the Cretaceous in Gondwana.

Keywords

Ichnofossils · Crocodyliforms · Cretaceous · Paraná basin

1 Introduction

The first historical record of vertebrate ichnofossils in Brazil dates from the beginning of the twentieth century, with reference to coprolites (Ruedemann, 1929), footprints (Huene, 1931) and eggs (Price, 1951). After these studies, almost nothing was published. However, in the 1990s, new discoveries and several theses examining vertebrate ichnofossils have popularized the study of paleontology and its diverse lines of research.

The Paraná basin is located in southwest Brazil and contains vertebrate remains preserved in the Adamantina Formation, specifically of different crocodyliforms that coexisted at the same time during the Late Cretaceous. The inferences described here are on the coprolite and egg remains of two distinct groups of crocodyliforms: *Baurusuchus salgadoensis* and *Mariliasuchus amarali*, which were collected respectively in outcrops from the General Salgado and the Marilia counties in Sao Paulo State (Carvalho & Bertini, 1999; Carvalho et al., 2005). The Adamantina Formation is characterized by fine siliciclastic sediments with flat-parallel structures, deposited in a distal braided fluvial system, with alluvial plains and intermittent small lakes in a semi-arid climate. On the basis of paleontology and geochronology, this formation has a Turonian-Maastrichtian age (Fernandes & Coimbra, 1996).

2 Materials and Methods

The egg clutches and coprolites, reported here, were preserved in a channel bar (sand flat), of the same level and associated with crocodyliforms. A morphological analysis was carried out using binocular stereoscope, polarization microscope, and scanning electron microscope. Coprolite composition was analyzed using powdered extract from the coprolite matrix and host sediment, and submitted for x-ray diffraction and x-ray fluorescence analysis.

3 Results

Most of the coprolites are characterized by a cylindrical shape with some specimens whole, preserved in isolated or agglomerated units, with a length of 2–5 mm, with autochthone and para-alochthone after the features of taphonomic preservation. The coprolite mineral composition by x-ray diffraction revealed partial or total replacement by apatite

P. R. de Figueiredo Souto (✉)
Universidade Federal do Estado do Rio de Janeiro, Instituto de Biociências, Rio de Janeiro, RJ 22290-240, Brazil
e-mail: prfsouto15@gmail.com

W. Nava
Museu de Paleontologia de Marília, Marilia, SP, Brazil

F. Lucci et al. (eds.), *Selected Studies in Geomorphology, Sedimentology, and Geochemistry*,
Advances in Science, Technology & Innovation, https://doi.org/10.1007/978-3-031-43744-1_39

Table 1 The results of chemical composition (%) by x-ray fluorescence analysis of host coprolites (left side) and sediment (right side) of crocodyliforms from the Adamantina Formation, in 1–General Salgado County (UFRJ/DG 353-IcV) and 2–Marília County (UFRJ/DG 129-IcV)

Elements	1	2
SiO_2	2.10/72.57	44.71/77.02
TiO_2	0.00/1.61	0.01/0.63
Al_2O_3	0.76/6.40	2.54/9.15
Fe_2O_3	0.39/4.51	0.64/3.87
MnO_2	2.36/0.00	3.34/0.01
MgO	0.41/2.92	0.01/1.84
CaO	46.02/4.66	21.91/0.73
K_2O	0.09/3.80	0.96/3.05
P_2O_5	33.91/0.00	16.65/0.01
SO_3	0.30/0.00	0.43/0.01
SrO	0.00/0.00	0.31/0.01
ZrO_2	0.09/0.00	0.00/0.00

hydroxyapatite or calcite and silica minerals. The chemical composition by x-ray fluorescence shows high percentages of phosphor and calcium elements (up to 20%). Within the coprolites, a small tooth and a bone fragment of epiphysis has been identified (Table 1).

The fossil eggs of crocodyliforms from Marilia have an ellipsoid form, the eggshells have thickness of 1.3–1.7 mm. The analysis of microscopic aspects shows that the eggshells consist of polycrystalline units of calcite in the form of compact wedges. The width of these units varies (0.18–0.32 mm) along the shell, and their height is 0.2 to 0.3 mm. The eggs from General Salgado County are bigger, they are more numerous and disposed in a typical clutch of modern crocodilians, while the eggs from Marília County are smaller and preserved side by side. However, the parasistematic classification of these specimens indicates the same oofamily and structural morphotype (Krokolithidae oofamily and crocodiloid morphotype), possibly differing in the oospecies (Fig. 1).

Fig. 1 Microstructure by transversal thin section of eggshell from the crocodyliform, with C1 units in wedge shaped and C2: group of basal blades

4 Discussion

The large amount of coprolites in both localities from the Adamantina Formation can be attributed to the fact that these structures were deposited in sand flats of seasonal lacustrine environments, which resulted in a long term exposure and, consequently, a reduced chance of preservation. This preservation feature of crocodilian coprolites can be better inside the lake bodies, as verified in coprolites from Eocene and deposited in a perennial lake environment [8]. Other strong evidence of nesting areas is the presence of young individuals in these populations, as a tooth found inside of a large coprolite produced by an adult crocodyliform, due to the size of the tooth, belong to the very young crocodyliform; and second, the absence of enamel which is a typical characteristic of teeth ingested by crocodilians, can be probably associated to cannibalism behavior, which has been related to extant crocodilians, due to disputes for food and territory in nest areas. The eggs from General Salgado County were not preserved in nests but in clutches. The difference between clutches of eggs and the built nests has been observed in groups of modern crocodilians and birds. The alligator females covered the soil with vegetation in laying eggs or building shallow nests, while some birds build a nest in the soil, with the shape of a well-defined depression.

5 Conclusion

The morphology, position and chemical composition of the fossilized excrement, as well as the position and microstructure of the fossil eggshells with calcite arranged in two different layers, leave no doubt that they are fossilized structures similar to modern crocodilians. In the case of the Marília County, where abundant bone fragments were discovered, the egg-laying and fossilized excrement related to *Mariliasuchus amarali* indicate endemism with nesting in specific areas. On the other hand, in General Salgado County, the presence of a population with larger individuals of baurusuchids justified the coprolites' and eggs' bigger presence, as a clutch of extant crocodilians.

In conclusion, the association of the coprolites and eggs studied and comparison with the excrement and eggs of extant crocodiles suggests development of nesting area in a fluvial environmental, and evidence of specific reproductive strategy and adaptation of these reptilian groups during continental conditions of the Late Cretaceous.

References

Carvalho, I. S., & Bertini, J. R. (1999). *Mariliasuchus*: Um novo Crocodylomorpha (Notosuchia) do Cretáceo da Bacia Bauru, Brasil. *Geologia Colombiana, 24*, 83–105.

Carvalho, I. S., Campos, A. C. A., & Nobre, P. H. (2005). *Baurusuchus salgadoensis* is a new baurusuchid crocodylomorph from Bauru Basin (Cretaceous), Brazil. *Gondwana Research, 8*, 87–100.

Fernandes, L. A., & Coimbra, A. M. A. (1996). Bacia Bauru (Cretáceo Superior, Brasil). *Anais Da Academia Brasileira De Ciências, 68*, 195–205.

Price, L. I. (1951). Um ovo de dinossáurio na Formação Bauru, do cretácico do Estado de Minas Gerais. *Rio De Janeiro, Departamento Nacional Da Produção Mineral/divisão De Geologia e Mineralogia, 53*, 1–9.

Ruedemann, R. (1929). Fossils from the Permian Tillite of São Paulo, Brazil and their bearing on the origin of the tillite. *Geological Society of America Bulletin, 40*, 417–426.

von Huene, F. (1931). Verschiedene Mesozoische Wierbeltierreste aus Südamerika. *Neues Jahrbuch Für Mineralogie, Geologie, Paläontologie Monastshefte, 66*, 181–198.